Biopower

Biopower

Foucault and Beyond

Edited by

VERNON W. CISNEY
NICOLAE MORAR

The University of Chicago Press
Chicago and London

Vernon W. Cisney is visiting assistant professor of philosophy at Gettysburg College. He is the author of *Derrida's "Voice and Phenomenon": An Edinburgh Philosophical Guide,* as well as coeditor or cotranslator of several other books. **Nicolae Morar** is assistant professor of philosophy and environmental studies and an associate member with the Institute of Ecology and Evolution at the University of Oregon. He is coeditor or cotranslator of several books, including *Perspectives in Bioethics, Science, and Public Policy.*

The University of Chicago Press, Chicago 60637
The University of Chicago Press, Ltd., London

Printed in the United States of America

25 24 23 22 21 20 19 18 17 16   1 2 3 4 5

ISBN-13: 978-0-226-22659-0 (cloth)
ISBN-13: 978-0-226-22662-0 (paper)
ISBN-13: 978-0-226-22676-7 (e-book)
DOI: 10.7208/chicago/9780226226767.001.0001

Library of Congress Cataloging-in-Publication Data

Biopower : Foucault and beyond / edited by Vernon W. Cisney and Nicolae Morar.
pages cm
Includes bibliographical references and index.
ISBN 978-0-226-22659-0 (cloth : alk. paper)—ISBN 978-0-226-22662-0 (pbk. : alk. paper)—ISBN 978-0-226-22676-7 (e-book) 1. Foucault, Michel, 1926–1984. 2. Philosophy, French—20th century. 3. Biopolitics I. Cisney, Vernon W. II. Morar, Nicolae, 1979–
B2430.F724B483 2015
194—dc23
2015011975

♾ This paper meets the requirements of ANSI/NISO Z39.48-1992 (Permanence of Paper).

CONTENTS

Acknowledgments / ix

VERNON W. CISNEY AND NICOLAE MORAR
Introduction: Why Biopower? Why Now? / 1

PART I : ORIGINS OF BIOPOWER

JUDITH REVEL
ONE / The Literary Birth of Biopolitics (translated by Christopher Penfield) / 29

ANTONIO NEGRI
TWO / At the Origins of Biopolitics (translated by Diana Garvin) / 48

IAN HACKING
THREE / Biopower and the Avalanche of Printed Numbers / 65

CATHERINE MILLS
FOUR / Biopolitics and the Concept of Life / 82

PAUL PATTON
FIVE / Power and Biopower in Foucault / 102

PART II : THE QUESTION OF LIFE

MARY BETH MADER
SIX / Foucault, Cuvier, and the Science of Life / 121

JEFFREY T. NEALON
SEVEN / The Archaeology of Biopower: From Plant to Animal Life in *The Order of Things* / 138

EDUARDO MENDIETA
EIGHT / The Biotechnological *Scala Naturae* and Interspecies Cosmopolitanism: Patricia Piccinini, Jane Alexander, and Guillermo Gómez-Peña / 158

PART III : MEDICINE AND SEXUALITY: THE QUESTION OF THE BODY

CARLOS NOVAS
NINE / Patient Activism and Biopolitics: Thinking through Rare Diseases and Orphan Drugs / 183

DAVID M. HALPERIN
TEN / The Biopolitics of HIV Prevention Discourse / 199

JANA SAWICKI
ELEVEN / Precarious Life: Butler and Foucault on Biopolitics / 228

PART IV : NEOLIBERALISM AND GOVERNMENTALITY: THE QUESTION OF THE POPULATION

TODD MAY AND LADELLE McWHORTER
TWELVE / Who's Being Disciplined Now? Operations of Power in a Neoliberal World / 245

FRÉDÉRIC GROS
THIRTEEN / Is There a Biopolitical Subject? Foucault and the Birth of Biopolitics (translated by Samantha Bankston) / 259

MARTINA TAZZIOLI
FOURTEEN / Discordant Practices of Freedom and Power of/over Lives: Three Snapshots on the Bank Effects of the Arab Uprisings / 274

PART V : BIOPOWER TODAY

PAUL RABINOW AND NIKOLAS ROSE
FIFTEEN / Biopower Today / 297

ANN LAURA STOLER
SIXTEEN / A Colonial Reading of Foucault: Bourgeois Bodies and Racial Selves / 326

ROBERTO ESPOSITO
SEVENTEEN / Totalitarianism or Biopolitics? Concerning a Philosophical Interpretation of the Twentieth Century (translated by Timothy Campbell) / 348

Contributors / 361
Index / 369

ACKNOWLEDGMENTS

Nicolae and Vernon would like to express their gratitude to the following people, without whom this project would never have been possible.

First, we would like to thank our contributors for their excellent essays. They have continually supported our ambitious project and truly shaped this volume. We would also like to thank the two anonymous reviewers. Their generous advice and suggestions for topics and contributors were indispensable in strengthening this work into a solid and rounded collection.

In addition, there are three people to whom we are deeply indebted. To Arnold Davidson, we greatly appreciate your support. You expressed faith in this project from its inception and helped us secure outstanding contributors whom, let's face it, we otherwise would not have gotten. Thank you. Next, Daniel W. Smith, you have been with us since this project's inception. In addition to your guidance, you have been and remain a genuine friend and an inspiration. Thank you. To Leonard Lawlor, few words can capture the gratitude that we feel for your constant support and friendship. You introduced Vernon to the concept of biopower and have advised us through the development of this project. Thank you.

Thank you to the editorial staff at the University of Chicago Press. To Elizabeth Branch Dyson, thank you for your willingness in 2010 to take a chance on two graduate students from Purdue University. Thank you also to Nora Devlin, Kelly Finefrock-Creed, Lisa Wehrle, Ryo Yamaguchi, and the countless typesetters, designers, marketers, and folks behind the scenes who have helped us bring this project to fruition.

Nicolae would like to thank Ted Toadvine, Colin Koopman, and all the members of the Critical Genealogies Collaboratory for their constant encouragements and feedback on early drafts of this book project. Special thanks also to Sebastian Big for having initiated me into reading Michel Foucault.

Vernon would like to thank Steve Gimbel, Kerry Walters, Lisa Portmess, Gary Mullen, Dan DeNicola, and Carol Priest for their support, encouragement, and friendship.

Finally, we would like to take this opportunity to thank our families. Vernon would like to thank his wife, Jody, and their two children, Jacob and Hayley. Jody, for your strength, your unwavering encouragement, love, and support, I thank you; I would not be me without you. Jacob and Hayley, in addition to being two amazing people, you give me countless reasons to smile and are my constant reminders as to why I do what I do; thank you.

Nicolae is always reminded that few things in his life would have been possible without Anca's love and endless support. Words are often poor and bounded to capture my wholehearted thank you.

Introduction: Why Biopower? Why Now?

VERNON W. CISNEY AND NICOLAE MORAR

"Biopower," a phrase coined by Michel Foucault,[1] is timely in the sense that it characterizes what Foucault calls the "history of the present"[2] (which is always, at the same time, a thought of the future). Biopower exposes the structures, relations, and practices by which political subjects are constituted and deployed, along with the forces that have shaped and continue to shape modernity. But it is untimely in that its relevance is necessarily dissimulated and masked—the mechanisms of power always have a way of covering their tracks. Before we can elaborate on this concept of biopower—the very etymology of which already points us toward the emergence of life into politics—it would behoove us to look at what power itself is, or what we typically *think* power itself is. For the traditional model of power is precisely what Foucault's concept of biopower assimilates and ultimately surpasses.

An Analytics of Power

What comes to mind when we think of power? Traditionally power was conceived as a commodity or a badge of honor supervening on life and the living, something one either has or lacks. Operating in a top-down manner, the bearer of power dictates, on possible penalty of death, what those *not* in power may and may *not* do. In other words, power is strictly delimiting, the conceptual model being that of the sovereign who rules over his (or her) subjects with greater and lesser degrees of legitimacy and severity. To guarantee its legitimacy, power must produce its own bodies of knowledge, its *truths*. "Power," Foucault claims, "cannot be exercised unless a certain economy of discourses of truth functions in, on the basis of, and thanks to, that power."[3] During the mid- to late middle ages, as tensions between the limits of secular authority and those of religious authority began to escalate, the rediscovery of the *Corpus Iurus Civilis* in about 1070 CE reanimated the Roman codes of juridicality and right, and served to adjudicate matters regarding the expanses and limitations of sovereign power. But whether the concepts of *law* and *right* were employed for the purposes of justifying the absolute power of the sovereign or drawing strict limits to it, and whether the sovereign is *one,* as in a monarchy, or *many,* as in a representational government, what is never in question is the *nature* of power relations themselves as a form of delimitation or "deduction."

On this model, the relation between the sovereign and the life of his subjects is a dissymmetrical one of permissiveness and seizure. The sovereign is in a position to *endanger* the lives of his subjects—in cases when society is threatened, he may put them in harm's way to defend its (or his) security; and he is also in a position to *terminate* their lives—in extreme cases when they blatantly transgress the laws of the sovereign or directly (or indirectly) threaten his life and the lives of his subjects. The sovereign's power over life is thus the power to *let* live or to *make* die: "The right which was formulated as the 'power of life and death' was in reality the right to *take* life or *let* live ['de *faire* mourir ou de *laisser* vivre']. Its symbol, after all, was the sword."[4] All of this operates under a perceived economy of subtraction, where the most visible manifestation of authority resides in the sovereign's power to take whatever possessions he wants or needs from his subjects, up to and including their very lives: "Power in this instance was essentially a right of seizure: of things, time, bodies, and ultimately life itself; it culminated in the privilege to seize hold of life in order to suppress it."[5] The question of *right,* on this model, is thus always a question of *how far* this deductive power extends and *what,* if anything, constitute the rights of the subjects. However, it is always in relation to the delimiting sovereign that this question is posed: "The system of right is completely centered on the king; it is, in other words, ultimately an elimination of domination and its consequences."[6] Thus the question of power is supplemental to or supervenient on, rather than constitutive of, the question of life: "For millennia, man remained what he was for Aristotle: a living animal with the additional capacity for a political existence."[7]

The question of *power,* then, is traditionally one of domination, or the overextension of power's reach. However, as Foucault rightly notes, domination occurs in myriad ways, at all levels of society: not just between de facto political rulers and subjects, but between lovers, spouses, parents and children, teachers and students, ministers and congregations, managers and employees, store clerks and customers—in short, in all relations of life. To truly get at the heart of power, therefore, requires not a general and totalizing interrogation of established systems of power, but rather close and particular analyses of the fundamental, constitutive relations and mechanisms at work in localized settings, which make the establishments of domination possible in the first place. Foucault writes, "As I see it, we have to bypass or get around the problem of sovereignty . . . and the obedience of individuals who submit to it, and to reveal the problem of domination and subjugation instead of sovereignty and obedience."[8]

Against the traditional understanding of power, Foucault demonstrates that relations of power are not homogeneous commodities that one either simply has or lacks; on the contrary, power pulses and reverberates through all areas of life; he who is *dominant* in one situation (as perhaps a father over his children) is, in a different situation, *subject* (as employee and citizen). Thus, rather than power being an external force that *impinges* on the lives of individuals (conceived as static and atomistic elements in the system), power flows through the lives of human beings, constituting the individuals themselves: "Power is exercised through networks, and individuals do not simply circulate in those networks; they are in a position to both submit to and exercise this power. They are never the inert or consenting targets of power; they are always its relays. In other words, power passes through individuals. It is not applied to them."[9] In question, then, are the mechanisms through which domination and subjectivation, in all their various forms, take place.

The Emergence of Biopower

At this level of analysis emerges a remarkable shift, according to Foucault. The sovereign, deductive model of power becomes "no longer the major form of power but merely one element among others."[10] A more insidious and *expansionary* model appears (or is invented): "a new mechanism of power which had very specific procedures, completely new instruments, and very different equipment. It was, I believe, absolutely incompatible with relations of sovereignty.[11] This new mechanism of power applies directly to bodies and what they do rather than to the land and what it produces."[12] Power now appears not to limit but to provoke, purify, and disseminate force for the purposes of management and control, ramified throughout all areas of life, the expansion of which is now its raison d'être. This new form of power is "working to incite, reinforce, control, monitor, optimize, and organize the forces under it: a power bent on generating forces, making them grow, and ordering them, rather than one dedicated to impeding them, making them submit, or destroying them."[13]

In this model of power, the expansion and efficiency of life, not its deduction, becomes the primary function of power, and deduction is merely an instrument in the service of expansion. Death, in and of itself, is anathema to a system of life: "death is power's limit, the moment that escapes it."[14] Death therefore must become strictly instrumental, a sometime necessity in the service of life, a purgation of impurity or threat. Foucault calls this new form of power *bio-power*: "one would have to speak of *bio-power* to designate what brought life and its mechanisms into the realm of explicit calculations and made knowledge-power an agent of transformation of human life."[15] And if, for Aristotle, human beings fundamentally are political animals, Foucault shows us that an important shift occurs with the emergence of biopower; the modern man becomes "an animal whose politics places his existence as a living being in question."[16]

Biopower operates around two poles, according to Foucault. The first, arising in the seventeenth century, functions at the micro level, what Foucault calls "an *anatomo-politics of the human body*."[17] It manifests in a host of disciplinary mechanisms and institutions: militarily, pedagogically, medically, and at the level of labor. The human body comes to be seen as a machine, complete with functions and utilities, inputs and outputs, predictabilities and precisions. "Disciplinary power," as Foucault already calls it in 1975 in *Discipline and Punish*,[18] demands and guarantees of the body "its disciplining, the optimization of its capabilities, the extortion of its forces, the parallel increase of its usefulness and its docility, its integration into systems of efficient and economic controls."[19] No longer does power emphasize the *law* as the product of an arbitrary dictate of the sovereign. Rather it functions under a different type of rule, one located in the natural realm, a *norm*, legitimated by the sciences.

The birth of this disciplinary form of power thus coincides historically with the multiplication and expansion of the human sciences, which are made to serve as the legitimating discourses of this new form of power: "They are extraordinarily inventive when it comes to creating apparatuses to shape knowledge and expertise, and they do support a discourse, but it is a discourse that cannot be the discourse of right or a juridical discourse . . . a discourse about a natural rule, or in other words, a norm. Disciplines will define not a code of law, but a code of normalization, and they will necessarily refer to a theoretical horizon that is not the edifice of law, but the field of the human sciences."[20] Beneath the banner of the anatomo-political, the concepts of *law* and *right* do not disappear or cease to function; rather, the law itself increasingly serves the role of controlling, regulating, correcting, disciplining—in short, *normalizing*.[21]

The other pole around which biopower is organized arises later, "emerging in the second half of the eighteenth century: a new technology of power, but this time it is not disciplinary."[22] This second pole incorporates the elements of disciplinary power but uses them in a new way, and to a slightly different end, "not to the extent that they are nothing more than their individual bodies, but to the extent that they form, on the contrary, a global mass that is affected by overall processes characteristic of birth, death, production, illness, and so on."[23] During this point in history, Foucault claims, emerges a host of disciplines and bodies of knowledge whose task is to calculate, interpret, and predict the overall *health* of the society writ large. The regulation and tracking of birthrates, death rates, fertility rates, economic and poverty statistics, infant mortality, average longevity, and disease, as well as all of the various factors that influence these aspects, operate within a power centered not on the *individual* living body but on the *species-body*. This Foucault identifies as a "*bio-politics of the population*."[24] These two aspects, according to Foucault, the disciplinary power mechanisms of the body and the regulatory mechanisms of the population, constitute the modern incarnation of power relations, labeled as *biopower*.

It is for this reason that sexuality, situated at the juncture of these two domains, becomes such a politicized issue in the eighteenth and nineteenth centuries (and, arguably, the twentieth and twenty-first centuries as well). It is something of a nexus in which these two, the health of the body and the health of society, discipline and regulation, fuse into one: "whence the four great lines of attack along which the politics of sex advanced for two centuries."[25] The first two lines of attack are summed up with the dictum *in the name of regulation, discipline.* The premature sexualization of children is seen to corrupt their development and thus their ability to healthily integrate into the world; therefore, it harms the very health of society itself. Dr. John Harvey Kellogg was one of the more well-known and impassioned critics of childhood masturbation, famously recommending all manners of psychological manipulation and genital mutilation for both boys and girls as processes of normalization. Likewise, in the name of the overall health of the familial institution (as the cornerstone of a healthy society) came a massive increase in diagnoses of female hysteria, a broad designation assigned to nearly *any* ailment that troubled a woman, especially one that was considered detrimental to the female reproductive capacities. This diagnosis brought with it an invasive medicalization of the woman's body and sexuality, including physician-prescribed "pelvic massage" as a regular treatment and, in some cases, a forced hysterectomy if the woman failed to be "cured." The other two lines of attack accord with the inverse of the above declaration, *in the name of discipline, regulation*: "here the intervention was regulatory in nature, but it had to rely on the demand for individual disciplines and constraints."[26] Birth control became a highly controversial topic of public discourse in the late seventeenth century, continuing into the twentieth, as women more widely and effectively began to use barrier methods of birth control. The dissemination of information about birth control as well as the legality of its methods fluctuates in tandem with societal attitudes toward population as a whole. For the majority in this debate, a robust population (for reasons such as economic or moral) is seen as integral to the overall health of society; hence, numerous legal restrictions regulated birth control information or means during this period.[27] Finally, the psychiatric "pathologization" of sexual perversions emerged as the study of sex in the form of a psychological phenomenon, but this brings the individual sexual practices of human beings directly into the domain of abnormality and "corrective" intervention.

These two poles of biopower (discipline and regulation), however, remain somewhat distinct throughout the eighteenth century, Foucault claims. When they finally unite, it is not in the form of a totalizing theory "but in the form of concrete arrangements that would go to make up the great technology of power in the nineteenth century."[28] A new form of political and social organization emerges in the nineteenth century, one dependent on the pushing to the limits of all human capacities. It demands social hierarchy, the machinization of bodies, and a vast crop of willing subjects. Its name—*industrial capitalism*: "the latter would not have been possible without the controlled insertion of bodies into the machinery of production and the adjustment of the phenomena of population to economic processes. But this was not all it required; it also needed the growth of both of these factors, their reinforcement as well as their availability and docility; it had to have methods of power capable of optimizing forces, aptitudes, and life in general without at the same time making them more difficult to govern."[29] Biopower seeks the consistent and ongoing increase in the forces of life without thereby suffering the loss of control over these forces—power in the service of vitality. All the societal institutions, therefore—the family, the church, the education system, the university, the military, and so on—normalize, structure, optimize, and subordinate the forces of individuals to enter them into the machine of the economic system, to make them *productive members of society* who will happily defend it to the death if necessary. Life, as both subject and object, has thereby emerged into the political.

Population, Security, and Governmentality

Interestingly enough, it is Foucault's more in-depth engagement with the concept of "population" that will ultimately result in a significant terminological shift on his part. It is important to note—given the tremendous dissemination of the concepts of biopower and biopolitics by theorists following in the footsteps of Foucault—the durational brevity of Foucault's own engagement with and employment of these terms specifically. By 1978 he drops the language of biopower and of biopolitics, or rather resituates their conceptual apparatuses within the context of the broader notion of what he calls "governmentality." This concept he understands primarily as "the ensemble formed by institutions, procedures, analyses and reflections, calculations, and tactics that allow the exercise of this very specific, albeit very complex, power that has the population as its target, political economy as its major form of knowledge, and apparatuses of security as it essential technical instrument."[30] As noted above,[31] Foucault introduces the concept of "biopolitics" in a 1974 lecture in Rio de Janeiro, and the concept of "biopower" first appears in the context of his final lecture of the Spring semester, 1976, delivered at the Collège de France on March 17; this is followed by the somewhat extended treatment of the concepts of "biopolitics" and of "biopower" in part 5 of *The History of Sexuality*, volume 1. Foucault then took a sabbatical in the 1976–77 academic year, returning to lecturing on January 11, 1978, and giving to his lectures for the 1977–78 year the title of *Security, Territory, Population*. The very first line of the opening lecture begins in the following way: "This year I would like to begin studying something that I have called, somewhat vaguely, biopower."[32] This announcement therefore heralds explicitly the reopening of the question of biopower, which he had launched for the first time in his last lecture almost two years prior as if to suggest that the lecture series on which he was about to embark was conceived as a direct continuation of a problematic that had been merely announced and whose substance had yet to be plumbed. This suspicion is further provoked by the fact that he takes up, point by point, what he considers to be five guiding intentions of the analyses to follow, many of which circulate around his earlier articulation of the concept of biopower itself: (1) the analysis does not propose to offer a general theory of what power *is*, as such; (2) relations of power are not supervenient upon relations of production, relations of family, sexual relations, etc.; rather, "mechanisms of power are an intrinsic part of all these relations, and, in a circular way, are both their effect and cause"[33]; (3) given the *extent* and domains of these analyses (economics, historical formations, etc.), they may branch into the framework of a general analysis of society, but ought not be *confused* with history or with economics, as Foucault's analyses concern, he says, only "the politics of truth,"[34] which, he claims, is the primary recognizable definition of "philosophy" for him; (4) the analyses are meant to offer only a "conditional imperative"[35]—if resistance is sought, here are the trajectories along which it may be pursued; that is, the analysis is meant to propose merely avenues, not particular directives, for political resistance; (5) the one and only "categorical" imperative he offers—never engage in theoretical polemics, as this serves only to severely exhaust the relation between struggle and truth. These five threads, specifically when they speak of the pervasive relationality of power that would subvert any theorization of power as such, echo quite closely the earlier formulation of the concept of biopower.

It is in the context of these five general threads that Foucault introduces in the lectures the notion of "security," which he understands as the structured efforts to rigorously study and manipulate the probabilities and statistics having to do with the phenomena detrimental to the overall "health" of a society. Generally speaking, these apparatuses determine what are tolerable and intolerable levels of these phenomena, to assess what are the "tipping points" at which the costs of restricting the phenomena outweigh the benefits, and ultimately, to stabilize these probabilities within acceptable ranges. In short, the strategies of "security," Foucault says, aim to manage forces and circulations (pathological, economic, sexual, pedagogical, disciplinary, etc.) in a society, by way of direct intervention on the given milieu occupied by the individuals of the society. In this way, "instead of a binary division between the permitted and the prohibited, one establishes an average considered as optimal on the one hand, and, on the other, a bandwidth of the acceptable that must not be exceeded."[36] Foucault articulates the distinction between the three specific models of power relations: "let's say then that sovereignty capitalizes a territory, raising the major problem of the seat of government, whereas discipline structures a space and addresses the essential problem of a hierarchical and functional distribution of elements, and security will try to plan a milieu in terms of events or series of events or possible elements, of series that will have to be regulated within a multivalent and transformable framework."[37]

By all appearances, therefore, in the inaugural lecture of Foucault's returning semester, he is on track to continue with a more thorough explication of the radical concept of biopower that he had, by this time, merely introduced, doing so within the framework of an analysis of "security" (which at this time sounds quite close to the description of, or at least a component of, the *biopolitics of the population,* discussed above). Yet, following the opening declaration of intent of the 1978 lectures, the terms "biopower" and "biopolitics" are almost entirely absent from the remainder of the lectures. Furthermore, just three weeks after the opening lecture, on February 1, 1978, Foucault makes the following startling claim: "Basically, if I had wanted to give the lectures I am giving this year a more exact title, I certainly would not have chosen 'security, territory, population.' What I would really like to undertake is something that I would call a history of 'governmentality.' "[38] What then, we should ask, happens in these interim weeks? Why does Foucault announce on January 11 that he intends to study "biopower," but then suggest on February 1 that the title for the lectures he is currently giving is no longer appropriate and should be replaced with an emphasis on "governmentality," a concept that he has only just introduced? There are a couple of important points to make in response to these inquiries.

A tremendously significant, albeit subtle, shift occurs in Foucault's terminology between the first and second lectures from the 1977–78 year, regarding the relation between the individual, multiplicity, and population. In the context of cursory comments on the *spatiality* of the three specific models of power relations, Foucault suggests that one might perhaps be tempted to differentiate the spatiality of the three in the following way: sovereignty acts on territory, discipline acts on bodies, and security acts on populations. This schematization, though tidy and neat, does not quite hold together, and Foucault dismisses it at the end. While sovereignty, it is true, is a model that is through and through inscribed with territoriality, it is nevertheless the case, Foucault thinks, that "the effective, real, daily operations of the actual exercise of sovereignty point to a certain multiplicity, but one which is treated as a multiplicity of subjects, or [as] the multiplicity of a people."[39] Sovereignty therefore, though it is essentially inseparable from a notion of territory, nevertheless operates on the *multiplicity* of the sovereign's subjects. In an analogous way, the same can be said, Foucault thinks, of the disciplinary mechanisms discussed above. Discipline, it is true, intervenes directly on the forces of the body itself to challenge and optimize them. At the same time, it does so precisely to the end of situating that particular body within a hierarchy that precedes it. Thus the disciplinary model as well operates on a *group* of individuals as on a *multiplicity:* "The individual is much more a particular way of dividing up the multiplicity for a discipline than the raw material from which it is constructed."[40] Hence, the basic spatial distinction that would say that sovereignty acts on territory while discipline acts on individuals simply does not hold up, as both, Foucault argues, act on multiplicities—a multiplicity of subjects in the case of the sovereign and a multiplicity of hierarchically organized individuals in the case of discipline. What then of security?

It is here that the major shift between the lectures of January 11 and January 18 takes place. On January 11, Foucault claims, "So sovereignty and discipline, as well as security, can only be concerned with multiplicities."[41] Therefore, the population, which is the object of security, is in this first lecture considered a "multiplicity," or an organization of the multiple *as* the multiple. By the following week of January 18, however, Foucault comes to characterize the population as precisely *not* a multiplicity—the multiplicity in the model of security comes to be that inconsequential group that falls *outside* the population, while the population itself will be seen as a newly emergent political subject unto itself, an *individual* we might say, of a different sort. Foucault makes this discovery in the context of an analysis of the phenomenon of, and of the discursive engagements surrounding, "scarcity."

In the January 18 lecture, Foucault writes, "I would like now to resume this analysis of apparatuses of security with another example in order to pick out something that is no longer the relationship to space and the milieu, but the relationship of government to the event. I will take straightaway the example of scarcity."[42] This mention of "government" announces the avalanche of terminological usages of concepts having to do with "government," "governing," and ultimately, "governmentality." The phenomenon of scarcity, the insufficiency of food resources to meet basic subsistence needs, is problematic in the context of the discussion of population, for a number of reasons. First, scarcity is a phenomenon such that it tends toward self-perpetuation. As it worsens, the phenomenon of scarcity becomes also a problem for the strategies of power, inasmuch as these heightened moments of crisis and need tend to produce the inevitable outcome of revolt: "So it is the scourge of the population on one side, and, on the other, catastrophe, crisis if you like, for government."[43] Thus, scarcity comes to be seen as an event that the government of population must prepare for and avoid. However, to do so, government through the seventeenth and eighteenth centuries will employ "a system of legality and a system of regulations, which was basically intended to prevent food shortage, that is to say, not just halt it or eradicate it when it occurs, but literally to prevent it and ensure that it cannot take place at all."[44] The strategies and measures employed impose strict regulations on the prices of grain, on the amount of land that can be cultivated, on the exportation of grain, and on its storage. The hope in the use of these strategies is to produce enough food to feed the townspeople (but only enough), such that the peasants (who produce the grain) are paid as little as possible (but not so little that they can no longer afford to plant a sufficient amount of grain) and the townspeople can be fed for as low a cost as possible, thereby staving off in advance the phenomenon of scarcity and, more important, the revolt that would be sure to follow.

This is the *theory* behind the system, but unfortunately, the contingency and intervention of natural forces will ensure, historically, that the theory does not go as planned. The actual *effect* of the implementation of these strategies is that the minimization of prices in fact severely impedes the planting of grain on the part of the peasantry. When there is a surplus of grain, the already-low prices are driven further down—a contingency that the power structure cannot anticipate or prevent—thus, the peasantry who produce the grain do not break even financially, which thereby further limits the amount of grain they can afford to plant the following season. This effect is intensified when, due to unpredictable climatic variations, however seemingly slight they may be, the harvest yields less than was predicted. In short, the very limitations and restrictions designed to *prevent* scarcity constantly run the risk of *producing* the very scarcity it sought to prevent.

But the very reason that scarcity was seen as the intolerable event that must be prevented in advance by legislative means was the governmental fear of revolt in the face of widespread famine. With the emergence of political economists of the eighteenth century, the tendency toward revolt will come to be seen as engendered precisely by the abrupt widespread recognition of pervasive scarcity: "it was precisely this kind of immediate solidarity, the massiveness of the event, that constituted its character as a scourge."[45] Therefore, with the freedom of commerce, proposed by these political economists, there will no doubt, indeed necessarily, be periods of scarcity. But the scarcity that occurs will not be like the scarcity that arises, severely and abruptly, under the conditions of a strictly regulated market. Rather, the scarcity that will arise under the freedom of commerce will be more gradual, gaining slowly in intensity and severity over time, as opposed to the abrupt sort of events that produce mob mentality. Before it gets too extreme, the scarcity will be ameliorated by the very forces that helped produce it. The loss of the unlucky few who succumb to the scarcity is natural and necessary, and it secures the avoidance of scarcity by the population writ large. By allowing these "naturally occurring" periods of scarcity to take hold of a small number of individuals, the scarcity for the population itself can be, by and large, drastically reduced. Foucault writes, "So there will no longer be any scarcity in general, on condition that for a whole series of people, in a whole series of markets, there was some scarcity, some dearness, some difficulty in buying wheat, and consequently some hunger, and it may well be that some people die of hunger after all. But by letting these people die of hunger one will be able to make scarcity a chimera and prevent it occurring in this massive form of the scourge typical of previous systems."[46]

Hence the terminological shift—where in the previous week, Foucault had suspected that the power models of sovereignty, of discipline, and of security had all operated on multiplicities, with the birth of the "population," Foucault now thinks, a new political subject/object is born. It is no longer a multiplicity because it is not at all the individual, nor is it *groups* of individuals, who matters under the model of security; rather it is the population itself. The multiplicity is precisely the name given to those individuals who fall *outside* the population, the unlucky ones who are its by-products. Hence under the model of security, we

> have two levels of phenomena therefore. Not a level of the collective and a level of the individual, for after all it is not just an individual who will die, or at any rate suffer, from this scarcity. But we will have an absolutely fundamental caesura between a level that is pertinent for the government's economic-political action, and this is the level of the population, and a different level, which will be that of the series, the multiplicity of individuals, who will not be pertinent, or rather who will only be pertinent to the extent that, properly managed, maintained, and encouraged, it will make possible what one wants to obtain at the level that is pertinent. The multiplicity of individuals is no longer pertinent, the population is.[47]

Therefore the population emerges as a political being unto itself, and it is at the level of the population that the strategies of security intervene, while the multiplicity, though a necessary component of the strategies of security (inasmuch as its "exclusion" makes possible the security of the population), is not its direct object.

We could perhaps suggest (and this will be one of the topics addressed in this volume), that when Foucault comes to in fact analyze the *biopolitics* of the population, which had previously been subsumed beneath the general banner of biopower (along with its companion pole of the anatomo-politics of the body), that what he discovers is that there is a more drastic difference between the objects of the disciplinary model and the model of security than he had perhaps previously thought. Hence Foucault perhaps no longer believed it obvious that the models of discipline and security fit so tidily and comfortably together beneath the general banner of "biopower," that the object of anatomo-politics and the object of biopolitics were so distinct as to justify a terminological shift such that, while the disciplinary model is no doubt still operative (as we should remind ourselves that for Foucault, a new regime of power always subsumes and incorporates the strategies of previous models), it is so only in support of the health and well-being of the population, now understood as its own sort of political animal.

The language of "governmentality" to which Foucault turns at this point is discovered in what he characterizes as a "flourishing development of a significant series of treatises that do not exactly present themselves as advice to the prince, nor yet as political science, but which, between advice to the prince and treatises of political science, are presented as arts of government."[48] It is in the sixteenth century, Foucault claims, in the shift between the sovereign model and the model of security, that the general problem of "government" arises, and this in a multitude of forms: (1) the government of oneself, in the form of a return to an emphasis on Stoic thought;[49] (2) the government of souls and of human conduct, in the form of pastoral power; (3) the government of children, apparent by an explosion of texts centered on pedagogical strategies; (4) finally, the government of the state. "How to govern oneself, how to be governed, by whom should we accept to be governed, how to be the best possible governor?"[50] It is important to note that the literature on government from this period does not treat the concept of government from the standpoint of sovereignty, according to Foucault. Even when it is ostensibly *discussing* the administration of the sovereign, it does so nevertheless in terms of the *right arrangement* of things—resources, wealth, the people, and so on. The art of government, therefore, presupposes a multitude of various ends, each suited specifically to their unique objects: "the government will have to ensure that the greatest possible amount of wealth is produced, that the people are provided with sufficient means of subsistence, and that the population can increase."[51] These discourses on governmentality operate within a framework that accords quite naturally with the delimitations of the sovereign model of power that had helped shape the language of "biopower" for Foucault. In addition, the areas over which "government" functions (economy, resources, health, security, population, crime, etc.) are well suited to the analyses of the statistical model, burgeoning in the eighteenth century and central to Foucault's earlier articulation of the *biopolitics of the population*. Thus, even as Foucault continues to promise an analysis of biopolitics and biopower through the 1977–78 and 1978–79 lectures, his terminology shifts progressively to the language of governmentality.

Objections and Replies

At the same time, the discoveries that Foucault makes with the concept of biopower have resulted in conceptual apparatuses that occupy his work for the remainder of his life. Some of these discoveries are as follows: (1) a model of power relations that is essentially expansionary of the forces of life, rather than delimiting; (2) the ubiquity of power relations throughout all other modes and types of relations; (3) the persistence with which new models of power employ the fear of sovereign power for the purposes of maintaining insidious control. All in all, these conceptual apparatuses, as the diversity of contributions in this volume attests, have not gone away—they continue to operate to this day throughout all areas of life.

Obvious objections present themselves to us. First, why does sovereign power, now for the most part relegated to a subordinate role, continue to guide our understandings of the political? Put otherwise, if biopower has indeed become the pervasive modern model of power, as Foucault claims that it has, why is it not recognizable as such? Foucault provides two responses. The first is that the very persistence of the conceptual model of sovereignty continues to serve as a critical tool *against* the reemergence of sovereign power itself. In other words, if biopower, first in its disciplinary form and subsequently in its regulatory form, is indeed to supplant sovereign power, it needs a critical apparatus to ever keep sovereignty at bay, and this apparatus is the very concept of sovereignty itself, serving as the constant reminder of the "dangers" of the excesses of power. Second, above we stated that the relevance of biopower is necessarily masked, for the sake of its preservation. In modern, "liberal" forms of government, the understanding of power as "sovereignty" persists, but in a dispersed form. The state is a "sovereign," but one in which its subjects, by virtue of their being "free" citizens or self-governing agents of the polis, are themselves understood to be "sovereign." Sovereignty as the perceived model of power continues, but it is democratized such that each participant is rational, reasonable, responsible, and capable of making good decisions, of exercising one's own "sovereignty," so long as she does not impose her will on another "sovereign" individual within the system. Thus, believing ourselves free, or sovereign, in this manner, the *de facto* mechanisms of domination elude us, and we do not recognize our own subjugation; or, put otherwise, we do not recognize the ways in which our desires and our very *choices* are constituted by the relations within the system itself.

Second objection: if this model of power is bent on generating forces, expanding and monitoring life, how is it that the nineteenth and twentieth centuries, the period in which the two poles of biopower were actively concretizing into an organic whole, exemplifies the most extreme nationalistic discourses that would produce the most totalizing slaughters the world has ever seen? Auschwitz, Buchenwald, Dachau, Dresden, Hiroshima, Nagasaki, and so on are all embodiments of this so-called modern form of power focused on "life." Is this not problematic for Foucault's argument? On the contrary, Foucault will claim that this economy of death is merely the dark underside of a power over life, its logical extension. With the emergence of the modern nation-state following the Peace of Westphalia in the twilight of the middle ages, the bellicosity that had permeated the daily relations of society is relatively marginalized. Warfare becomes more centralized, and a permanent, "professional" military comes to be formed. Furthermore, the state apparatus itself provides a more consistent level of stability, such that bellicose violence, when it *does* occur, does so at the limits or frontiers of the state. Paradoxically, this marginalization of the role of warfare in daily life is contemporaneous, Foucault claims, with the emergence of a new form of discourse, one that sees *war* as the constitutive element of established institutions: "we have to interpret the war that is going on beneath peace; peace itself is a coded war."[52] In the throes of this war, one must choose sides: "A binary structure runs through society."[53] This war is believed to be a permanent one, stretching back to a social organization's origins, extending through the modern age in an insidious and subterranean form, and culminating in a decisive final battle that is yet to come and for which we must prepare. This discourse is thus motivated by mythical elements, imbued with glorious notions of good and evil, and heroic notions of sacrifice, bloodshed, and ultimate triumph. The binary structure of society is one in which the "good," or the "pure" is threatened from within by the "bad" or "impure," and in this struggle there can be only one victor. It is not difficult to see, then, where this line of reasoning leads Foucault: "The war that is going on beneath order and peace, the war that undermines our society and divides it in a binary mode is, basically, a race war."[54]

Whether in its explicitly biological form or its socially (or even culturally or religiously) motivated form, the interpretation of the whole of a society's history in terms of internal struggle, between the *true* and the *impostor*, who would seek to usurp the birthright of the "genuine" descendants, is the dark underbelly of a model of power intimately invested in the discourse of life. In the interest of preserving *one* way of life, another must be annihilated—there is no middle ground or compromise possible: "But this formidable power of death . . . now presents itself as the counterpart of a power that exerts a positive influence on life, that endeavors to administer, optimize, and multiply it, subjecting it to precise controls and comprehensive regulations. Wars are no longer waged in the name of a sovereign who must be defended; they are waged on behalf of the existence of everyone; entire populations are mobilized for the purposes of wholesale slaughter in the name of life necessity: massacres have become vital."[55] Biopower does not operate anymore at the level of the individual, but it is waged at the level of the population, and, more important, exercised in the name of *the biological existence of a population*. It is "situated and exercised at the level of life, the species, the race, and the large-scale phenomena of population."[56]

Finally, the third objection: what of the death penalty? How can a system of power, focused on the multiplication and fostering of life, ever deliberately execute an individual citizen? First, we must note the overall decrease in the use of capital punishment, especially in the Western world, throughout the twentieth century (the United States being the last Western nation where it is still commonly practiced): "As soon as power gave itself the function of administering life, its reason for being and the logic of its exercise—and not the awakening of humanitarian feelings—made it more and more difficult to apply the death penalty."[57] Here the terminological apparatuses of failed normalization and pathology, and of threat, intervene. Whereas capital punishment used to be merely an exercise of justice, a punishment carried out for one of those crimes falling under the class of "capital offense," it now becomes something of a proverbial "last resort" of the state against a criminal *nature*, a monster, a *thing* so beyond the pale of humanity as to be irredeemable, someone who, were they to ever escape or be released, would pose a severe and imminent threat to the safety and security of society as a whole.

Foucault thus presents us with a vision of power relations far more insidious and pervasive than any model of power hitherto conceptualized. The fecundity of his concept is salient in the groundbreaking works of later European political theorists, whether in the form of critical expansion (Giorgio Agamben)[58] or in the form of inheritance and application (Michael Hardt and Antonio Negri,[59] as well as Roberto Esposito[60]). At the outset, we claimed that biopower is a timely concept. It is our conviction that the conceptual mechanisms characterized by Foucault beneath the banner of "biopower" can go a long way toward helping us understand who we are, where we have come from, who we are becoming, and who we *can be*. It is a paradoxical fact of our culture that we can become so emotionally and psychologically invested in preserving the life of a single individual who has existed in a bare, vegetative state for fifteen years,[61] while simultaneously shrugging off the excess of 100,000 unprovoked Iraqi civilian deaths[62] caused by the Multi-National Force's military intervention since 2003. Or that we can stage protests to protect the rights of US fetuses, while not flinching at the fact that in 2010, 7.6 million children in the world died from preventable causes.[63] We enter the twenty-first century engrossed in a seemingly permanent "war on terror," where no one and everyone is "the enemy," welcoming the disciplinary structures that accompany this war. Here the executive power of the United States can, at will and without trial, order (1) the execution of anyone on foreign soil, including US citizens, deemed to be *hostile* to the interests of the US population, or; (2) the indefinite detention of insubordinate US citizens arrested on US soil.[64] Meanwhile the debates that populate our public sphere center on issues of health insurance requirements, birth control, gay marriage, dietary restrictions for families, immigration laws, stem cell research, future possibilities for radical genetic alterations and enhancements, and a multitude of other issues centered around life, the body, and the subject as a living citizen in the polis.

In This Volume

Biopower: Foucault and Beyond critically engages with Foucault's concept of biopower while also reaching into the future, addressing today's problems but with an eye toward tomorrow's. As we have shown, biopower is a multifaceted concept. Thus, as editors, we have chosen each section to approach this concept from a different perspective to ultimately provide a general understanding of biopower.

Part 1, "Origins of Biopower," focuses on the heterogeneous genesis of the concept of biopower, both in Foucault's works and in the works of other major biopolitical thinkers such as Giorgio Agamben and Antonio Negri. Judith Revel opens this part with an enigmatic concern: how can one raise the question of a literary (and linguistic) birth of biopolitics when the literary moment in Foucault's oeuvre largely precedes his biopolitical concerns? Revel reconsiders here the different statutes of language in Foucault's writings and shows how the linguistic period makes possible the process of historicization of the power of words. Ultimately she reveals the slow transformation from *knowledge about the world* to *power over the things and subjects in the world.* And, Revel argues, subjectivation processes can be resisted and overcome only through a series of inventive processes, similar to the creative literary ones, that enable human beings to reappropriate themselves.

In his contribution, Antonio Negri argues that the origins of biopolitics trace both to Foucault's work and, more broadly, to a series of heterodox currents in Western Marxism, as they have developed in Italy and France. Negri's point is that we cannot fully understand the concept of biopolitics without reinscribing it within a series of events in the 1960s and 1970s; only then can we fully appreciate the political problem for which the concept was supposed to stand as a new strategy of intervention. Understanding this new form of power, which is exerted on populations, on multiplicities, on *the multitude*, creates a new space for creative subjects and makes possible the exercise of their freedom.

In a similar vein, Ian Hacking argues that understanding the origins of biopower requires an even broader historical timescale. The study of biopolitics needs a historical perspective that highlights how the political body has become the object of numerical manipulations. Statistical technologies have made possible the study of populations and have transformed them into objects of knowledge. The gathering of massive quantities of data was a necessary condition for recognizing patterns and for defining norms within a multiplicity and, ultimately, for controlling and altering social practices.

The recent diversification of views about biopower and biopolitics—from Foucault to Agamben, Negri, Hardt, and Esposito—has not always clarified these concepts; as a consequence, some scholars have rejected these notions altogether.[65] Catherine Mills's contribution further clarifies the two concepts by considering the prefix "bio" and the ways in which different conceptions of biopolitics (Foucault, Agamben, Esposito) have appealed to various aspects of the manifold concept—*life*. Ultimately, Mills argues that these biopolitical theories all share a significant difficulty: namely, they are unable to identify the limits of biopower and, consequently, are unable to conceptualize life independently of its biopolitical production. Paul Patton's essay reinforces this critical perspective. Along with Mills, Patton identifies some of the confusions involved in Foucault's early definitions of biopower (and biopolitics). Patton spells out very clearly some of the important reasons that have led Foucault to abandon these conceptual tools in his subsequent work.

The second part, "The Question of Life," takes seriously the challenge of formulating the concept of life. Mary Beth Mader invites us to reconsider Foucault's engagement with the concept of life and the ways in which it has first emerged as the object of sciences of life. By highlighting the relations between classical natural history (as developed in *Les Mots et les Choses*) and Georges Cuvier's functionalist notion of life, this essay exposes the conceptual basis that underlines Foucault's understanding of life and problematizes the very object of biopolitics.

Building also on Foucault's broader conception of life as the one emerging from *The Order of Things* and from the *History of Madness*, Jeffrey Nealon finds a way not simply to answer to Donna Haraway's charge of "species chauvinism" against Foucault, but also to show how, in Foucault's early archaeological work, a whole series of neglected formulations of animality are directly linked to the emergence of biopower. The notion of animality is supposed to decenter the focus of biopolitics from human biological processes to a broader concern for species or to a new zoology. This broader view of animality also plays a key role in Eduardo Mendieta's chapter. He explores a new space between biology (as the realm of necessity) and politics (as the realm of possible) to call into question the Judeo-Christian *scala naturae* and to show how, through art, we can gain a better understanding of a (hermeneutical) middle between becoming animal and ceasing to be human.

Foucault's bipolar conception of biopower, as developed in *La Volonté de Savoir*, serves as a matrix for parts 3 and 4, which emphasize the *anatomo-politics of the body* and the *biopolitics of the population*. Part 3, "Medicine and Sexuality: The Question of the Body," focuses on two very significant technologies of normalization in Foucault's works: medicine and sexuality. In his contribution, Carlos Novas examines the ways in which new forms of political activism and biopolitical resistance have emerged to address the problems of rare diseases and orphan drugs. Patients' associations, Novas argues, have formed effective coalitions to support specific legislation requiring pharmaceutical companies to pay sufficient attention to individuals affected by rare conditions. With respect to sexuality, David Halperin raises the question of gay subjectivity and how this notion of subjectivity has been constantly pathologized under numerous labels like "moral insanity," "sexual perversion," "personality disorder," "mental illness," and so on. Keeping *the queer project alive*, he argues, requires a dual process.

First, it demands a radical separation of the question of subjectivity from any medical psychological, psychoanalytical, and "scientific" explanations of homosexuality since those explanations will always tie the queer culture back to normative taxonomy (e.g., unsafe sex, HIV/AIDS risks) whose biopolitical function is to produce "normal" subjects. Second, the notion of queer subjectivity has to be replaced with an entire political program grounded in a notion of gayness as a quasi-ethnic social identity as a way to escape the psychological explanation of queer subjects.

Jana Sawicki's essay closes part 3 on the body by exposing the role that Foucault's analysis of biopolitics has played in Judith Butler's work. Sawicki addresses three important ways in which Butler incorporates and responds to Foucault's projects: first, through an analysis of the concept of normalization in relation to subjection and resistance; second, through the notion of precarious life since it brings back the question of the very object of biopolitics; and ultimately, through the ethical turn, which stresses the need for a new ethic.

Part 4, "Neoliberalism and Governmentality: The Question of the Population," focuses on the second pole of biopower and takes into consideration Foucault's treatment of the question of the population through the lenses of his later engagements with the concepts of governmentality and neoliberalism. Todd May and Ladelle McWhorter invite us to think of how things have changed even for those of us who, following in the path of Foucault, used to see *discipline* everywhere. Hence, their question: how do we understand our present situation? Do we live in a completely new world where Foucault's concepts are not useful anymore to map out our present? May and McWhorter's thesis is that a new form of power has arisen in the past forty years, a *neoliberal power*, and their argument is that *The Birth of Biopolitics* lectures show us how Foucault has already anticipated this new form of governmentality.

The *Birth of Biopolitics* lectures (1978–79) represent a unique moment in Foucault's works. On the one hand, Foucault produces new hypotheses about neoliberalism that are not taken up elsewhere (in either his books or interviews). So, it is difficult to predict to what extent they represent Foucault's ultimate view on the matter. Second, within the series of lectures at Collège de France, *The Birth of Biopolitics* is the place where Foucault first analyzes immediately contemporary facts and current political events. Frédéric Gros argues that Foucault's engagement with neoliberalism, which seems initially foreign to biopolitics, actually makes visible a new form of *neoliberal biopolitics*. The last contribution on biopolitics and populations follows Foucault's engagement with immediate political events and considers very recent revolutionary movements such as the *Arab Uprisings.* Martina Tazzioli focuses on a politics of/over life that relates to forms of governmentality that manage and control migrant populations. Those migrant populations have destabilized the strategies of what she calls "b/ordering" through their increased mobility and have created modes of resistance and of transgression by redefining the meaning of space and by inventing new forms of political struggle and survival strategies.

Part 5, "Biopower Today," captures how the notion of biopower has evolved *beyond* Foucault. In other words, it highlights the ways in which this concept has gained in clarity and thereby provokes a certain academic traction to be used in future work.[66] Through a process of conceptual clarification, Paul Rabinow and Nikolas Rose's argument extracts and builds a model for recognizing biopolitical events in the future. Those events entail a number of specific features such as truth discourses about living beings, forms of authority who are deemed competent to speak in the name of that truth, strategies for collective intervention, and ultimately, modes of subjectification. In the light of this model, Rabinow and Rose consider also a series of recent biopolitical developments in race, population, and genomic medicine.

In her contribution, "A Colonial Reading of Foucault," Ann Laura Stoler calls into question the idea that sexuality was "originally, historically bourgeois." Foucault's assertion does not stand the test of colonial archives since it fails to capture the issue of race. Stoler's incentive is to explain why Foucault is so elusive on this problem and to suggest that the biopolitical treatment of race should occupy a significantly more important place in colonial studies concerning sexuality and the education of desire.

Roberto Esposito closes this volume by raising a significant question of how we should philosophically understand the twentieth century: through the lenses of totalitarianism or biopolitics? Rather than subordinating the movement of history to the logic of a given philosophy, Esposito sees events as consisting of elements that are themselves philosophical. Meaning is no longer stamped from outside, that is, from a point that coincides with the philosophical perspective of the person looking at the world. Esposito's response focuses on how meaning originates and is constituted by facts themselves. These two modalities for understanding contemporary history—that of the more traditional philosophy *of* history and that of history *as* philosophy—are often confused and superimposed. In Esposito's view, they are mutually exclusive in their presuppositions and effects on meaning. This is why he emphasizes the importance of properly construing these two political paradigms of totalitarianism and of biopolitics.

Every single contribution in this volume takes up some aspect of the concerns we have discussed about the concepts of biopower and biopolitics. Whether connecting Foucault's literary texts with the applied analyses of the biopolitical, or problematizing common understandings of the totalitarian face of the twentieth century, or contextualizing the biopolitical within the larger discourse of governmentality, each piece in this volume directly contributes in an indispensable way to the animation of thought with respect to this concept of biopower, the concept of life in and as the political. The editors invite the reader to engage with these contributions in an effort to, in the spirit of Foucault, think the history of the present, which is always by necessity a thought of the future.

Notes

1. Some commentators have pointed out that this domain of inquiry might have preceded Foucault's work. Mauro Bertani points out that the notion of "*biocratie*" has already been present in Auguste Comte's writings as well as in those of French psychiatrist Edouard Toulouse. See Mauro Bertani, "Sur la généalogie du biopouvoir," in *Lectures de Michel Foucault,* vol.1 , *A propos de "Il faut défendre la société,"* ed. Mauro Bertani, Daniel Defert, Alessandro Fontana, Thomas C. Holt (Lyon: ENS Editions, 2001), 19 Antonela Cutro makes a similar point in *Biopolitica Storia e attualità di un concetto* (Verona: Ombre Corte, 2005), 8. In his book, *Biopolitics: An Advanced Introduction* (New York: New York University Press, 2011), Thomas Lemke provides another interesting story about the birth (and the evolution) of the concept of biopolitics (see 9–33).
2. In *Discipline and Punish,* Foucault says "I would like to write the history of this prison with all the political investments of the body that it gathers together in its closed architecture. Why? Simply, because I am interested in the past? No, if one means by that writing a history of the past in terms of the present. Yes, if one means writing the history of the present." In Michel Foucault, *Surveiller et Punir—Naissance de la Prison* (Paris: Gallimard, 1975), 35, published in English as *Discipline and Punish—The Birth of the Prison,* trans. Alan Sheridan (New York: Vintage, 1977), 31. See also Michel Foucault, "What Is Enlightenment?," in *The Foucault Reader,* ed. Paul Rabinow, 32–50 (New York: Pantheon, 1984).
3. Michel Foucault, *Il faut défendre la société* (Paris: Seuil/ Gallimard, 1997), 22, published in English as *Society Must Be Defended: Lectures at the Collège de France, 1975–1976,* trans. David Macey (New York: Picador, 2003), 24. Hereafter, we will provide the English pagination first, followed by the French in square brackets.
4. Michel Foucault, *La volonté de savoir* (Paris: Gallimard, 1976), published in English as *The History of Sexuality,* vol. 1, *An Introduction,* trans. Robert Hurley (New York: Random House, 1978), 136 [178].
5. Ibid.
6. Foucault, *Society Must Be Defended,* 26 [24].

7. Foucault, *History of Sexuality*, vol. 1, 143 [188].
8. Foucault, *Society Must Be Defended*, 27 [25].
9. Ibid., 29 [26].
10. Foucault, *History of Sexuality*, vol. 1, 136 [179].
11. This quote, though coming explicitly from Foucault, is quite controversial due to its hyperbolic use of the words "absolutely incompatible." Foucault changes his mind a number of times over the years as to the extent to which the strategies of sovereign power are incompatible with biopower. Nevertheless we include it precisely because it demonstrates just how active Foucault's thinking is at this point in his life.
12. Foucault, *Society Must Be Defended*, 35 [32].
13. Foucault, *History of Sexuality*, vol. 1, 136 [179].
14. Ibid., 138 [182].
15. Foucault, *History of Sexuality*, vol. 1, 143 [188—"il faudrait parler de 'bio-politique' pour désigner ce qui fait entrer la vie et ses mécanismes dans le domaine des calculs explicites et fait du pouvoir-savoir un agent de transformation de la vie humaine."]. Preceded by the notion of biohistory, the concept of biopolitics appears first in the second lecture Foucault delivered at the State University of Rio de Janeiro in October 1974, two years prior to the concept of biopower in La Volonté de Savoir (1976). See "The Birth of Social Medicine," in *Power*, ed. James Faubion (New York: New Press, 2000), 137. Between 1978–79, Foucault dedicates a whole course to this concept, *Naissance de la Biopolitique*, published in 2004 (Paris: Gallimard/Seuil), and published in English as *The Birth of Biopolitics*, trans. Graham Burchell (New York: Palgrave Macmillan, 2008).
16. Foucault, *History of Sexuality*, vol. 1, 143 [188].
17. Ibid., 139 [183].
18. Foucault, *Discipline and Punish*, 170 [172].
19. Ibid.
20. Foucault, *Society Must Be Defended*, 38 [34].
21. Foucault dedicates a whole collective volume to study the technologies of normalization. See *Généalogie des Equipements de Normalisation*, ed. Michel Foucault (Paris: CERFI, 1976), including some important sections (esp. January 8, 1975, lecture) in his 1974–75 lectures at Collège de France, *Les Anormaux* (Paris: Gallimard/ Seuil, 1999), published in English as *Abnormal*, trans. Graham Burchell (New York: Picador, 2004).
22. Foucault, *Society Must Be Defended*, 242 [215].
23. Ibid., 242–43 [216].
24. Foucault, *History of Sexuality*, vol. 1, 139 [183].
25. Ibid., 146 [193].
26. Ibid., 147 [193].
27. One of the more famous of these is the Comstock Act passed in the United States in 1873, which outlawed the selling of all methods of birth control, as well as all posting of advertisements for birth control.
28. Foucault, *History of Sexuality*, vol. 1, 140 [185].
29. Ibid., 141 [185].
30. Michel Foucault, *Sécurité, territoire, population: Cours au Collège de France (1977–1978)* (Paris: Gallimard, 2004), 111; published in English as *Security, Territory, Population: Lectures at the Collège de France, 1977–1978*, trans. Graham Burchell (New York: Picador, 2007), 108.
31. See note 14 above.

32. Foucault, *Security, Territory, Population,* 1 [3].
33. Ibid., 2 [4].
34. Ibid., 3 [5].
35. Ibid.
36. Ibid., 6 [8].
37. Ibid., 20 [22].
38. Ibid., 108 [111].
39. Ibid., 11 [13].
40. Ibid., 12 [14].
41. Ibid.
42. Ibid., 30 [32].
43. Ibid.
44. Ibid., 31 [33].
45. Ibid., 41 [43].
46. Ibid., 42 [43].
47. Ibid., 42 [44].
48. Ibid., 88 [91].
49. This might arguably provide the link between Foucault's thought on the genealogy of power structures and his later works centered on the concept of the "care of the self."
50. Foucault, *Security, Territory, Population,* 88 [92].
51. Ibid., 99 [101].
52. Foucault, *Society Must Be Defended,* 51 [44].
53. Ibid.
54. Ibid., 59–60 [51].
55. Foucault, *The History of Sexuality, Volume I,* 137 [180].
56. Ibid.
57. Ibid., 138 [181].
58. Giorgio Agamben, *Homo sacer: Il potere sovrano e la nuda vita* (Turin: Giulio Einaudi editore, 1995), published in English as *Homo Sacer: Sovereign Power and Bare Life,* trans. Daniel Heller-Roazen (Stanford, CA: Stanford University Press, 1998).
59. Michael Hardt and Antonio Negri, *Empire* (Cambridge, MA: Harvard University Press, 2000).
60. Roberto Esposito, *Bios: Biopolitica e filosofia* (Turin: Giulio Einauid editore, 2004), published in English as *Bios: Biopolitics and Philosophy,* trans. Timothy Campbell (Minneapolis: University of Minnesota Press, 2008).
61. The sensationalized Terri Schiavo case in Florida from 1998 to 2005. For an illuminating analysis from a biopolitical perspective, see John Protevi, "The Terri Schiavo Case: Biopolitics, Biopower, and Privacy as Singularity," in *Deleuze and Law: Forensic Futures,* ed. Rosi Braidotti, Claire Colebrook, and Patrick Hanafin, 59–72 (New York: Palgrave Macmillan, 2009).
62. The exact figure is uncertain. This statistic comes from Iraq Body Count, accessed February 12, 2015, http://www.iraqbodycount.org.
63. "Young Child Survival and Development: Introduction," UNICEF, accessed February 12, 2015, http://www.unicef.org/childsurvival/index.html.
64. The National Defense Authorization Act, signed into law by President Barack Obama on December 31, 2011. See "President Signs Indefinite Detention Bill into Law," press release, ACLU, December 31, 2011, http://www.aclu.org/national-security/president-obama-signs-indefinite-detention-bill-law.

65. A significant example is Paul Patton's work both in this volume and in *Giorgio Agamben: Sovereignty and Life*, ed. S. DeCaroli and M. Calarco, 203–18 (Stanford, CA: Stanford University Press, 2007).
66. An interesting recent development beyond the notion of biopower is exemplified in Colin Koopman's work. See Koopman, "The Age of Infopolitics," New York Times, January 26, 2014, http://opinionator.blogs.nytimes.com//2014/01/26/the-age-of-infopolitics.

PART ONE

Origins of Biopower

ONE

The Literary Birth of Biopolitics

JUDITH REVEL
TRANSLATED BY CHRISTOPHER PENFIELD

It is no doubt paradoxical to want to anchor the Foucauldian concept of biopolitics—known to have appeared relatively late in Michel Foucault's life, in the second half of the 1970s—to the core of "literary" and linguistic writings from the previous decade. To state this in reverse, it is paradoxical to make the interest in language, speech, and writing, which characterizes much of Foucault's work in the 1960s, the real medium of culture for what will emerge ten years later, within a formidable analytic of power, as a new problematization of the relations between subjectivity, power, and practices of freedom.

This is paradoxical because two real difficulties must be confronted: first, to challenge the traditional division to which Foucault's oeuvre is generally subjected, which neatly separates the periods and themes of research in an effort to nonetheless draw the difficult figure of an interrogation that is complex but coherent even in its apparent discontinuities—in sum, to link "literature" to "politics"; second, to save Foucault from a simplistic reduction to the philosophical "spirit of the times," a reduction that claims that regardless of the disciplinary domain to which they belong, all French thinkers to some extent took up the question of language during the 1960s and the question of power during the 1970s.

In the first case, then, it is a matter of locating in Foucault not the hint of a system, of a unitary and teleological project or a secret design discernable from the origin—concepts from which Foucault is known to have been far removed—but, on the contrary, the echo of a continuous and ceaselessly relaunched work, the movement of which renders precisely necessary the changes in the methodological paradigms, conceptual tools, and fields of inquiry: a coherence that thus in no way claims to erase the discontinuities, the turnabouts, and the changes—which are clearly numerous and important—but aims to read them as stages of a reasoning that proceeds by successive reformulations and displacements or, more exactly, as ever larger revolutions of a spiral winding around itself as it advances.

In the second case, it is a matter of thinking what may be singular in Foucault's thought, independently of the "generational" identification with a certain "milieu" (journals like *Critique* or *Tel Quel*, institutions like Vincennes or the Collège de France, collectives like the Prison Information Group, etc.) or with a certain time period (pre-1968, structuralism, the events of May, the militancy of the 1970s, etc.). Clearly, nothing would be less Foucauldian than to read Foucault's work independently of the historical and material conditions of its production; and turning Foucault into an author "outside of his time" runs such risk of ridicule that only certain hagiographic commentaries are willing to do so unrestrainedly. But it is equally true that Foucault cannot be reduced to the simple epiphenomenon of a more general set of currents that would have been largely derived from prevailing intellectual modes—whether one thinks of this strange, second linguistic turn that traversed France from the middle of the 1950s onward, or of the progressive politicization of intellectual discourse and, more generally, of the work of thought—in the ten years that followed 1968, with its procession of debates on the relations between theory and practice, on the forms of engagement, and on the role of intellectuals in the struggles that were then underway or still to come. If Foucault was always a thinker *à la mode*—and without question he was—it would nevertheless be unfair to consider him as only a modal phenomenon; there is nothing ignominious in public approval if not the risk of becoming static and endlessly repeating oneself, and we know that Foucault was anything but this.

The literary birth of biopolitics, then: a deep-rootedness in language and style of what still remains one of the most complex Foucauldian concepts, for biopolitics plays a pivotal role between the analytic of power and a number of themes that characterize the last years of Foucault's work—the production of subjectivity and the invention of the self, ethics and aesthetics, the relation to the body, and more generally the relation to life. Before venturing forth on this brief genealogy of biopolitics, however, it is necessary to momentarily retrace our steps so as to specify what was truly Foucault's interest in discourse and the extent to which literature is both a subcategory of discourse and its strange blind-spot. To speak of a *literary* birth of biopolitics is precisely not to speak of a linguistic or discursive birth, and it is on this distinction that we would like to focus.

There are two different statutes of language in Foucault. The first, which largely sustained the identification—at first carefully maintained, then severed in just as clean a manner by the interested party—with structuralism, is really the product of a double displacement: on the side of the materiality of the sign (that is, against a hegemony of sense then identified as being essentially the phenomenological matrix) and on the side of the historicization of the very structures of the discourses of knowledge. In the first case, this leads Foucault to focus closely on linguistics, on general grammar, and more broadly on a number of formal inquiries regarding the functioning of language; in the second, this pushes him in the direction of an archeological reading of discourse. Very quickly, the notion of discourse thus refers in Foucault to a set of statements that, though they may well belong to different fields, nonetheless obey common operating rules—rules that are not only linguistic or formal, but that reproduce a certain number of historically determined divisions (for example, the great reason-unreason division highlighted with respect to madness in the classical age). The "order of discourse" proper to a given period becomes, from the publication of *The Order of Things* in 1966 onward, a quasi-synonym for what Foucault understood at that time by *episteme*: a system of traces, of which the differences, variations, heterogeneities, and singularities nevertheless form a great isomorphism, endowed with a normative and regulative function and producing mechanisms for organizing the real through the production of knowledge and the induction of strategies and practices.

For Foucault, it is therefore a matter of analyzing discursive traces by searching to isolate the laws of their operation independently of their nature and conditions of enunciation: as he explains clearly, "it was original and important to say and to show that what was done with language—poetry, literature, philosophy, discourse in general—obeyed a certain number of internal laws or regularities: the laws and regularities of language. The linguistic character of language facts was an important discovery for a certain period."[1] But at the same time, it is also a matter of describing the transformation of the typology of discourse between the seventeenth and eighteenth centuries, which is to say, of historicizing the procedures of identification and classification proper to this period—all the more so, no doubt, since what is in question is a configuration to which we still belong. In this sense, Foucauldian archaeology is a linguistic analysis reinvested within a vaster interrogation of the conditions of emergence of certain discursive apparatuses that sometimes support practices (as in *History of Madness*) and other times engender them (as in *The Order of Things* or *The Archaeology of Knowledge*). It is in this context that Foucault thus simultaneously engages in the analysis of what he calls the "discursive masses" and an archaeology of the human sciences.

Yet there is equally another statute of language, one that Foucault sometimes refers to as a veritable "structural esotericism," and whose existence he takes particular note of at the time of the work on Raymond Roussel,[2] before the Rousselean model is in turn applied more generally to a number of writers—or rather: acts of writing—in the years to follow. By contrast to the first configuration we considered, this esoteric literature is a practice of language that seems to frustrate both linguistic analysis and archaeology: an experience of speech that, precisely because it is *speech*, indicates the "outside" of all language and denounces both the internal economy and the foundational divisions of language.

Now, at the end of the 1940s, Foucault attended the courses at the École Normale Supérieure that Maurice Merleau-Ponty had dedicated to Saussurean linguistics. We may recall that for Ferdinand de Saussure, language is the object of linguistics, whereas speech, being entirely invested in the subjective dimension of the speaker, is by contrast an impossible *object* of general description: in sum, while linguistics, to the degree that it is the science of the functioning of language, is also the science of historical languages, speech, in turn, is the object of no science. Speech is thus the outside of any linguistic approach.

One finds in the works of Foucault, fifteen years later, an important echo of this division. Even if only at the beginning of the 1960s, Foucault substitutes for the traditional Saussurean language-speech distinction two oppositions that he makes alternate use of: the discourse-language couple, where discourse is paradoxically what is restive toward the order of language in general—that is, what seems not to conform to a strictly linguistic analysis (this is the case, for example, in the production of Raymond Roussel); and the discourse-speech couple, where discourse becomes by contrast the linguistic echo of the articulation between knowledge and power such as it is described through the mechanisms of identification, distribution, and taxonomization, and where speech, insofar as it is subjective, seems by contrast to embody a practice of resistance in the face of the major, rigorously ordered and codified objectivation of discursive production.

In one case as in the other, it is surprising that under the appellation "discourse" and then under that of "speech," Foucault returned to Saussure's idea of a limit of linguistic analysis. And even though Foucault will quickly abandon the discourse-speech identification to definitively place discourse on the side of language—his inaugural lecture in 1971 at the Collège de France, "The Discourse on Language,"[3] is the most patent example of this: discourse becomes henceforth a space of order *par excellence*—the fact remains that this *limit*, this *outside*, and this *impossible object* are three figures that are rejoined at the same moment in Foucault in a number of texts on literature and literary writers. Everything thus transpires as though Saussurean *speech* had been crystallized by Foucault around this specific writing that is literature—or rather: with regard to a certain literature that indiscriminately mixes authors of the past (e.g., Rousseau, Sade, Hölderlin, Nerval, Flaubert, Mallarmé, and Roussel) and contemporary writers (e.g., Artaud, Bataille, Blanchot, Klossowski, and numerous writers associated with the drafting of the journal *Tel Quel* with whom Foucault was quite close in the early 1960s) and that groups them together precisely on the basis of this "exteriority" of language.

We must therefore return to the three major matrices of this "exteriority" that appear in Foucault between 1962 and 1966: "limit," "outside," and "impossible object." So, too, must we understand the extent to which Foucault's analyses thereby present an absolutely paradoxical situation, for though the books from the same period[4] affirm the absolute impossibility of an exteriority to the *episteme* of a given period—because one is never *outside of one's own time,* and accordingly one must be resolved to undergo its determinations—the numerous "peripheral" texts on literature—generally published in journals under the form of reviews or critiques, and today included in volume 1 of *Dits et Écrits*[5]—by contrast underscore this exteriority to linguistic description and search for its model and key.

The first major matrix of exteriority appears in Foucault as an obvious homage to Bataille in a formidable text of 1963 that is dedicated to him,[6] which is to say, precisely at the moment when Foucault is writing on Raymond Roussel: it is, namely, the model of transgression. The functioning of this model is simple: there is no limit that does not refer to transgression, which means that there is no space that does not also, immediately, designate its own exteriority. Because a limit is always an act of passage, this act sketches, contrary to its proper function of constraint, the possibility of the gesture that denies it: it is thus at once what refuses and founds the "passage to the limit." And if despite everything the notion of transgression is abandoned fairly rapidly by Foucault, this is because the dialectical link that seems to unite the limit with its own transgression can also be read in the other direction: transgression is also, always, the inevitable reaffirmation of a limit.

The second matrix then intervenes: it, too, is placed under the tutelary shadow of a protective figure—that of Maurice Blanchot—on the occasion of a text that is once again an homage,[7] and which seeks to think exteriority without falling into the dialectical trap. Now, this exteriority is not a metaphysical entity; it is an *experience.* And it brings into play a theme that at that time was central for Foucault: that of the subject, or rather, its disappearance.[8] Foucault defines the "experience of the outside" in effect as the dissociation of the "I think" and the "I speak," to the degree that language must confront the disappearance of the subject that speaks and register that empty place as the source of its own indefinite outpouring. Language thereby "escapes the mode of being of discourse—in other words, the dynasty of representation—and literary speech develops from itself, forming a network in which each point is distinct, distant from even its closest neighbors, and has a position in relation to every other point in a space that simultaneously holds and separates them all."[9]

We can thus understand both what links Foucault to Saussure—the idea that a certain experience of language denies linguistic analysis, or objectivating discourse—and what radically separates them; for where the subjective investiture of speech was for Saussure the criterion of its exclusion from the field of scientific analysis, for Foucault, by contrast, it is precisely the disappearance of the subject that marks the specificity of the experience of an exteriority to the discursive order. The order of discourse and disorder of speech are thus both opposed and rendered symmetrical by this absolute choice to think language as structure without any reference to the one who carries it out. This explains, then, why Foucault still tries to account for the exteriority of language in linguistic terms—to speak of a "structural esotericism" is not the least of the paradoxes since it amounts to attempting the description of the outside of linguistic structures in terms of structure.[10] However, this probably explains less the evident fascination that he experiences for the figures that embody this outside: for Foucault may say that he is not interested in the personage of Raymond Roussel,[11] but the fact remains that all of his work is permeated by its strangeness, just as, many years later, Foucault will confess to having been reduced to silence by the personage of Pierre Rivière, "the parricide with the reddish-brown eyes."[12]

In the first case, the pursuit of a structure of what paradoxically escapes linguistic structures owes perhaps more than is believed to Jacques Lacan's work—to the extent that to affirm, as Lacan did, that the unconscious is *structured like a language* is both to accept the principle of a structural description of the unconscious and to assert that there is not *language* but *languages,* that is, to explode the principle of a unity of the linguistic field. The lesson of Lacan, from which Foucault once again perhaps greatly benefited, is thus that where linguistics functions from a double presupposition of the unity of language and the differentiation of historical languages, the psychoanalyst must henceforth think the structural rupture of the unity of language, which is to say, radical difference as foundation. In sum, we must think difference—exteriority, heterogeneity—at the very base of structural analysis[13]: not one but several structural systems, several regimes of signs, several orders. In Foucault, the opposition between the order of discourse and the order of speech is thus less the discovery of the power of subjective disorder (in the manner of Saussurean *speech*) than the affirmation that several orders exist and that they are incommensurable. No possibility, therefore, of redirecting different orders toward one alone that would be hegemonic and foundational—in the way that reason redirected unreason back into itself and reabsorbed the other under the reassuring figure of the same; no possibility of thinking the unconscious starting from consciousness (Lacan); no possibility of integrating the structural "esotericism" under anything other than the structures that it gives to itself (Foucault).

It is thus a matter of discovering these other structures, that is, a mode of thought whose "fundamental forms and categories must one day be defined," and which Foucault very quickly locates according to a sort of lineage in the margins of Western culture, from Sade to Hölderlin, from Nietzsche to Mallarmé, from Artaud to Bataille and Klossowski. It is everywhere a matter of understanding the functioning of the experience of the outside, which is to say, both the fracturing of the experience of interiority and the decentering of language toward its proper limit. In this sense, according to Foucault, Blanchot, who remains the emblem of this thought, seems in effect to have succeeded in dislodging the reflexivity of consciousness from language and in transforming fiction into a dissolution of narration that highlights what had hitherto been rendered invisible: "the interstice of images."[14] The paradox of this rootless and groundless speech, which discloses itself as seepage and murmur, as gap and dispersion, is that it represents an advance toward what has never been received by language—language itself, which is neither reflection nor fiction but infinite streaming—that is, the "infinite oscillation between the origin and death."[15]

Although the text dedicated to Blanchot was published the same year as *The Order of Things*, in 1966, it is rather surprising that the account given there by Foucault is still inflected by quite marked phenomenological accents—starting precisely with this reference to the origin and the limit-experience of death. Understanding what aspects of phenomenology Foucault retains would make for a different discussion.[16] Let us simply emphasize here that a strangely present figure—whose name, however, never appears[17]—visibly haunts this literary Foucault intent on giving voice to what is invisible in the interstices of the world, the dispersion of signs, at once the hollow and the material of a language claiming to be inaugural and not the repetition of a single order, an experience and not an instrument of reference: namely, Maurice Merleau-Ponty, by virtue of whom Foucault had begun reading Saussure from 1947–48, and whose title, *The Prose of the World,* will become that of the first chapter of *The Order of Things*.

Exteriority, then: with the risk, always incurred, of redirecting the experience of the outside toward the dimension of interiority, and the problem of endowing it with a faithful language; with the difficulty of making the space of radical exteriority that this experience seems to sketch coexist with an archaeological description of discursive apparatuses as presented in *The Order of Things,* and in light of which evidently no outside is possible. And as Foucault will conclude some years later, "We are always on the inside. The margin is a myth. Speech from the outside is a dream that we keep renewing."[18]

In truth, the possibility of thinking an absolute difference (and not, once again, *the other of the same*) seems trapped in advance by the recourse taken to spatial metaphors[19]; and if Foucault had believed that in affirming the radical heterogeneity of the experience of the outside, he had freed himself from the dialectical trap of thinking the transgression of limits, it seems he could have done so only by taking recourse to one of two subterfuges: either anchoring the possibility of the heterogeneity of the outside in an experience that is by definition incommensurable (the origin, death), or pretending that the outside can be thought without ever alluding to its inverse, the inside. Strangely, in the text on Blanchot, Foucault uses a term that is hardly philosophical, one that, before him, Merleau-Ponty had already made use of: the "fold." And perhaps he senses that the figure of the fold, whose philosophical consistency Deleuze will demonstrate much later on, is the only one capable of canceling the vicious circle of inside-outside—which is really just as dialectical as the limit-transgression couple—for the fold is literally the inside of the outside and the outside of the inside, which is to say, a kind of linguistic Mobius strip: the inside *and* the outside, both separately and simultaneously.

The third major matrix is one that takes shape from the disjunction between language and thought. Here again, it is a question, on the one hand, of recovering a philosophical theme that is not new, and on the other, of a unique twist imposed by Foucault. It is one of the recurrent points in contemporary thought that language and thought share no relation, or, more precisely, that language does not clothe thought as its intimate reflection, which is to say, that the function of language is not to say out loud or *externalize* what consciousness thinks to itself in private (objects): indeed, the linguistic turn in philosophy largely takes as its starting point the recognition of this disjunction. This could also already be understood in the Saussurean theorizations since the affirmation of the arbitrary double character of the sign proceeded exactly in this direction.

In Foucault, the irreparably widened gap between words and things is precisely what enables the archaeological analysis of the way in which, beginning with the modern period, words purported to govern and bring order to the world: in sum, the way in which the construction of a principle of intelligibility was ceaselessly sought in the activity of naming, as though language itself could be the sole guarantor of the possibility of shared meaning and objective representation. From here, it became possible for Foucault to historicize this relationship to the power of words, that is, to read between the lines the history of this slow transformation from *knowledge about the world* to *power over the things of the world*; for to name is to transform things—be they *subjects*—into *objects*, and to make truth the effect of a nominal distancing.

Except that certain problems remain. The first consists precisely in whether there is something that can escape the power of words—and the question has not ceased to rebound in contemporary philosophical thought to the extent that its weightiness has been dramatically increased by the trauma of the Holocaust. Can one give voice to what is unspeakable? This is another way of asking: are there words for everything? Does the power of language have no limits? The second problem, on the other hand, returns to a reconsideration of this particular object that is thought itself. If the power of words consists in giving voice to the world on the basis of a historically determined structure of order (a "partitioning," Foucault will say, before he borrows from Deleuze that other word "apparatus"), the articulation and functioning of which are precisely in question, then what about this part of the world called thought? In dissociating thought and language—or rather, the act of thinking from that of naming—are we not inevitably forced to admit that thought thinks outside all language? And would it not be better to think that thought is always already language, which means that how we think is structured in advance on the basis of an order of which language is only one of the faces? In brief: just as we must conduct the archaeology of words and of how they organize the world to construct forms of knowledge about it, must we not also conduct a history of systems of thought, that is, of the way in which thought is always already ordered by discourse?

Most Foucauldian analyses seem to go in this direction, and the fact that Foucault, several years later, took over a chaired position at the Collège de France with precisely the title of "Chair of the History of Systems of Thought" only confirms the point. However, a strange twist of theme occurs from the beginning of *The Order of Things*. One doubtless recalls the extraordinary opening of the book—the great laughter that seizes Foucault while reading Borges's Chinese encyclopedia. One perhaps remembers less the expression that concludes the paragraph, after the reading of Borgesian taxonomy has been completed: for what provokes the laughter is less the bizarreness of the quoted text than the paradoxical observation of a "naked impossibility of thinking that." "That" is Borges's Chinese encyclopedia, which, because it is unthinkable, can receive no other qualifier than this pure demonstrative designation; "that" is the overthrow of the question of whether everything can be said by a new question, infinitely more dangerous, of whether everything can be thought. Because, as Foucault subtlety remarks, this unthinkable Chinese encyclopedia is paradoxically perfectly sayable: put into words without thereby agreeing to be constituted as an object—which means there is a usage of words (an esoteric structure) that does not conform to the common rules of thought and that, because it provokes a crisis thereof, opens a breach in the order that had been believed absolute. Objectivating thought has a limit because it cannot constitute itself there where it must confront words in the form of impossible objects. It is no longer things that resist words, but words that undermine the very possibility of things.

Now, where all philosophy since Descartes had endorsed the double idea that thought was both a cogito and the thought *of something*—since there was no thought without subject or object—the Chinese encyclopedia becomes, through Foucault's reading of it, the deconstruction of these two planes: for no thinking subject is needed to undergo the experience of language—in fact, Borges's text mentions none—just as it happens that the object refuses to appear behind the words that collapse its space in advance. The 1960s go by. The linguistic-literary decade of Foucault—with its procession of paradoxes, beginning with this strange subversive privilege of literature that claims to oppose itself to the order of language while maintaining the dissolution of the subject and the possibility of structural analysis—closes with Foucault's lecture at the Collège de France, "The Discourse on Language." Here, Foucault returns a final time to the "danger" that can be felt arising beneath the surface of words[20]; then he definitively stops taking an interest in it.

What is interesting is that this apparent abandonment from 1971 onward of the theme of discourse for an analysis of power relations, practices, and strategies actually corresponds to what Foucault describes as the passage from an archaeology to a "dynastic of knowledge": no longer just the description of a regime of discursivity and its eventual transgression, but the analysis "of the relation that obtains between these major kinds of discourse and the historical, economic, and political conditions of their appearance."[21]

Yet this displacement, which actually amounts to problematizing the methodological shift from archaeology to genealogy, equally poses the problem of the conditions for the *disappearance* of a given discursive regime and, more generally, of any historical determination. Where the end of *The Order of Things* confined itself to mentioning the possibility of a hypothetical effacement of the discursive economy that the book had so meticulously described, genealogical analysis pushes Foucault to openly confront the concrete theme of practices of resistance, which became ubiquitous during the 1970s.

However, it is difficult to extend the theme of discourse to that of practices without taking into account the subject of these practices—especially, perhaps, since in the meantime there begins a brief but decisive experience for Foucault, that of the *Prison Information Group*, which confronts him with precisely this other resistance to objectivation that consists in the reappropriation of subjectivity.

Here is probably the real passage from the literary to the political: the discovery that resistance to objectivation plays out not only on the side of impossible objects—in the manner of Borges—but on the side of resubjectivation. And it is on this terrain that Foucault will slowly construct, at the heart of the analytic of power that will occupy him throughout much of the 1970s, that other axis of research, which is that of biopolitics.

Let us return for a moment to this analytic of power. Foucault never treats power as a coherent, unitary, and stable entity but as "relations of power" that presuppose complex historical conditions of emergence and imply multiple effects, including those outside what philosophical analysis traditionally identifies as the field of power. The Foucauldian analyses effectuate two remarkable displacements: if it is true that power is only ever exercised by some on others—the "some" and the "others" never being fixed in a role but occupying by turns, even simultaneously, each of the poles in the relation—then a genealogy of power is indissociable from a history of subjectivity; and if power exists only in action, then the question of "how" will guide the analysis of its modalities of exercise, which is to say, just as much the historical emergence of its modes of application as the instruments that it gives itself, the fields where it intervenes, the network that it outlines, and the effects that it implies in a given period. It is thus never a matter of describing a fundamental first principle of power, but an assemblage where discourses, practices, forms of knowledge, and institutions intersect, and where the type of objective pursued not only is irreducible to domination but belongs to no one and itself varies historically.

By characterizing relations of power as complex modes of action on the action of others, Foucault also includes freedom in his description to the degree that power is exercised only on subjects—individual or collective—"who are faced with a field of possibilities in which several kinds of conduct . . . are available. Where the determining factors are exhaustive, there is no relationship of power."[22] The Foucauldian analysis thus destroys the idea of a face-off between power and freedom: it is precisely in rendering them indissociable that Foucault can recognize the role of power as not only repressive but productive (of truth effects, of subjectivities, of struggles) and, conversely, that he can establish the phenomena of resistance at the very interior of the power they seek to contest, rather than in an improbable "outside."

The genealogy of power that Foucault outlines has both constants and variables. If, beginning with Plato, all Western thought thinks in effect that there is an antinomy between knowledge and power ("where knowledge and science are found in their pure truth, there can no longer be any political power"[23]), Foucault, in the wake of Nietzsche, will seek by contrast to dissolve this myth and to reconstruct how, in every period, political power is interwoven with knowledge: how the former gives rise to truth effects and, conversely, how games of truth make a practice or a discourse into a challenge for power. But while, in the Middle Ages, power functions roughly through the recognition of signs of fealty and the collection of goods, starting from the sixteenth and seventeenth centuries, it will organize itself around the idea of production and provision.

To obtain productive benefits from individuals means above all to exceed the traditional juridical framework of power—that of sovereignty—so as to integrate the bodies of individuals, their gestures, their very life: this is what Foucault will describe as the birth of the "disciplines," that is, as a type of governmentality whose rationality is in fact a political economy. This disciplinarization, in turn, undergoes a modification to the extent that the government of individuals is completed by the control of "populations" through a series of "biopowers" that manage life (hygiene, sexuality, demography, etc.) in a global manner to enable a maximization of the reproduction of value (i.e., a less expensive management of production). "There is a sort of schematism that needs to be avoided here . . . that consists of locating power in the State apparatus, making this into the major, privileged, capital and almost unique instrument of the power of one class over another"[24]: just as the juridical model of sovereignty cannot account for the emergence of a political economy, so the political critique of the state cannot shed light on the circulation of power in the social body as a whole and the diversity of its applications, which is also to say, the variability of the phenomena of subjection and, paradoxically, of subjectivation to which it gives rise.

Now, it is precisely this theme of subjectivation or, as he will later say, of the *production of subjectivity* that allows Foucault to move from the old theme of the exteriority of the structure of order, of the outside of objectivity or the refusal of procedures of objectivation, toward a nonspatialized, political formulation that is that of resistance taking form as a practice of freedom. Subjectivation in no way means a return to the old figures of the Cartesian subject: on the contrary, it designates in Foucault a process by which a subject is constituted, the creative process by which a subjectivity is affirmed not only negatively—against order, against relations of power whose rule it collapses—but positively in the world: in brief, a constitutive movement.

The "modes of subjectivation" or "processes of subjectivation" of human being actually correspond to two types of analysis: on the one hand, as we have seen, the analysis of the modes of objectivation that transform human beings into subjects—which means that for power, there are only ever objectivated subjects, and that modes of subjectivation are in this sense practices of objectivation (this is the direction that Foucault takes in *The Order of Things*, for example, when he reads the presence of man into the heart of the edifice of the human sciences and offers the hypothesis of his disappearance); and, on the other hand, the analysis of how the relation to oneself, through a certain number of techniques of the self, enables one to constitute oneself as subject of one's own existence—and this is, of course, the direction taken to carve out, at the very core of power relations, a space of freedom.

Foucault initially locates three principal modes of objectivation of the subject: "the modes of inquiry that try to give themselves that status of sciences,"[25] such as the objectivation of the speaking subject in grammar or linguistics, or again that of the productive subject in economics and the analysis of wealth; the "dividing practices" that divide the subject within itself (or in relation to other subjects) to classify it and make it an object, such as the divisions between the mad and the sane, the sick and the healthy, the courageous and the criminal, and so on; finally, the way in which power invests the subject not only through the modes of subjectivation already mentioned but also by inventing others—such are the stakes of techniques of governmentality. However, Foucault's question then seems to reverse itself: if it is true that modes of subjectivation produce something like subjects by objectivating them, how do these subjects form relations to themselves? What procedures does the individual employ to appropriate or reappropriate his or her proper self-relation?

The problem of the historical production of subjectivities thus pertains at the same time to the archaeological description of the constitution of certain forms of knowledge on the subject; to the genealogical description of the practices of domination and strategies of government to which individuals are subjected; and to the analysis of the techniques through which human beings, in cultivating the relation that binds them to themselves, produce and transform themselves: as Foucault finishes by recognizing, "in the course of their history, men have never ceased to construct themselves, that is, to continually displace their subjectivity, to constitute themselves in an infinite, multiple series of different subjectivities that will never have an end and never bring us in the presence of something that would be 'man.'"[26] This unassignable place of subjectivity in motion, in perpetual "dispossession" in relation to itself, remains for Foucault the product of historical determinations, but it is also the new space of a work taken on oneself, the modalities of which are themselves historical, but that gives rise to the dimension of the outside (creation, invention, the new, the pre-objective) at the very core of relations of power; and from this double anchorage the possibility of the subjective resistance of singularities takes form. Ten years after the first literary explorations of resistance—a resistance, once again, thought on the model of what it must free itself from: as both gap and structure, as refusal of the objective grid of knowledge and confirmation of the dissolution of the subject—the place of self-invention is no longer sought at the exterior of the power-knowledge grid, but is lodged instead in its inner torsion.

Let us now turn to the final element of this short excursus: namely, the theme of biopolitics. Biopolitics is an ambiguous concept insofar as it is at first completely assimilated to that of biopower. "Biopolitics" in effect refers initially in Foucault to the way in which power tends to transform itself, between the end of the eighteenth and beginning of the nineteenth centuries, to govern not only individuals through a number of disciplinary procedures, but the set of living beings constituted as *populations*: biopolitics—by way of local biopowers—therefore deals with the management of health, hygiene, nutrition, sexuality, birth, and so on, to the extent that these become political challenges.

In this context, Foucault means by "population" a set of living, coexistent beings who present particular biological and pathological traits, and whose very lives are susceptible to being controlled to assure a better management of the labor force: "The discovery of population is, alongside the discovery of the individual and the body amenable to dressage, the other great technological core around which the political procedures of the West transformed themselves. At this moment, what I will call 'bio-politics,' in opposition to the anatomo-politics I mentioned a moment ago, was invented."[27] Whereas discipline was given as the anatomo-politics of bodies and applied itself essentially to individuals, biopolitics represents that great "social medicine" that was applied to the population to govern life: life is henceforth part of the field of power. Biopowers are thus literally powers over the *bios*, over life.

However, insofar as the concept of biopolitics is assimilated to that of biopowers, it raises a real problem. Is biopolitics actually to be thought of as a set of biopowers, or, to the extent that saying power has invested life also means that life is in turn a power, can we localize in life itself—which is to say, of course, in labor and language, but also in bodies, affects, desires, and sexuality—the site of emergence of a radical resistance, the site for a production of subjectivity that would constitute itself as a moment of desubjection? This is the direction that Foucault seemingly wants to take at the very end of the 1970s since he privileges both the dimension of ethics (that is, a problematization of the relation to oneself) and that of aesthetics (that is, an analysis of the conditions of possibility for the production of subjectivity). Biopolitics represents precisely the moment of passage from the political to the ethics-aesthetics of the self. As Foucault confirms in 1982, "the analysis, elaboration, and bringing into question of power relations and the 'agonism' between power relations and the intransitivity of freedom is an increasingly political task—even, the political task that is inherent in all social existence."[28]

It was therefore necessary for Foucault to return to the figure of the subject—this figure whose presence the work on language and literature had seemingly wanted to radically exclude. Let us be clear: not the subject as self-sufficient and ahistorical consciousness, nor even the subject as entity or *thing*—that is, as matter to objectivate—but rather, the subject as constitutive process. Subjectivity is nothing other than this: a pure moment of invention within the very historical determinations that it depends on while escaping from; a production of being without end or goal, and yet that opens again the possibility of an ontological event; an outside that does not need exteriority to mark out its difference, since instead of searching for it *elsewhere,* in another space, it inaugurates it *here* in an *actuality* that has again become creative. From space to time, Foucault finally found the means to escape the specters of the dialectic of the limit—without, of course, thereby falling back into the aporias of time considered as a dimension of phenomenological consciousness: here, time is that of productive ontology, a creative *kairos*.[29]

To speak of the literary birth of biopolitics thus means tying back together the threads of two inquiries that never ceased to operate within Foucauldian analyses: on the one hand, the reconciliation of a radical critique of the subject with a reinvestment of subjectivity free of prior objectivation; and on the other, the passage from a spatial analysis of difference to a temporal ontology of creation. The illusion induced by the radical critique of the classical subject (that is, of that subject that was both transcendent condition of possibility of its own thought and ultimate object of its discourse) had pushed Foucault toward the hypothesis of a total desubjectivation of the experience of "esoteric" literature as well as the economy of discourses in general: it was thus to a hypothetical "outside" that fell the task of escaping from all processes of objectivation while reaffirming once again the dissolution of the subject—and this outside, which could be neither object nor subject, remained analyzable nonetheless as *another* linguistic structure. In addition, the spatialization of the metaphors that voiced this desire for an exteriority to the discursive system of knowledge never ceased to recreate the form—oh how spatial!—of the dialectical circle, which is to say, that ultimate ruse of reason that claims that every other is necessarily redirected to the same, and that every counterpower is only the symmetrical inverse of what it critiques. Here again, time had to be rediscovered: a time no longer reduced to the present—if the present is taken to mean that temporality of average duration, which is determined precisely by the concept of *episteme*—but returned to its proper topicality; a time that, because it rediscovers that only an act of invention can enable human beings to reappropriate themselves, will therefore be defined as actuality.[30]

Still, it remains the case that from the beginning, Foucault's interest in literature no doubt anticipated much of what will enable the "biopolitical response" to biopowers. For Foucault, what was fascinating about the literary experience that he had detailed for more than ten years, but for which he was unable to fully account, was the possibility of an invention of new linguistic forms, different structures, unforeseen codes. In sum: a creation. What fascinates Foucault ten years later by the idea of biopolitics is that the power (*puissance*) of life can respond to a power (*pouvoir*) over life: which is another way of saying that one never finishes creating, and that if man is a figure in the sand destined to progressively efface itself, that production of being is there to introduce us to other shores. Biopolitics, production of subjectivity, actuality—all names that, we will wager, would probably not have displeased Raymond Roussel.

Notes

1. Michel Foucault, "Truth and Juridical Forms," in *Power: Essential Works of Foucault, 1954–1984*, vol. 3, ed. James D. Faubion, trans. Robert Hurley (New York: New Press, 2001), 2.
2. Michel Foucault, *Death and the Labyrinth: The World of Raymond Roussel* (London: Athlone Press, 2001); trans. of *Raymond Roussel* (Paris: Gallimard, 1963).
3. Michel Foucault, "The Discourse on Language," in *The Archaeology of Knowledge & The Discourse on Language* (New York: Pantheon, 1972).
4. If we, of course, set aside *Death and the Labyrinth* for obvious reasons.
5. For the English translations of some of these texts, see Michel Foucault, *Aesthetics, Method, and Epistemology: Essential Works of Foucault, 1954–1984*, vol. 2, ed. James Faubion, trans. Robert Hurley et al. (New York: Free Press, 1998). [Translator's note.]
6. Michel Foucault, "A Preface to Transgression," trans. Donald Bouchard and Sherry Simon, in Foucault, *Aesthetics, Method, and Epistemology*, 69–87.
7. Michel Foucault, "The Thought of the Outside," trans. Brian Massumi, in *Aesthetics, Method, and Epistemology*, 147–69.
8. One need only recall the end of *The Order of Things*: "If those arrangements were to disappear as they appeared, if some event of which we can at the moment do no more than sense the possibility—without knowing either what its form will be or what it promises—were to cause them to crumble, as the ground of Classical thought did, at the end of the eighteenth century, then one can certainly wager that man would be erased, like a face drawn in the sand at the edge of the sea." Michel Foucault, *The Order of Things: An Archaeology of the Human Sciences* (New York: Vintage, 1994), 387.
9. Foucault, "Thought of the Outside," 148–49.
10. This is an idea that will remain for many years in France one of the most accredited hypotheses of the "linguistic reading" of resistance to the established order. See, for example, Gilles Deleuze and Félix Guattari, "On Several Regimes of Signs," chap. 5 in *A Thousand Plateaus: Capitalism and Schizophrenia*, trans. Brian Massumi (Minneapolis: University of Minnesota Press, 1987).
11. Michel Foucault, "Pourquoi réédite-t-on l'œuvre de Raymond Roussel? Un précurseur de notre littérature moderne," *Le Monde*, August 22, 1964, n6097, republished in *Dits et Ecrits*, vol. 1 (Paris: Gallimard, 1994), text n26: "But a text which hardly advances us in knowing that Roussel was mad, that he had the premonition of beautiful obsessional symptoms, that Janet had treated but not cured him" (422). It is precisely starting from this indifference to the speaker and this refusal of a psychologizing and individualizing biography that the Foucauldian critique of the concept of the author will be developed several years later.

12. *I, Pierre Rivière, having slaughtered my mother, my sister and my brother . . . A Case of Parricide in the 19*th *Century*, ed. Michel Foucault, trans. Frank Jellinek (New York: Random House, 1975): "We fell under the spell of the parricide with the reddish-brown eyes" (xiii).
13. This primacy of difference must be emphasized: if a language is a system of signs, each sign possesses in itself no positivity; it is in reality nothing other than the location it occupies in relation to all the others within the linguistic chain. The diacritical dimension of the production of meaning is here fundamentally tied to the affirmation that all language is in essence differential—which is paradoxically to say that it has no *essence*, but a structure.
14. On this point, the link between Foucault and the later Merleau-Ponty seems more consistent than could at first be imagined. Let us remember, for example, what Merleau-Ponty wrote in one of his "Working Notes" in *The Visible and the Invisible*: "Meaning is *invisible*, but the invisible is not the contradictory of the visible: the visible itself has an invisible inner framework (*membrure*), and the in-visible is the secret counterpart of the visible." Maurice Merleau-Ponty, "Working Notes," in *The Visible and the Invisible*, trans. Alphonso Lingus (Chicago: Northwestern University Press, 1968), 215. And in regard to the theme of the interstice and how it is tied to that of the differential matrix of language: "The separation [*écart*] which, in first approximation, forms meaning, is not a no I affect *myself* with, a lack which I constitute as a lack by the upsurge of an *end* which I give myself—it is a *natural* negativity, a first intuition, always already there" (ibid., 216).
15. Foucault, "Thought of the Outside," 218.
16. See, e.g., Pierre Macherey, "Aux sources de *l'Histoire de la folie*," *Critique* 471–72 (1986): 753–74.
17. As Foucault explains in a fairly surprising manner at the end of his life, "I think it is important to have a small number of authors with whom one thinks, with whom one works, but about whom one does not write. Perhaps I will write about them one day, but at such a time they will no longer be instruments of thought for me. In the end, for me there are three categories of philosophers: the philosophers that I don't know; the philosophers I know and of whom I have spoken; and the philosophers I know and about whom I don't speak." Michel Foucault, "The Return of Morality," trans. Thomas Levin and Isabelle Lorenz, in *Politics, Philosophy, Culture: Interviews and Other Writings, 1977–1984*, ed. Lawrence Kritzman, trans. Alan Sheridan (New York: Routledge, 1988), 250. On "the philosophers I know and about whom I don't speak," see my "Les grands absents: une bibliographie par le vide," in *Cahier de L'Herne: Foucault* (Paris: Éd. de l'Herne, 2011).
18. Michel Foucault, "The Social Extension of the Norm," in *Foucault Live: Collected Interviews, 1961–1984*, 2nd rev. ed., ed. Sylvère Lotringer, trans. Lysa Hochroth (New York: Semiotext(e), 1996), 198, trans. modified.
19. On this subject, see Michel Foucault, "Questions on Geography," in *Power/Knowledge: Selected Interviews and Other Writings, 1972–1977*, trans. Colin Gordon, Leo Marshall, John Mepham, and Kate Soper (New York: Pantheon, 1980). There, Foucault develops the idea that to try "to decipher discourse through the use of spatial, strategic metaphors enables one to grasp precisely the points at which discourses are transformed in, through and on the basis of relations of power" (70). Except that this choice of method—which is opposed to an entire register of temporal metaphors that for Foucault are tied to the privilege of consciousness—cuts both ways: when the register of spatiality is used for describing the possibility of practices of

resistance, is it not on the contrary the surest means of bringing these practices back under the categories of discourse?

20. Foucault, "Discourse on Language," trans. Rupert Swyer: "What is so perilous, then, in the fact that people speak, and that their discourse indefinitely proliferates? Where is the danger in that?" (216, trans. modified).
21. Michel Foucault, "De l'archéologie à la dynastique," in *Dits et Ecrits*, vol. 3, text n119.
22. Michel Foucault, "The Subject and Power," in *Power*, 342.
23. Foucault, "Truth and Juridical Forms," 32.
24. Foucault, "Questions on Geography," 72.
25. Foucault, "Subject and Power," 326.
26. Michel Foucault, "Interview with Michel Foucault" (interview with D. Trombadori), in *Power*, 276.
27. Michel Foucault, "The Meshes of Power," trans. Gerald Moore, in *Space, Knowledge and Power: Foucault and Geography*, ed. Jeremy Crampton and Stuart Elden (Aldershot: Ashgate, 2007), 161.
28. Foucault, "Subject and Power," 343.
29. On this subject, see the development given it several years ago by the Italian philosopher Antonio Negri in "Kairòs, Alma Venus, Multitudo," in *Time for Revolution*, trans. Matteo Mandarini (New York: Continuum, 2004).
30. The theme of actuality is, of course, central in a number of Foucauldian texts from the final years, particularly in those that he dedicated to a commentary on Kant's text "Was ist Aufklärung" (1784). Refer, for example, to "What Is Enlightenment?" in *Ethics: Subjectivity and Truth: Essential Works of Foucault, 1954–1984*, vol. 1, ed. Paul Rabinow (New York: New Press, 1998); or also "Qu'est-ce que les Lumières," *Le Magazine Littéraire*, May 1984, n207, republished in *Dits et Ecrits*, vol. 4, text n351.

TWO

At the Origins of Biopolitics

ANTONIO NEGRI
TRANSLATED BY DIANA GARVIN

Marx in Italy and France: The "Break" between Traditional and Critical Marxism in the 1960s and 1970s

This chapter analyzes the *propagation* of the concept of "biopolitics" primarily from the standpoint of its genesis in Michel Foucault and the philosophical development that can be charted in his works; however, it also tries to stay out of the history of philosophy field and instead follow a series of heterodox currents in Western Marxism, developed in Italy and France in particular. By doing so, it illustrates how, while belonging to and breathing in the aura of events that were contemporaneous to the genesis of the concept, these currents developed biopolitics along the lines of strategies of intervention that, in the 1960s and 1970s, bore its mark and expressed its potential in turn. To do so, a longer story must be told to recall the premises that made this development possible.

To start, recall what "Italian *operaismo*" has been, not so much to dwell on the value of the theory, which will somehow emerge if the reader is patient, but to clarify how the scientific innovation of the "biopolitical" developed in the 1960s, a decade characterized by the demise of the experience of European resistance and the spread of the economic development of Fordism. Essentially, this innovation traversed an experience that sought to bring together the thought and practice of a new working-class politics at a particular time—the end of the 1950s, the aftermath of the 20th PCUS Congress and the workers' revolt in Budapest, but more important, the defeat of FIOM (Antonio Gramsci's metal workers' union in Turin and FIAT Mirafiori) and in a determinate space—the modern factory. At stake was the reconstruction of political organization in the factory. Now, allow me to present my firsthand experience of Italian *operaismo*. We were looking for a *strong subject*, the working class, capable of fighting against and bringing about a crisis of the mechanism of capitalist production, and it worked: this strong subject was a movement that, in the large factories, led the struggles until 1968 as an exercise of workers' counterpower against the bosses, and often did so against the official unions; a movement that grew to become an autonomous power and produce hegemonic forms of political comportment amongst the workers. In Italy, 1968 evokes both 1968 and 1969, with the youth protests and the workers' "hot autumn," marking a significant change in the power relations between workers and capital, whereby wages directly affected profits. One might say that 1968[1] lasted until 1977.

This was possible because of *operaismo* and its call for the centrality of the factory and the political centrality of the working class in general social relations. As Tronti puts it, "One must turn the question on its head, change the sign, and go back to the beginning: and the beginning is the class struggle of the working class. At the level of socially developed capital, capitalist development is subordinated to the workers' struggles. It comes after them and to them it must make the political mechanism of its own reproduction correspond."

The point was to give a new form, both theoretical and practical, to this fundamental contradiction. Theoretically, the fundamental contradiction was identified in the capital relation itself, and thus in the relations of production, or what at the time we called "the scientific concept of the factory." Here the collective worker potentially had, *when* fighting and *if* autonomously organizing its own struggle, *a sort of sovereignty* over production; then he was, or rather could become, a revolutionary subject. Karl Marx writes, "Labor not as an object, but as activity; not as itself *value*, but as the *living source* of value. [Namely,] it is general wealth (in contrast to capital in which it exists objectively as reality) as the general possibility of the same, which proves itself as such in action. Thus, it is not at all contradictory or, rather, the in-every-way mutually contradictory statements that labor is *absolute poverty as object*, on one side, and is, on the other side, the *general possibility* of wealth as subject and as activity."[2]

The analysis and militant activity of organization were based on the dominant "mode of production" in the Fordist era: the line worker,[3] the worker on the assembly line in the Taylorist organization of work. Tronti notes that the alienation of the worker reached its apex here, but so did the maximum level of resistance. The worker with whom we built organization and struggle not only loved but also hated his work. The *refusal of work* thus became a deadly weapon against capital. Labor power, as an inner part of capital (the variable capital distinguished from constant capital), once autonomous, removed itself from the function of productive work, and in so doing it placed a threat right at the heart of capitalist productive relations.

What more can be said? That this militant standpoint, which represented a shift from the analysis of the laws of motion of capitalist society toward an examination of the laws of motion of labor and workers' resistance, soon became dominant not only in Italy but wherever there were struggles in the Fordist factories of the mass worker. This period was violent but also full of hope. Marx's statement seemed correct: by emancipating itself, the proletariat will free the whole of humanity, or, better, its emancipation will abolish class society.

How did the capitalist powers react to this attack? They developed a counterrevolution, pure and simple. By now we are in the 1970s. To respond to the threat of workers' centrality, capital decided to bring down the centrality of industry and abandon, or revolutionize, the industrial society that had been both the reason for and the means of its own birth and development. This it did to the extent that it turned itself from industrial into financial capital.

But let us examine the transitions determined by this counterrevolution more closely. First of all, as we have seen, there was a transformation of the mode of production. The Toyotist "workstation" provisionally substituted the assembly line. Then, in a continuous and structural way, came the contraptions of automation; what was left of direct production[4] started being "put out" of the factories, processes of "outsourcing" proliferated, and gradually and eventually, the company became computerized and placed under the control of financial capital. Enter *post-Fordism*. But, to come to our second point, labor changed too: capitalist socialization was now occurring on the basis and by means of processes of exploitation that had become *social*. The wage was no longer that monetary quantity the worker negotiated in the factory. Instead, it was reconfigured so as to become a machine that followed the reproduction and the formation of labor power at the level of society as a whole, and throughout the time of life.

At this stage, the question we were paradoxically forced to ask was: is there still a working class, as a *subject* central to the critique of capitalism, as a political subject, rather than a *sociological object*? And with the transformations of work and of the figure of the worker, the rise of the service industry, the shift from employment to self-employment, from material to cognitive labor, from security to precariousness, from the refusal of work to the lack of work, we had to question the political significance of these transformations. Well, faced with these questions, following the shift in reality that was implicit in its analysis, *operaismo* (or at least its most active currents) had the intelligence of turning a historical archaeology (that of workers' struggle) into a new genealogy, that of the *social worker*,[5] thus creating a device for the future.

Here the *biopolitical* entered the scene: biopolitical as life put to work, and therefore as politics mobilized to organize the conditions and control of the social exploitation of all realms of life. As we said in Marxian terms, capital "subsumed" the whole of society. The Frankfurt school had described the actuality and violence of this subsumption but failed to grasp the fundamental aspect of the process: the transformation of the figure of class, the metamorphosis and continuity of resistance. In other words, the biopolitical became central in political discourse at a time when the nature of labor power had changed and social activity had come to replace industrial labor as the source of productivity. At the end of the 1970s and especially during the 1980s and 1990s, this process became more widely acknowledged. Political thought and the critique of sovereignty had to adapt to this new ontology, and *biopower* and *biopolitics* came to describe respectively a new figure of sovereignty and financial command over labor, and the terrain where labor power exerted both its productive capacity and its resistance, where it suffered its alienation and expressed new forms of refusal of work in the shape of "exodus."

So far we have charted, through the knowledge that relates to militancy and workers' struggles, a deepening of the analysis of political control and its transformations. The story we have just told took place in Italy in the years that concern us. At that time, analogous processes were unfolding in France, albeit with differences, as they were often but not always linked to militancy. These led to the discovery of a new field of criticism and thus of a new subversive voluntarism. Let us start from the beginning again here and contextualize the political debate and the experience of the movement of the same period in France.

Where was critical Marxism in France at the time? We certainly would not find it in the French Communist Party, so let me put forward a different hypothesis.

In the 1950s and 1960s, the most attentive side of French political philosophy, highly influenced by the underground currents of communist thought, was invested in what might be defined as the question of "reproduction." Under question was the way in which concepts, knowledge, and ideology influence the reproduction of social systems and how, going through the ideological consistency of knowledge, social action and social being can self-perpetuate or change, be interrupted or subverted. The question was no longer one of continuity/discontinuity, of historical being, but rather of the *dispositif* of change.

In classical economics and Marxism, reproduction refers to the economic field where the modes of production are regenerated. Extending the concept, one might say that reproduction represents a constant renewal of the conditions of exploitation of labor power in the system of capital. Why does capitalism manage to reproduce itself, thus reproducing, or even augmenting, the relations of exploitation? How can one break this process or cycle of production and circulation, commodities and knowledge?

In the 1960s, around Louis Althusser, Claude Lévi-Strauss, Jacques Lacan, Michel Foucault, and later Jacques Derrida, the question of reproduction was raised in a radical way. These thinkers construed it as the search for a break, a rupture. They wished to comprehend what their critical consciousness found hard to believe: that is, why does the capitalist reproduction of the world seem to flow continuously, without breaks, when it is actually always the result of struggles, and thus of discontinuity, excess, and innovation?

One could confront the same issue from a different perspective, from the question Gramsci and critical Marxism were asking in the 1920s and 1930s. In that context, the debate critically focused on the relationship between structure and superstructure: where dogmatic Marxism claimed that the structure was economic and the superstructure ideological, Gramsci (and many others) denied the effectiveness of the distinction and affirmed that relations of domination became real when ideology was implicated in production. The affirmation of hegemony was a power that made it possible for the relations of production and those of ideology to mutually influence one another. This issue later assumed primary relevance for the philosophers of *rue d'Ulm* at the École Normale Supérieure in the 1950s and 1960s; then they faced the problem of reproduction by tracing the relations of production back to a series of anthropological equivalents, namely to the claim that *everything, in society, is productive* and thus there is no longer a realm "outside" production. Whether it critiqued or denied the centrality of workers' labor, theirs was not an anti-Marxist stance; quite the contrary, because it emphasized the importance of labor understood as social activity.

We thus came, in the 1950s and 1960s, in one of Europe's intellectual centers, Paris, to finally understand what Gramsci, Gyorgy Lukàcs, Walter Benjamin, and others had been saying in the 1920s and 1930s: there is no longer an "outside"; production and reproduction are one, a whole. A refusal, contra the tradition of orthodox Marxism, of any possibility of mediation that is external to the movements, of any recourse to a dualist model, including the claim to truth of the Party, thus became possible.

If there is no "outside" of production, and knowledge, ideology, and the concept are found in the processes of reproduction, then this whole of powers is organized autonomously, or, rather, structurally. But what is a structure; what is structuralism? Gilles Deleuze identified five aspects of structuralism: (1) the overcoming of the static-dialectical relation between real and imaginary; (2) the topological definition of conceptual space; (3) the recognition of structure through the identification of a differential relation of symbols; (4) the recognition of the unconscious (and conscious) character of the structural relation; (5) the serial or multiserial movement of structure itself, that is, its internal self-regulation.[6] According to Althusser, the structure is a "process without a subject," a completely closed logical space.[7] Thus, the real is framed as a synchronic section of this whole. Every relation must be understood with reference to its position in the system. The philosophy of history, positivism, and teleology were thus eliminated.

Yet all these definitions would be completely irrelevant had we not been able to single out one unique and solid result of such a formidable period of research: namely, the severance of all *transcendental* concepts from our approach to history and to the world. What had been visible in the past was an internal criticism of the doctrine of Marxism whose foundational philosophy was a dualistic epistemology of structures and superstructures, and whose political project consisted of a dualism of party (leadership) and movements (spontaneity). This critical standpoint was now reemerging but transformed, thanks to the definition of an ontological fabric where such dualisms could *no longer* find the possibility or conditions for their existence.

The new question in France, then, was one that had already been addressed in Italy albeit in a more experimental and practical way. To confront it, one needed to move forward because the structuralist standpoint was untenable: how was a "process without a subject" conceivable? Ultimately, this premise had to be evacuated and replaced by a different project: how to *reestablish subjectivity* and situate it *within* a new framework that was solidly and fully immanent?

To this challenge rose Foucault's thought, which confronted it by turning the structuralist perspective into a *biopolitical* one.

When Foucault began his work, a set of conditions had matured. First of all, structuralism had successfully attacked the "autonomy of the political" and any ideology that isolated the function of the political from economic and social struggles. In structuralism, the political had already taken on a biopolitical semblance: the social was subsumed under capital, even more so when it came to the expressions of the imaginary; through these the political was practiced. The most extreme example was the narrative that demanded that madness and its disciplining be recognized as political, economic, and social.

Second, the biopolitical framework was set against the "isolation of the social," that is, the French tradition originating in Durkheim that regarded the social as a sort of independent realm capable of affecting other aspects of life. No, no category of the social that rules over the rest exists; all there is is a rich social reality that is an economic and political whole—and would later become also one of libido, passions, and fantasy.

Third, the situation is such that the structure started to come to life anyway. Foucault's shift to the biopolitical was not a translation of the positions of the Frankfort school (Marcuse's *One-dimensional man*[8]) where the human was still reunited around an ontology of power, albeit in alienation and generalized despotism. In Foucault, we are dealing with *the exact opposite,* because the human agitates, moves, and changes. In other words, in so far as the structuralist context is traversed by subjectivity, it opens up to multiple dimensions.

Here, the encounter with the positions of Italian *operaismo,* which had always seen capital as a relation of command, and thus as a unity *split into two*—variable capital *against* constant capital, resistance and command, *puissance* and *pouvoir—finally the encounter took place.*

To conclude, from the beginning, in Foucault there is an insuppressible tendency to break the functional nexus of the structure, and what Foucault did not develop, Deleuze would take on, as in a relay.

Deleuze too came from the experience of structuralism, but by the 1960s, he had already turned both structuralism and its realm into the rigorous construction of a "field of immanence." He did so by immediately reinterpreting the biopolitical field as a terrain of constitutive *dispositifs.* It was Félix Guattari who helped him in this crucial operation. The *dispositif* was not only an epistemological operation but also an ontological one. It reconstructed the real from below, in the situation, following an oriented pragmatics. Here crucial was the reference to Baruch Spinoza—a new reading of Spinoza that removed all static effects from pantheism to reveal, on the contrary and in all its riches, a creative drive. Thus biopolitics was traversed by "cupiditas," desire, the *puissance* of action, and Friedrich Nietzsche's thought, with Spinoza's, put at the service of a materialist and emancipatory philosophy.

Another aspect of Italian *operaismo* was here taken on: the recognition that, in so far as society is structured and completely subsumed under capital and an "outside" no longer exists, "inside" class struggle unfolds *everywhere*; the recognition, then, that class struggle constituted the real, and revolutionary militancy interpreted it.

In fact, from the 1930s and late surrealism to the 1960s and Guy Debord, French philosophy had developed a singular view of the "real subsumption" of society under capital because this subsumption was immediately seen as a totality of being. What I am trying to say is that in poststructuralism the characters of real subsumption were partly recovered from a tradition. But this recovery went beyond all ideologies and was thus really operative because, in the meantime, a set of *historical conditions*, both economic and social, had come into being. What were these conditions? The first, from a historical standpoint, was that (as we have already underlined) around 1968 a shift occurred from the Fordist society of production to the post-Fordist society of communication. Anticipated by the reflections of Debord, there was a heightened awareness of this process, and the world of production was interpreted in this light.

Second, the transition from "disciplinary society" ("government") to the so-called society of control (*governance*) was being registered. An analysis developed to recognize that, in the society of control, production and resistance are organized into "modes of life." This operation amounted to a total reversal of the structural field and thus to an articulation of the "field of immanence" as a biopolitical terrain. There is no "outside," *dehors*; the *bios* is that "inside" wherein each one is entirely enveloped. Resistance thus exemplifies acting in this contradiction, but the contradiction one is immersed in is a biopolitical reality. The collective body lives there because it produces everything, because it works, but most of all because it resists, and in this resistance it configures reality.

Third, labor power was redefined. In post-Fordism it became and had to be recognized as being ever more socially active and cooperative, all the more immaterial the more valorization was realized in cognitive services.

At this point we need a second definition of the *field of immanence*. We have already partly characterized it as a creative *biopolitical* terrain. In so far as the poststructuralist ontology of "real subsumption" constitutes an ontology of the biopolitical, the field of immanence must reveal a creative dimension. But what is this creativity?

It is not easy to say. This philosophical path, constantly driven by a sort of revisionism (with respect to Marxist orthodoxy) and yet (in my view) always revolutionary, had determined a specific "topos": the field of immanence. Now that our poststructuralist authors had ruled out the possibility of a transcendental approach to reality, how was it possible to set an engine of creativity into motion starting from this topology? Would we not become trapped in an idealistic imagination?

Once again we must go back to the analysis of capital. As stated earlier, capital is a *relation,* but in this relation, in struggle, one must construct a materialist "telos." What will this "telos" be?

Here allow me to simplify matters and present my reasoning in synthesis. In the field of immanence, human activity tends toward (or rather, desire and will subjectively tend toward) the construction of a world where one can freely live and build happiness. The intellectual, cognitive, and immaterial labor power of today that produces all wealth will therefore want to destroy any force that is contrary to and prevents this happiness. To return to Marx: labor is general wealth as possibility, the living source of value. Therefore the capital relation is subject to an enormous pressure that can cause it to explode.

In other words, where is *class struggle* today? How does critical Marxism work as a movement practice rather than a philosophy? There are two possibilities that follow from what has been said so far. By the end of the 1970s, evidently dogmatic Marxism was over, but it also seemed obvious that historical materialism had invaded the *entire* field of political thought. One can no longer escape class antagonism. Second, and this is very important, the concept of class, without losing its antagonistic characters, had profoundly changed as a social subject: the working class had changed its *technical* composition via a process that it itself had set into motion—*from the factory to society*. Against the ontological backdrop of these transformations of the relations of production and political struggle, the working class thus reappeared as a *multitude,* as a collection of singularities that built the common.

Although many problems are still open, our path in search of the political genesis of the "biopolitical" ends here. We are going to take a break and open the discussion to new issues and aspects of the "biopolitical." In any case, it will be difficult to draw definitive conclusions.

Discussion

B1. We return now to the issue of the biopolitical foundation of the political and open another parenthesis in this line of inquiry. A question necessarily arises: when we put forward the *topos-telos* relation from a materialist standpoint, do we perhaps do so in terms of an uninterrupted flow? That is to say, could "vitalism" be the philosophy that nourishes and expresses this particular development of subversive thought in France? *Is vitalism the trace of the biopolitical?*

The answer to this question is resolutely negative; the authors under consideration have nothing to do with the vitalist tradition, despite its greatness. The three names of the vitalism of the early twentieth century, Georg Simmel, Henri Bergson, and Giovanni Gentile, touched on the issues proposed by poststructuralism but always understood flow as "form." Forms could be social and constitutive (Simmel), spiritual and fluid (Bergson), or disciplinary and dialectic (Gentile), but none of these perspectives accounted for what is essential today; they lacked an interpretation of the course of history as constituted by events and weaved together by singularities. Today, in the philosophical experiences we are examining, there are no "forms," and were there any, they would present themselves as singular and evenemential. If a source of this so-called poststructuralist vitalism was to be found, it would not be in the early twentieth century but in the great tradition that goes from Machiavelli to Nietzsche through Spinoza and Marx. In these authors, vitalism is a *philosophy of power* (puissance).

Therefore, we here face the ontological consistency of the elements of flow. The difference between a classic vitalist conception and the current definition of the dynamic context of philosophical analysis is that, while in the vitalism of 1800 to 1900 the process of life presented itself as a metaphysical flow that is intercepted, separated, and configured by forms, in the work of our authors the agents that structure, constitute, and express this flow are events and singularities. From this standpoint, a truly crucial transformation became manifest because here the *possible* of life expressed itself as *power* (puissance). Put another way, the standpoint of vitalist metaphysics is translated here into one of the *ontology of practice.* The subjectivity that had been expelled as phenomenological-transcendental here returns as a practical subjectivity, as the capacity to act, as a materiality that is constitutive of the process. This doing can be uncontrolled, unconscious, but is always irreducible, strong, and real. At the beginning of this affair, in structuralism, there was a reversal of consciousness toward matter that in poststructuralism became a return of matter to consciousness. The ontological plane was paradoxically fixed, determined, and normalized by the exceptionality of the innovation, of the event, of the singular, and the problem of the *topos-telos* relationship became inserted in the tension of the possible, situated in the *dispositif* of the spread of power.

B2. If we now take up the thread of the argument that led us from the critique of modernity to the immanentist solution of the question of reproduction and the break with the structuralist framework, and eventually brought us to the emergence of a new subjectivity in the biopolitical approach—we can ask what, in the contemporary philosophical approach we share, is defined as "politics." In short, having so heavily insisted on the "bios," we can now try to provide a definition of the "political." First and foremost, note that, in the period that interests us here, the political lies at the center of philosophizing. Indeed, when the philosophical terrain is radically defined as a "field of immanence," and when language and bodies already represent the only matter in this immanence, then the ontological interrelation of subjects, the logical constitution of the common, that is, an ever-renewed genesis of the city, become the heart of philosophical analysis and the latter must increasingly become oriented by the political (*philosophia ancilla politicae*).

How to place a definition of the political in contemporaneity?

The political can be interpreted and defined along the following lines: (1) *synchrony*, (2) *diachrony*, (3) and the figure of the *relation of the political with life*. It could be assumed that, from the synchronic point of view, the political adheres to the surface of ontology and can only be represented within the ontological realm. It is within this ideal hegemony of the "inside" of ontological interiority that the political is determined.

On this premise, some contemporary theories, and the *deconstructionist perspective* in particular, nonetheless seek an alternative to the density of the ontological field, opening up to a notion of the political that is, so to speak, *excessive* or *disseminative*.[9] To bring about this operation, the field of immanence must be "shaken" and made to react to a diachronic impulse. According to "deconstruction," the "inside," assumed as an existential and political totality, becomes dynamic and temporal by means of different meanings of the possible that present themselves as if by exceeding it. This exceeding is something placed at the margins: something that concerns the forms that lie at the edge of ontological totality and defines them as "disseminative," "rhizomatic," and so on. From the standpoint of immanence, it could be suggested that here, albeit tenuously, one might see a recourse to transcendence—but this would be unfair. In fact, the analytically negative and deconstructionist perspective, construed on this flat and full being, in this world "without an outside," acts on the possibilities that are revealed and called forth from the *edge*, from the margins, and seeks to reopen, from these margins, an ethical and/or political development capable of shaking up the fullness of a seemingly static being. In fact, at work here is a moral drive, an ethical urge, as if the field of postmodern immanence, by virtue of its being ontological, had eliminated the possibility of making value judgments about it. Deconstructionism seeks to rediscover and realize the value of judgment, and it is because of this that political judgment starts at the border, on the edge, suspended over nothingness. This condition is best represented by the Derrida of Levinas inspiration, who develops the philosophy of deconstruction.

Other authors, and Jacques Rancière in particular, have more radically tried to escape the political grip of the ontological "inside." To them, the political, a "kind of paradoxical action," has nothing to do with the material dimension of power structures and what the synchronic and diachronic conditions of its actual effects impose, nor does it concern historically determined power relations and regimes, that is to say, biopolitics and biopowers. The paradoxical nature of the political consists of opposing, to the reality of power, the self-determination of "supplementary parts," parts empty of power, "without parts"[10] in the overall social "*partage.*" With respect to Derrida's definitions, the marginality of the political subject here becomes extreme and can no longer flow back toward the inside of the system, a possibility that deconstruction, on the contrary, had permitted. Clearly, though paradoxically, here transcendence and a sort of absolute purity of judgment are called on to testify to the definition of the political, which suggests that the specter of dialectics might reappear, or hover between the fullness of reality and the absolutely "different" of the political. Alain Badiou pushes this paradoxical dualism to the extreme and denies to the political any ontological reality.

This scenario can be summed up by emphasizing that whenever the definition of the political is sought in the *exceeding* of being *at the margins*, the notion that the "field of immanence" can be perceived as an insurmountable horizon for such definition is denied.

B3. It must be noted that the ontological "inside," if assumed as the exclusive realm of political experience, must be seen as "overabundant"; in other words, this reality is a being that contains the presence of a "beyond," or rather, it is an expression of *innovation* as *consistency* or substance of the field of immanence itself. Deleuze first expressed this concept, and from a diachronic standpoint he also proposed the possible as a "fold," *pli*, as a constant reopening of innovative tensions, of events, on the flat and powerful terrain that the field of immanence had revealed. Here it is no longer necessary to define a disseminating margin and, from there, construe a development of value: instead, it is the center of this being that is expressive, not through deconstruction but through the constitution of a power, *puissance*, in the constant sequence of the folds and (tenuous yet strong) movements of being.

At this point, let us return to an analysis of the field of immanence and ask whether in the alternative definitions we have presented one might already find a different appreciation of the field of immanence itself and its ontological consistency. As far as the authors of deconstruction are concerned, the field of immanence they operate on does indeed seem *anahistoric,* and therefore flat and hard. By contrast, Deleuze's "smooth field" is uneven, full of caverns, folds of being: it is a determination of being (and historicity) as a plurality of events and interweaving. But this perspective is also not exempt from criticism. Indeed, if one assumes that, in Deleuze's work, the mutation of all terms of reference is continuous, and the basis of each of their substance and/or desiring drives aleatory, then it will still be hard to define an idea of politics and/or power in his definition of "field of immanence." We are in a position where, if the *political* is given, it is given *without power*. The expression of freedom on this aleatory field seems to exclude the very possibility of power. One might object that power exists anyway (there are courts, prisons, taxes, armies, etc.); but the philosopher would then reply that these forms have no value, that they do not represent an ontological reality. And he or she would be right: in the field of biopolitical immanence, the negative cannot be a transcendental condition. At best, it could be an absence of being, which is to say that the negative is not there. In fact, if power (*pouvoir*) presents itself as a *dispositif* of total and full constitutiveness, if the ontological constitution is power (*puissance*), then the political is configured not so much as resistance but as *generation*, no longer as "being against" but as "being for." The negative that opposes the "being for" and the power that negates "generation" are not there; all there is is their negativity, their "nonbeing."

But to follow this path one must add concrete historicity to immanence, chase the *res gestae* in the field of immanence, because only thus can resistance positively restitute the negative to power and generation return power to nonbeing *tout-court.* What is missing in Deleuze is a full reduction of immanence to historicity. A convincing definition of the field of immanence and the biopolitical calls instead for the coincidence of immanence and history.

On this premise we can see that not only have we reached the end of the Platonic tradition that turned the political into transcendence, imposing it as force and order; not only have we surpassed the Aristotelian tradition that regards generation and corruption as elements that nourish one another in a reciprocal act. With Deleuze, despite the limits of his exposition, we find ourselves in a totally new position: that is, the definition of a politics of the eternal, if one sees as "eternal," irreversible, and irreducible, that material constitution on which the strategies of *cupiditas* rest like matrixes of being, as in Spinoza.

B4. Having examined how the political is defined within the development of the synchronic and the diachronic, of being and history, we must now consider the political in relation to life. Now, between Derrida and Deleuze, between "dissemination" and "generation," lies the Foucauldian experience of power and life: an experience that tends toward the Deleuzian alternative, internal to poststructuralism, and opens toward the vitality of being and generation rather than deconstruction and dissemination, moving forward with reference to the determinations of historicity. So as Foucault understands the centrality of Deleuzian immanentism, he also underlines its concrete limits. Here biopolitics becomes a full experience: life reveals the political conditions of its production and reproduction, and philosophy, with sociology and the other human sciences, manifest the extent of the depth and intimacy of this interrelation. *The field of immanence is biopolitical.*

But the biopolitical, the expression of the vital desire of subjects, conflicts with *biopower.* Theirs is not a polar or molar conflict, but rather a microphysical and molecular dynamic that the biopolitical expresses by colliding with and passing through biopower. The latter seeks to dominate each and every expression of life, and to present itself as the dissolution of the biopolitical fabric. The exercise of power wants to resolve the differences of the biopolitical within itself, subsuming the singularity of their acts and unifying them into a subject. By contrast, the experiences of life that constitute the field of biopolitical immanence give substance to *dispositifs* that are different from those the biopolitical wishes to establish. They are not binary, bipolar, or molar oppositions, by all means. They are infinite lines of flight, each driven to build new realms and terrains of biopolitical being. Both the *reductio ad unum* and the operation that tries to fix this movement in transcendence are not only inadequate but also impossible. The emperor has no clothes. The falsification or the covering of his nakedness is the product of parasitic speculations. Thus the field of immanence, perfected, is restored to us—at the end of a long path running from the debate on "reproduction" in the postwar years to today. Foucault is the author who best managed to provide a logic and a form to this long process of inquiry, but the biopolitical would be incomprehensible without Deleuze because it is not only a physical structure, a corporeal possibility, or a singular device; it is, above all, power (*puissance*). *The biopolitical, with power, produces new subjectivities.*

B5. Here allow me a last parenthesis on the concept of the *multitude.*

Paradoxically, one of the most important aspects of French philosophy (poststructuralist and post-1968) is that with its insistence on the field of immanence it opened itself to the perspectives and problems of globalization. This thought has not simply interpreted historic events like an owl of Minerva; it has also anticipated their development. The formation of empire, the end of national sovereignty, the deterritorialization of the concepts and categories of political science, and thus the shift from modernity to postmodernity, were often anticipated by the political thought that took its breath from "bios." It is sufficient to recall the pages of *Anti-Oedipus* on globalization, the writings of Derrida on the nation-state, but also more banal contributions such as Jean-François Lyotard's on value, Jean Baudrillard's or Paul Virilio's on communication: all of these theoretical elements have intervened to provide a rich definition of the shift to globalization.

In any case, what seems most important to us in this shift from the modern to the postmodern, or rather from the political to the biopolitical, is that the postmodern critically dissolved the moment it was defined, and this made it possible to open a breach through which the constituent flow of the biopolitical and its freedom could spread. We inhabit a field where constituent biopolitical processes develop in every way; diverse bodies couple, miscegenate, and hybridize; in immaterial labor one finds the cooperation of subjects that create ever-new services and relational goods; now, this multitude of bodies and activities, physical and intellectual, this multitude of souls functions as a creative subject. *This subject is multitude.* In modernity, both Bodin and Hobbes reduced the *multitudo* to the *vulgus*, Hegel to the *Poebel*. The great novelty is that here, this *multitudo*, even without unity, is power. *Multitudinis potentia.*

When modern philosophers deny the possibility that the *multitude* can be powerful, they do so based on the impediment (inherent in the concept) of being one. They thus replace the concept of *multitude* with that of nation, people, or race, where unity is imposed from the outside, or with that of sovereignty, which claims to unify the multitude from within. By contrast, here we find ourselves in a position where the *multitudo* does not call for unity, and yet it is productive. It is a body without organs, a *corps sans organs* in *Mille Plateaux*. That is to say: *every body is a multitude, but the multitude is not a body, it is a whole of bodies—a whole of freedom.*

Concluding Remarks

Our argument has shown the biopolitical to be an "experience of being." This reference to Jean-Paul Sartre is opportune because the experience of being—obviously experienced differently in postmodernity—can be referred to as a condition of reflection and action, of *engagement* and ontological constitution that Sartre, at the beginning of the postwar episode of French philosophy, dramatically hypothesized. From this reminiscence and renewal, the experience of the biopolitical leads once again to *praxis*.

And so we return to *operaismo*, to that notion of praxis that opened the analysis of this chapter. The journey of *operaismo* had barely begun. Now we must end it. For this, it will be useful to remember that this journey, so subversive (as French philosophy also was in the latter half of the twentieth century) always has to be nourished by struggles and organization. These must raise a specter—the resurrected specter of communism (as Derrida put it). This must be a communism that stands on new social legs—those of "mass intellectuality," of cognitive labor power, of the migrant proletariat—and on the new ability to know and imagine that this new work demands and generates. It must emerge from new experiences: those of a life subjected to the command and consumption of capital that—within this, and no other domination—*rebels*. The biopolitical experience must be invented with all the intensity that rebelling "inside" demands, be immersed in the geographic and temporal breadth that the global nature of biopower determines, and for this reason, be aware that there are no margins from which we can aleatorily defend our souls, no possible escapes for our bodies. The "inside" of biopower has a "heart": fighting against biopower is possible only if, from this heart, we remove all nourishment and circulation. Because power does not interest us, we have understood the *dissymmetry* between the biopolitical desire of democracy and the exercises of biopower; because the "inside" of the field of biopolitical immanence will eventually no longer have a center or a "heart," but only "love" that circulates with violence and construes freedom.

Notes

This text has been read and discussed frequently: first on February 3, 1997, at a seminar directed by Eric Alliez and Barbara Cassin at the Collége International de Philosophie, then on September 30, 2000, at a seminar at the Fondazione Lelio Basso in Rome, and finally on November 21, 2009, at the Volksbuehne in Berlin, for the Rosa Luxemburg Stiftung.

1. Negri cites the events of 1968 and 1969. This English rendering reflects the fact that the duration from 1968 to 1977 includes the year 1969. [Translator's note.]
2. Karl Marx, *Grundrisse: Foundations of the Critique of Political Economy, translated by Martin Nicolaus* (London: Penguin, 1973), 144 (available at http://www.marxists.org/archive/marx/works/download/Marx_Grundrisse.pdf [accessed January 6, 2015], 235).

3. *Operaio* is translated here as the general term "worker" to preserve Negri's linguistic consistency with reference to *operaismo* (workerism). While this does flatten the distinction between *operaio* and *lavoratore*, use of the alternative term "laborer" is not always applicable. This convention is followed throughout the essay. [Translator's note.]
4. *Produzione diretta* implies not only the direction production by labor but also a sense of manual production that accompanies assembly-line work. [Translator's note.]
5. "Social worker" is the conventional translation for *operaio-sociale.* See Antonio Negri, *Time for Revolution*, trans. Matteo Mandarini (New York: Continuum, 2003), 282. [Translator's note.]
6. See Gilles Deleuze, "How Do We Recognize Structuralism?," in *Desert Islands and Other Texts 1953–1974*, ed. David Lapoujade, trans. Michael Taormina (Cambridge: Semiotext(e), 2004), 170–92.
7. See, e.g., Louis Althusser, *For Marx* (London: Verso, 2005).
8. Herbert Marcuse, *One Dimensional Man* (New York: Beacon Press, 1964).
9. Negri's use of *dessiminazione* for the French term *dessimination* is obviously a direct translation. English translations of Derrida's term tend to follow this norm. See Jacques Derrida, *Dissemination*, trans. Barbara Johnson (Chicago: University of Chicago Press, 1981). [Translator's note.]
10. "The part without parts" is the conventional translation for *la part sans part.* Rancière and his translators often shorten the designation to *sans-part,* as Negri does here in referring to the *senza-parte.* For readers who may be unfamiliar with this tendency, I have translated Rancière's full phrase. See Jacques Rancière, *The Politics of Aesthetics: The Distribution of the Sensible*, trans. Gabriel Rockhill (London: Continuum, 2004). [Translator's note.]

THREE

Biopower and the Avalanche of Printed Numbers

IAN HACKING

Long ago, in *Les Mots et les choses*, Michel Foucault taught that life, labor, and language are not eternal objects of thought but arise as self-conscious topics only at the end of the eighteenth century. Of these three, it is of course life that is at the heart of his more recent work on sexuality, although the other two are curiously carried in train throughout his research of the past ten years. Life became not only an object of thought but an object of power: it was not merely individual living persons who might be subjected to the orders of the sovereign, but life itself, the life of the species, the size of the population, the modes of procreation. Such power, Foucault writes, "evolved in two basic forms" that constituted "two poles of development linked together by a whole intermediary cluster of relations." One pole centered on the body as a machine. It was disciplined by procedures of power that he calls *an anatomo-politics of the human body*. The second pole "focused on the species body" that serves as the "basis of the biological processes: propagation, births and mortality, the level of health, life expectancy and longevity." This was *a biopolitics of the population* "that gave rise to comprehensive measures, statistical assessments, and interventions aimed at the entire social body or at groups taken as a whole. . . . One may speak of *bio-power* to designate what brought life and its mechanisms into the realm of explicit calculation and made knowledge-power an agent of transformation of human life."[1] Biopolitics is, of course, less fun to study than anatamo-politics. The numerical manipulations of the body politic are and always were dusty, replete with dried up old books—the "Blue Books" of the British parliament, for example—books of ciphers. They offer no appeal to the voyeur. (John Fowles, who works Victorian prostitution statistics into *The French Lieutenant's Woman*, brought a little statistical voyeurism to the page and the screen, but he was actually faking the statistics.)

Yet these very interminable countings and tabulations beautifully illustrate the manifold ways in which knowledge is, and is not, power. They neatly display the *overt* and the *subversive* ramifications of knowledge.

Overt statistical study of populations comes to amass gigantic quantities of data that are seldom effective in controlling or altering the populations of study in the ways intended. Although data collectors often tell why they collect this or that bit of information, why fear the census taker, we might ask. The census will never be able to use what it finds out about you anyway. The fetishistic collection of overt statistical data about populations has as its motto "information and control," but it would more truly be "disinformation and mismanagement."

Yet there is a quite unintended effect of enumerating, and I call this subversive. Enumeration demands *kinds* of things or people to count. Counting is hungry for categories. Many of the categories we now use to describe people are byproducts of the needs of enumeration. What could be more inevitable than the class struggle about which Marx hectored us? Yet the social classes are not something into which a society is intrinsically sorted. On the contrary, it is the early nineteenth-century counting-bureaucracies that designed the class structure in terms of which we view society. After 1800, there was a radical shift in the ways in which people were classified in terms of occupation. You might say that came because new occupations arose with the new industrialization. Of course. But that they should be classified in the ways in which we group them was not some inevitable feature of the factories' organization. On the contrary, bureaucrats—even the very factory inspectors or people from the Board of Trade—designed easily countable classifications into which everybody had to fall—and thenceforth did.

Nothing was left untouched by the statistician. Nothing is as certain as death and taxes, and though you readily grant that tax brackets are an artifact, ways to die, you will think, are a part of nature. On the contrary. It is illegal to die, nowadays, of any cause except those prescribed in a long list drawn up by the World Health Organization (WHO).[2]

This is inherited from the nosology of William Farr, who throughout the middle of the nineteenth century ran the Registrar-General's Office for England and Wales, the dominant fact-gathering institution of its day. As for taxes: taxation was among the earliest motives for counting a people, but if individuals are not to be taxed equally by a poll tax, distinctions must be made. In the beginning, it was deemed impossible to number the people. Only the permanent fixtures of their dwellings would stay still long enough to be counted, and so hearths and windows were the basis of taxation—incidentally, a far more "progressive" mode of taxation than modern income tax, for the rich were soaked vastly more than the poor. Hearths and windows were for long not only the basis of taxation but also the most reliable estimator of the size of the population. One can tell the story of biopolitics as the transition from the counting of hearths to the counting of bodies. The subversive effect of this transition was to create new categories into which people had to fall, and so to create and to render rigid new conceptualizations of the human being.

There is a longer and a shorter story of biopolitics. The longer story gradually assumes a definite form in the mid-eighteenth century, and it continues today. Whereas Foucault's early books talked of sharp transformations, his research on sexuality directs itself not to mutation and revolution but to evolution in the longer term. There is no inconsistency in this: the world knows both revolution and evolution. At present, the dictates of fashion will encourage some of Foucault's admirers to mimic mindlessly each tum in the master's writings, so we shall expect that ruptures in the history of ideas will be less and less commonly brought to our attention, while smoother trajectories will be given to the development of ideas. So it will be salutary if some of us go on noticing mutations within the more gradual expansion of the biopolitical empire. I think there is a sharp change that I will locate in the two decades from 1820 to 1840. It is sufficiently abrupt that even the student of overt biopolitics will see it at once. For example, long ago Harald Westergaard, in a sober review of all statistical history, dubbed the period 1830 to 1850 the *era of enthusiasm* for statistical data collection.[3] Actually, I believe 1820 to 1840 is the time to look at, for by 1840 everything had been set in motion, and most of the impetus arose in the 1820s. It is that period that I call the avalanche of printed numbers. I am not disagreeing with Westergaard about dates: he is noticing the fulfillment of a fetishism for numbering, while I study its inception. Before we stop toying with dates, note that his "1850" is too neutral, too decimal a date to mark the end of the era of enthusiasm. The era of enthusiasm, as Westergaard labels it, began with the revolutions of 1830 and ended with those of 1848, which takes us back to overt biopolitics.

Statistical enthusiasm, then, bracketed the two years of revolution. It represented an overt political response by the state. Find out more about your citizens, cried the conservative enthusiasts, and you will ameliorate their conditions, diminish their restlessness, and strengthen their character. Statistics, in that period, was called moral science: its aim was information about and control of the moral tenor of the population. The motives were genuinely philanthropic, but that, as we have come to realize, means that they aimed at the preservation of the established state. As I began by saying, the overt political strategy of moral science—of bureaucratic statistics—did not work, and enthusiasm waned after the next wave of troubles, in 1848. But enthusiasm had put in place a large number of bureaucracies, which could go on functioning on their own, unenthusiastically perhaps, but with a certain inevitable growth. Equally important was the subtler, subversive influence of the new group of human categories coming from this avalanche of numbers.

During the years 1820 to 1840, the rate of increase in the printing of numbers appears to be exponential whereas the rate of increase in the printing of words was merely linear.[4] After the 1840s, both proceeded at about the same rate. However, an anecdote will better illustrate the avalanche of numbers than any tabulation of figures, for the avalanche is not merely a quantitative fact but a change in our feeling about the sort of world in which we live.

In England, as in many other industrializing countries, working men found themselves in need of mutual support so they formed small groups for purposes of self-insurance. Their weekly contributions gave some security against disease and death. These "friendly societies," as they were called, were regulated by a 1793 act of Parliament, an act that "encouraged persons to form and to establish one or more societies of good fellowship for mutual relief and maintenance of all and every member in old age, sickness and infirmity, or for the relief of widows and children of deceased members."[5]

By 1815, some 925,429 such societies had registered under the act. Evidently they were small and did not fare well. One reason was size: they were too small for any law of large numbers to be effective. They had more practical difficulties too, for they usually met in public houses; the secretary was commonly the landlord, who was not averse to seeing society funds spent on drink rather than savings. But the fundamental problem was simply that no one knew how much to charge by way of weekly contributions. The premium tables were usually signed by "petty schoolmasters and accountants whose opinion upon the probability of sickness and the duration of life, is not to be depended upon."[6] Yet the schoolmasters were hardly to be blamed: "It is remarkable that until a very few years ago no *data* were collected whereon a calculation of the average occurrence of sickness at the several ages of man could be formed with tolerable accuracy."[7] Thus runs the preamble to a report by a Select Committee of the House of Commons, first meeting in 1825 and continuing in 1827. It had other worries than actuarial rates, for it was widely supposed among the governing class that the friendly societies served as a cloak for illicit combinations (i.e., trade unions). The Select Committee, among other things, scotched that idea for lack of evidence. It was an excellent committee and treated its witnesses hard, besides trying to get information from abroad. "Have you any tables of sickness formed upon actual observation?" wrote the secretary on April 28, 1825, to Baron Delessert, in charge of the corresponding French *Sociétés des Prévoyances*. "No" was, in substance, the prompt reply. Similar negatives came from elsewhere—with one exception, Scotland.

The Scots taught a lesson that could not, for a moment, be learned: there are indeed regular laws of the occurrence of sickness. As evidence of resistance to this idea, consider the testimony of John Finlaison, chief actuary in the National Debt Office, given on March 11, 1825:

> FINLAISON: I conceive it is totally impossible to obtain authentic materials to reduce the average prevalence of sickness to any certain law.
>
> QUESTION: When you say that sickness is incapable of valuation, you mean that there are no data whereon any calculation can be made?
>
> FINLAISON: I mean that life and death are subject to a known law of nature but that sickness is not, so that the recurrence of one event may be foreseen but not so the other.[8]

Moreover, looking at "the extraordinary differences of sickness among the reports" of the friendly societies, Finlaison doubts that "any conclusion" could ever be drawn.

He was wrong. The first sign of this came from the Highland Society, a highly effective organ for agricultural reform. In 1820, it had appointed a committee to study the history of Scottish friendly societies to determine "the average rate of Sickness among mankind."[9] The Highland Society offered prizes for the best observations on the financing of friendly societies and was able to publish an important report by 1824—before the Select Committee of the House of Commons met in London. The report notes that "generally speaking, it would introduce a new idea among the member Friendly Societies, could a belief be implanted that the schemes of these institutions are in any degree susceptible of calculation."[10] Finlaison did not care to have this "new idea" implanted in *his* mind! The Select Committee presented him with the Scots report and told him to digest it and come back next day. That night he must have read expressions like this: "The Law of Sickness from 20 to 70 years of age . . . , the quantum of sickness which an individual on an average experiences each year, from 20 to 70 years of age."[11] Finlaison did not believe in such laws or such quanta. He would argue instead that "it would on the whole seem, that the army quartered in England ought to present Sickness at a minimum among mankind."[12] After all, the army contained young men in the prime of life, inducted only if healthy, with sergeants to prevent malingering and the feigning of illness. It stands to reason that the sickness rate among soldiers stationed in England (as opposed to the lethal barracks of the West Indies) must have been less than among indifferent and probably idle or dishonest Scotch workers. Yet he found that the sickness rate in the British Army stationed at home was *"more than thrice the quantum of Sickness among benefit societies"* in Scotland. Such an absurd anomaly showed simply that the Highland Society was living in fantasyland and that laws of sickness do not exist. Or so Finlaison thought.

On April 22, Finlaison was given a very bad time by the Select Committee. It thought he must be wrong and the Scots must be right. We now know that Finlaison had made a startling and important discovery: the best way for a young Englishman to sicken and die was to join the peacetime standing army and sicken in a barracks. This scandal went on for half a century. Everyone knows how Florence Nightingale joined the good fight against military incompetence. By the time she got into the act, it really was incompetence amounting to malice, for ten years after Finlaison had testified, the numbers were in and understood: everyone except the commissioned officers knew that the army barracks at home were deadly.

That is the first half of the anecdote. In 1820 began the first systematic study of sickness rates, and that only by one of the most advanced groups of intellectuals in Europe, the gentlemen farmers of the Highland Society. In 1825, the leading professional actuary was sure that there were no laws of sickness, not merely for lack of data, but because the every existence of such laws made no sense. For the second half of the anecdote, we need wait only ten years after the Select Committee had yielded up its 1827 report. In 1837, the hero was William Farr, about to be appointed compiler of abstracts to the newly founded Office of the Registrar-General of England and Wales. For forty years thereafter, he was the effective head of this Office (the titular head was a patronage appointment). The Office set the style of official statistics for many other nations.

At the start of his career, in 1837, Farr could publish papers with titles such as "On a method of determining the danger and duration of diseases at every period of their progress"[13] or "On the law of recovery and dying in Small Pox."[14] He was also able to produce tables for the various regions and occupations, providing just the laws of sickness that Finlaison had said could not possibly exist. Nor did he have to collect new data in every case.

Much of it was simply lying about in records that no one hitherto had thought of using. That is an important fact: in 1820, you might say, there was no point in analyzing available data because a law of sickness was well-nigh a conceptual impossibility. By 1837 the concept was unproblematic and laws were known. For example, the East India Company had long kept ledgers of the daily employment of its vast London workforce, showing the days in which each employee was absent for sickness. One could also extract information from other offices of government. Thus the numerous factories enquiries were not only the foodstuff of Marx's *Capital*. They could also provide, in Farr's words of 1837, items like this:

> A large volume containing a list of 2,461 labourers employed in the month of April 1823, with a statement of the number of days illness experienced by the labourers one by one, year by year, for the ten succeeding years [i.e. until 1833], also the date of every death and the date when any labourer ceased to be employed, by being superannuated and pensioned, dismissed, or by voluntarily leaving the service of the Company.[15]

Thus in the decade separating the Select Committee and the start of Farr's career, (1) sickness had become subject to statistical laws of nature, and (2) an immense amount of data not only was accumulating but was beginning to be understood. The extent of these data, even by 1840, is stupendous, a stunning contrast with 1820. Moreover, we could not have (2) without (3): many new bureaucracies were created to collect information about the people and to arrange populations into a well-organized data bank. The subversive element comes only with (4): the need to create classifications by means of which people would be counted in groups. "Checked in for work"; "Was off work for some reason or other"—those were natural-enough kinds in the factory (but not, note, in a farm, where those categories were imposed by the industrial model). The kinds of sickness for which a person might be absent were a function not only of the diseases of the time but also of the approved medical lore then current. But even a classification by illness was "natural" enough, given the "knowledge" of the day. The hardline sociological realist can tell us that it was germs, not doctors, that determined who would be classified as ill on April 1, 1833. But even such a hardliner may be moved to wonder at the sheer cascading of artifactual numbers after 1820. Let's look.

Laws of sickness furnish an anecdote that can be multiplied indefinitely. In England endless reports on the poor, the criminal, the infanticide, the mad, as well as questions of trade were churned out in the "Blue Books"—

Official Papers of Parliament, printed bound in blue figuring in most of the novels in which a member or a lord was in a country estate and had to get back to work ("He picked up his Blue Books," the novel goes, while the hero or villain is meditating). The annual reports by the Board of Trade are a morass of figures that never existed in print before.

It was not the English but the French who started the game of annual reports. Swedes had been doing it for a while, providing the "best" statistical information for late eighteenth-century Europe but always within the framework of the parish. The Lutheran minister always had some contact with any Swedish number. *Secular* numbers are a French achievement, doubtless a Bonapartic legacy. Statistical tables for Paris and the Seine department begin in 1821. Readers of Foucault will not be surprised at what goes on in these reports—admissions to, deaths in, and releases from the great hospitals and asylums, the Charentons and the Bicêtres. There are crimes against property and crimes against the person, and the first whisper of accumulated wisdom about the new category, recidivism.

On July 23, 1824, the celebrated mentalist Esquirol could address the French medical academy, asking: "Does there exist, in our day, a greater number of mad people than 40 years ago?"[16] Neither he nor anyone else knew. He was inclined to doubt rumors of *cette effrayante augmentation du nombre des aliénées qui menace la France*. The new and larger numbers were due, said this doctor-emperor, to the fact that better institutions and better diagnoses existed. When we leave his propaganda behind, we observe how few were his numbers in 1824. Within ten years there were so many numbers that every salon spoke in fear, and with full knowledge, of the *croissance effrayante* (buzzword of the day) for crime, insanity, prostitution, vagabondage, vagrancy, suicide. A chief source of the new terror was the annual set of sombre pages printed by the Ministry of Justice, begun for the year 1826. There had been ministerial minutes and abstracts of such information for some time, but they were fragmentary and essentially private. Often they were what we now call "classified" or "for your eyes only," and if they were printed, only a handful of exemplars was run off. After 1820 ministry reports were still supposed to have limited circulation, but the sheer fact of multiple printings put them in the public domain. They were ready to be reproduced or condensed in the mass circulation police gazettes and the like. Such was the crime wave consciousness that a Parisian in 1832 was much more worried than a comparable person in contemporary Manhattan. After all, the Parisian had already experienced one revolution from below and sensed another coming; revolution had to be capped, somehow or other. And it was—but not by statistics.

It is hard to say whether the 1832 cholera epidemic more accustomed Europe to numbers than the "crime wave," but between the two of them the years 1820 to 1840 turned the reading public from nonnumeracy to numeracy. There are pretty good grounds for speculation that the illiterate were, by gossip, made fairly numerate thirty years before general education made them literate. (Contrary to the usual claptrap of educational theorists, numeracy almost always, perhaps always, precedes literacy.)

Disease, madness, and the state of the threatening underworld, *les misérables,* created a morbid and fearful fascination for numbers on which the bureaucracies fed. (*Les misérables* is not merely the title of Hugo's masterpiece, but a standard set of pages in statistical reports, and when the first international scientific congresses commenced—they were statistical congresses, of course—*Les misérables* was a regular section at which learned papers would be presented.) In emphasizing the statistics of deviancy and alienation, I do not mean to underestimate the numbers by which the globe got about its business. Of course, commercial tables were also printed during this avalanche of numbers. We would be in error if we saw them simply as instruments for assessing profit and loss. Their purpose is more manifest and on the surface than the deviancy compilations, but even the businessmen look more at theft than trade. Social statistics, the stuff of moral science, everywhere meet the eye. In virtually every case it was the period 1820 to 1840 that got them moving.

Take education. The will to know the number of primary school pupils, their enrollments, their percentages of this and that, their spatial distribution: these were demanded in various French decrees of 1816, 1821, and so on. A full study was first made in 1829; the result was published in 1831 by the *Ministère de l'Instruction Publique* as *Statistique des Ecoles Primaires.*[17] Next step: correlate crime rates and education rates, using for data, if necessary, the third disciplinary element after education and deviancy: the army. Conscription rates for the army will tell you, department by department, educational levels. Match them with crime levels and you will get, by 1833, a lot of information, which is absolutely—useless.

Or consider the *Annales d'hygiène publique et de médicine légale.* Here we find a nonchance meeting of Foucault's anatomo-politics and his biopolitics. This journal is the source of the now notorious Pierre Rivière case, exhumed by Foucault and his collaborations, a fine example of disciplining the individual by the new medico-legal establishment. The *Annales* have plenty more reports like that: Foucault does not adequately convey the sense that his report is typical of the day. Yet, for all the docudrama of Pierre Rivière, the *Annales* are also a prime place for the new generation of

French social statisticians to publish their work. The *Annales*, commenced in 1829, are the culmination of occasional publications that started in 1819. So swift is the avalanche of numbers that almost immediately after formal publication began, the *Annales* would requires as many as six volumes a year to encompass their news.

These examples are French and British. Belgian ones would be even more compelling: one figure is dominant, Adolphe Quetelet, master statistical informant and creator of national bureaucracies and international congresses of the nineteenth century. But a phenomenon should not be illustrated by such a unique and phenomenal man who created a national discipline: better to take France and England where the greatest "hero" is Farr, who was never even promoted past the grade of compiler of abstracts.

Sweden, of course, is anomalous, for the state gathered demographic data long before anyone else. Yet despite the unequalled high quality of its records and even its analyses of data, it did not quite know what to do with this fortuitous diamond that is the hallmark of the modern state. Sweden is a parable. Here lie vast arrays of numbers garnered by icy pastors simply because the numbers were good in themselves.

As for the rest of Europe, the German states and principalities were a little slow to start but later set the pace. Petersburg was never more than a couple of years behind. The Italian countries took up statistics later but engaged in the greatest parody of modern Europe ever conceived. The Italians took English and French statistical know-how, combined it with Kantian *Anthropologie*, and created a criminal anthropology of which Sherlock Holmes addicts have heard a distant echo. Remember how Holmes at once identifies an ear in a box? I dare say that Conan Doyle had been looking at the pages of ears in Bertillon's *Bulletin Signaletique*.[18] The echo extended beyond fiction and pages of ears, for it was heard by Francis Galton. He invented theory of correlation and regression at the same time that he became the originator-champion of eugenics and invented—single-handedly, one might say—the practice of police fingerprinting.

The period from 1820 to 1840 is only an episode in biopolitics, a mutation within Foucault's evolution. I emphasize a specific avalanche of printed numbers with a strong element of the surveying of deviancy. It stands within a long-standing concern with war, taxation, and profit. In the early days, counting people was hard because no one knew how to do it—and anyway, there were no officers to do the counting if they did know what to do.

Characteristically the first complete census in the modern Western world took place in Canada in February and March 1666. (There had been partial censuses in some of the French colonies as early as 1605.) The Canadian census classified the 3,215 inhabitants according to sex, marital status, age, and profession. Note the time of year. There was a captive population—try leaving Quebec City on your snowshoes in early March. It would not have been possible to count, exactly, the inhabitants of Lyon or York that month. Why did they get counted at all? The object was clear, the same object as all biopolitics until the end of the nineteenth century, and the object of the Chinese census in the 1980s. How big is the population, and how might we influence its size? For the Canadians, the dominant fear was underpopulation in the face of English expansionism. In general, a census would tell the sovereign how to tax the subjects and how many would be available for war. The immediate effect of the Canadian census was a fantastic tax incentive for large families. Like every tax incentive in the subsequent 317 years, it did not have the desired effect. The intended, overt biopolitics never worked—*and that is the rule.* But the subversive biopolitics set the stage of categorization in which we still live.

The census became a hallmark of the enlightenment, but always for what at first sight seem good practical purposes. A bizarre debate occurred during the eighteenth century on the falling populations of the European countries. To lower the population was to damage efficacy in trade and war. The French reenacted this in the late nineteenth century. One of the chief products was the Durkheimian wing of sociology, a response to "pathologically" falling birthrates in France. (Need I add that most of the "data" on which these scare stories were founded were mistaken? Need I add that the governmental measures taken had no intended effect?) Malthus is the population analyst most celebrated of after-dinner speech mongers, but he is merely the most memorable, and perhaps the most charming, of ten thousand public voices. The first enlightenment "state" is, of course, the United States of America, whose very name was invented by Richard Price, publisher of the work of Thomas Bayes that we now call Bayesian statistics, and also publisher of the only tolerable actuarial tables available until well into the middle of the nineteenth century. The very name of the first enlightenment state was invented by a "statistician." I mention this not because he is such a famous figure, which he is not. I mention it because he is typical.

The idea of counting the people was so well entrenched by the time of Richard Price, and before the heyday of Malthus, that it was written into the American constitution in 1778. Ever after it would be *unconstitutional* not to count the American people. Was this some trifling bureaucratic aside? Not at all. It is written into article 1, section 2. If you were simple-minded, as I am, you could say that the second most important feature of the American dream was that the people should be counted.

The overt purpose of the constitutional counting was plain enough. It was to determine the size of congressional districts, so as to give equal representation to all free persons. People who have not recently read their constitution sometimes think that slaves did not count. O the contrary, all the other people, excluding "Indians not taxed," were figured in at the rate of 3/5 of a person. What we call the enlightenment was nothing if not exact.

Only four questions were asked in the 1790 US census. Americans were to lag behind Europeans for some time. No avalanche of printed numbers took place in America in 1820, and indeed the number of questions had not risen much above four even by 1870. But once the American bureaucrats caught on, they typically made "avalanche of printed numbers" a mild understatement. Should we speak rather of a volcanic eruption, a veritable St. Helens that covered America by 1880? In 1870, 156 questions in all were asked. In 1880, the number of questions was 13,010. The 1890 census asked only a handful of questions more but did something more important. It developed techniques for mechanical manipulation of data. A clever man saw that the "punch cards" invented by J. M. Jarquard in France, and later used in Lancashire cotton mills for the weaving of cloth, could also be used for storing and sorting other kinds of "information." So the 1890 census gave us the punch card, chief peripheral tool of early computing devices. By 1911 this census taker and statistician, Herman Hollerith, helped form the Computing Tabulating Recording Company. It later changed its name to IBM.

Nor is one to think that mechanical computation is an early twentieth-century production of counting the people. Charles Babbage is usually given credit for having invented the idea of a logic-processor, which would vastly speed up handling numbers. (Yes, he was one of the chief witnesses before the Select Committee of the House of Commons on the Friendly Societies.) Babbage's analytic engine cost a fortune—both his own and that of the British government—but for all the vast research and development money that went into it after 1832, it never worked. We owe the first useful and well-used computer to the Swedish inventor, Scheutz. Scheutz's original machine was widely exhibited in Europe and did print out logarithm tables. A Mr. Rathbone bought it for the Dudley Observatory in Albany, New York, where it caused endless trouble and never was a cost-effective tool for astronomy. But while it was on show at Somerset House in London, William Farr, of the Registrar-General's Office, had Bryan Donkin pirate the machine by making a copy. This was itself an amazing feat as the object had 4,320 pieces, 2,054 screws, 364 chains, and so forth.

> Here we had to do with the second calculating machine as it came from the designs of the constructor and the workshop of the engineer . . . the idea had been as beautifully embodied in metal by Mr. Bryan Donkin as it has been conceived by the genius of its inventors; but it was untried. So its work had to be watched with anxiety, and its arithmetical music had to be elicited by frequent tuning and skillful handling, in the quiet most congenial to such productions.[19]

This machine printed out numbers, and one of its great virtues was that it not only computed but also printed the results of computations and so avoided the errors of transcription common to clerks. Since it was real state-of-the-art technology, it is not surprising to find that "some of the pieces get deranged and print errors" or that "it approaches infallibility in certain respects but it is not infallible, except in very skillful hands." Farr used this machine to compute and print out the new annuity tables that were based on the census of 1851, and, of course, he worked in all his other amazing amounts of data. Annuity tables may be thought of as a rather minor aspect of numbering, but in fact the selling of annuities was a chief source of capital for the nineteenth-century state. During the Napoleonic wars, the British government had sold annuities using rates intended by Richard Price for the selling of insurance. What is good for the insurance company is bad for the seller of annuities, so not surprisingly Price's tables served in the 1820s to bankrupt successive British governments. But following the avalanche of printed numbers came the census of 1851 and a computing machine to devise annuity tables, which made the government safe in at least that respect. It is a happy parable that the avalanche of printed numbers ends with the first real use of a computing machine to provide printouts.

I have said that there is a longer- and a shorter-term account of biopolitics. The longer term has the traditional concern of taxation, the formation of armies, and the size of populations. I do not think these longer-term concerns had such subversive effects as occurred during the avalanche of printed numbers, from 1820 to 1840 or thereabouts. The reason is that the long-term countings had rather little to do with the formation of new categories, new ways of conceiving the person. Evolutionary biopolitics concerns itself with long-term populations, but it is revolutionary biopower that brought into being new concepts of the person. As Frege taught us, you cannot just print numbers. You must print numbers of objects falling under some concept or other. The avalanche of printed numbers brought with it a moraine of new concepts.

The prime "evolutionary" questions of a census have always been age, sex, marital status. We can use even Thomas Jefferson to introduce the new "revolutionary" categories. In 1800 he unsuccessfully urged that the US census should commence a breakdown by social class. His idea was that people fall into the following groups: first of all, educated people (lawyers, doctors, artists, writers, teachers, and clergymen); then businessmen; sailors; artisans; farm laborers; other laborers; domestic servants; the poor; and finally the rest, who live on private incomes.

That categorization, although a trifle odd, is more familiar than one thirty years earlier, in prerevolutionary France, which divided the people into the "productive class" and the "sterile but useful class." Farmers, fishermen, and miners make it into the former class, while merchants, carriers, artisans, and actors are sterile. Those words "productive" and "sterile" will not be lost on the readers of Foucault's biopolitics.

Of course, the fine categorization of people by what they "do" takes a while. By 1870 the French census would find you out in terms that Zola could only copy, never improve on. Do you run a bathhouse? Or perhaps you fall under the category "*acrobates, charlatans, monteurs de bêtes et curiosités*"? In France in 1872, there were 1,652 such men and 785 women who, in turn, supported as family members 1,090 males and 1,804 females. These acrobats had 168 male domestic servants and 87 female ones. That may make them seem a trifle underprivileged, but consider the fact that the 351,210 male and 197,071 female permanent farm workers had between them exactly 6 male domestics and 5 female ones. At any rate, for acrobats the *nombre d'individus que chaque profession fait vivre directement ou indirectement* was 5,586.

The idea of classifying people in minute divisions according to their "work" is already an instructive transformation that occurs in the nineteenth century. There is also a certain sort of back-bracketing, that is, a putting together of what a previous society had kept apart. Thus there is the group of *hommes de peine*, the people who do all the miserable work that no one else will do, such as cleaning toilets. Surprisingly these are counted together with two formerly honorable professions: embalming and grave digging (as opposed to *pompes funèbres*, a separate category)—*ensevelisseurs* and *fossoyeurs*. There must be a Foucaultian story to tell even here about this fugue of meldings and dividings in how to count the people who literally handle bodies. The cadaver industry seems to have provided far more support for the population (208,701) than the *Chefs et directeurs* in the category of transport, credits, banks, and commission work, despite the fact that embalmers (as opposed to funeral directors) had almost no servants and chiefs in transport, and bankers had servants roughly one on one (but only one on one, somewhat to my surprise). These statistics are endlessly amusing and often make one feel that progress is not yet around the corner. It looks as though, in 1872, 1 in every 11 *patrons* in heavy industry was a woman (47,368 to 4,216). Almost as many women as men were day-by-day employees in what the census calls *attachés aux usines et fabriquants*. Where did we get this fantasy that only in recent times have women joined the workforce?

It is not the categories of workers that I wish to emphasize, but rather the very idea of categorizing them—albeit eccentrically—according to their role in the workforce. But working is not the only basis for counting. I have mentioned the nosology of William Farr. He designed an increasingly countable list of diseases. No one would be allowed to die without a doctor signing a death certificate, but the "causes of death" had to be uniform and easily applicable so that the certificates could be enumerated to discover who was dying of what, in what numbers. Thus the aim was a standardized list of diseases. This nosology is ancestor of the international list of diseases published by the WHO and revised every few years. These categories are as responsible to the *need to count uniformly* as they are to any interest in "correct" diagnosis.

Should anyone suppose that I am telling out-of-date anecdotes about things that cannot happen today, consider DSM-III. This is the third in a sequence of classification principles issued by the American Psychiatric Association (APA). The acronym stands for *The Diagnostic and Statistical Manual of Mental Disorders* III. It is only newly out.[20] Its predecessor, DSM-II, appeared in 1968. Its aim is to produce a nationally standardized classification of mental disorders, not in terms of their causes but in terms of the ability of practitioners from different ideologies to agree on the classifications into which patients fall.

The APA is currently pressing the WHO to adopt DSM-III. Then everyone in the world will have to be classified in the same way. It is illegal to die in any way except those listed in the WHO nosology, and if the APA has its way, it will become illegal to go mad except according to the classifications of DSM-III. Then we will be able to count up all the mad people in the world, suitably grouped into categories designed for ease of classifications by doctors.

I think of DSM-III as the lineal descendant of that influential journal commenced in 1829, the *Annales d'hygiène publique et de médicine légale*. A diagnostic and statistical manual was just what the contributors wanted. They presented endless schemas of classification, and the people entering and leaving the French asylums were all properly checked off according to the weird array of complaints that they were judged to have. There are two vectors in the invention of categories of complaints. In this chapter I have emphasized one: the fetishism for counting, which brings with it the need for easily applied categories in terms of which to count. The other vector, of course, is the theoretical and practical reasoning of individual professionals—doctors, say—when confronted with the bodies of individual patients. The present chapter is a footnote to Foucault's remarks about biopower. A companion piece to the present one is a footnote discussing the other vector, anatamo-politics.[21] It shows how the medical profession invented split personalities in 1875.

Notes

This essay was initially published as "Biopower and the Avalanche of Printed Numbers," Humanities in Society *5 (1982): 279–95.*

1. Michel Foucault, *The History of Sexuality*, vol. 1, *An Introduction* (New York: Random House, 1978), 139, 143.
2. This phenomenon was drawn to my attention by Dr. Anne Fagot, in a draft of her dissertation "L'Explication causale de la mort" (1978). It will be published in English by Reidel in 1983. [Anne Fagot-Largeault's dissertation was published in French as *Les causes de la mort, histoire naturelle et facteurs de risque* (Paris: Vrin, 1989). Unfortunately, the book is still not published in English.—Editors' note.]
3. Harald Westergaard, *Contributions to the History of Statistics* (London: P. S. King, 1932).
4. I am here counting something like what philosophers call types rather than tokens. Thus the number of paupers in Devon in 1831, printed in a Blue Book issued by the parliament, counts as "one number" rather the number of tokens that appear, in this case, the number of Blue Books containing that number that were printed. Likewise the first word in *Great Expectations* counts as one printed word, even though there were innumerable printings of that first word as successive editions rolled off the press.
5. *Report of the Select Committee to Consider the Laws Respecting Friendly Societies*, July 5, 1825, London, 4.
6. Ibid., 13.
7. Ibid., 14.
8. Ibid., 45.
9. *Report of the Committee of the Highland Society of Scotland Appointed to Inquire into the State of the Friendly Societies in Prize of Essays and Transactions of the Highland Society of Scotland* 6 (1824): 271–560.
10. Ibid., 312.
11. Ibid., 420.

12. *Report of the Select Committee,* 139.
13. *British Annals of Medicine, Pharmacy, Vital Statistics and General Science* 1 (Jan.–June 1837): 72–79.
14. *British Annals of Medicine, Pharmacy, Vital Statistics and General Science* 2 (July–Dec. 1837): 134–43.
15. W. Farr, "Vital Statistics," in *A Statistical Account of the British Empire,* by J. R. Mc-Cullough (London, 1837).
16. J. E. D. Esquirol, "Address to Académie Royale de Médicine," *Annales,* July 23, 1824, 32–50.
17. Jacques Ozeuf, "Les Statistiques de l'enseignement primaire au XIXe siècle," *Pour une histoire de la statistique,* vol. 1, Institut National de la Statistique et des Etudes Economiques, Vaucresson, n.d. (report of a conference of 1976).
18. Alphone Bertillon, *Identification anthropométrique: Instructions, Signalétiques* (Melun, 1885). For the ear photographs, see the *Album* for this volume (Melun, 1893).
19. W. Farr, "On Scheutz's Calculating Machine," *Philosophical Transactions of the Royal Society* 140 (1859).
20. This essay was written just after DSM-III appeared in 1980. For my recent views about what happened since, see my review of DSM-V issued on May 18, 2013, in the *London Review of Books.*
21. Ian Hacking, "Multiple Personality and Its Hosts," *History of the Human Sciences* 5, no. 2 (1992): 3–31.

FOUR

Biopolitics and the Concept of Life

CATHERINE MILLS

In contrast to its relative neglect in the initial Anglo-American reception of the first volume of Michel Foucault's *History of Sexuality*, in recent years, "biopolitics" has come to appear as a key term for understanding the contemporary era, with a range of competing theories and approaches on offer. Unfortunately, the diversification and proliferation of biopolitics scholarship has often resulted in conceptual confusion rather than clarification; this has prompted some critics to reject the notion of biopolitics altogether.[1] The aim of this chapter is to make some headway toward the clarification requisite for retaining the concept, focusing in particular on the "bio" of biopolitics. As Eugene Thacker deftly summarizes, "in an era of biopolitics, it seems that life is everywhere at stake and yet it is nowhere the same."[2] Thus, one could point to the proliferation of notions of life, which exploit the manifold senses of the term—*nuda vita* or bare life,[3] creaturely life,[4] and surplus life,[5] to name but a few. However, while the equivocations of the concept of life have undoubtedly been productive, the referent of the "bio" in the term "biopolitics" largely remains undisclosed; it is the dark background on which the machinations of modern politics play out. Supposing for the sake of argument that "bio" and "life" can be used interchangeably, this chapter pauses in the midst of this proliferation, to ask what "life" means, what the prefix "bio" does, and what it is that one is committed to doing in analyzing life and its scientific, economic, and political mobilizations and manifestations.

I focus on the prefix "bio," which identifies biopolitics as a specific political rationale and form of organization, to consider how several influential approaches to biopolitics construe the relationship between life and politics. In the first section of the chapter, I discuss the contributions to a philosophy of life suggested by Giorgio Agamben in his work on biopolitics, especially the idea of an absolutely immanent "happy life," or "form-of-life." I show that Agamben is effectively unable to theorize biological life in anything other than a negative relation with biopolitics. Following this, I consider the empirically focused approach to biopower developed by Paul Rabinow and Nikolas Rose. While obviously valuable for its insights into the workings of contemporary bioscience and medicine, I suggest that the empirical strength of this work is also its weakness, insofar as it also eschews articulation of a conception of life independently of the discourses of biomedicine. In the next section, I turn to Roberto Esposito's work, in which he draws on Georges Canguilhem to initiate a theorization of life in which the norm is an immanent impulse of life. Finally, I turn to the relation of life and norms in Foucault's discussions of biopower in light of his essay on Canguilhem, in which he emphasizes the productive capacity for error internal to life. It remains a moot point whether errancy would have provided Foucault with a new way of thinking about life, as Agamben suggests. But in any case, this shift in perspective highlights the limits of biopower and the *reactivity* of the biopolitical state in relation to life. It allows for an understanding of life that is not outside or beyond biopower, but is nevertheless engaged in a productive or positive relation with it.

Agamben: Bare Life and Form-of-Life

Agamben's *Homo Sacer* project has done a great deal to focus attention on biopolitics and has contributed much to contemporary reflection on the concept of life. Agamben himself suggests a number of different formulations for thinking about life, most notably the category of "bare life," which he sees as the principal object of biopolitics, and its opposite, the postbiopolitical notion of "happy life" or "form-of-life." What is especially interesting to me in the *Homo Sacer* project are the ways that Agamben leverages the terms of *bios* and *zoē* to yield these notions, which map onto the problems of biopolitical domination and subsequent liberation (to momentarily use terms that are not part of Agamben's own vocabulary). In brief, bare life emerges from the fracturing of *bios* from *zoē* that takes place in the constitution of the state and the exceptional apparatus of law, while form-of-life emerges at the site of the "indetermination of life and law," best seen in Franciscan monasticism. Within this frame, then, the "bio" of biopolitics—the life conjoined to politics—appears as bare life, and life "beyond" biopolitical capture is a particular manifestation of *bios* in its indetermination with rule.

Agamben sets out his account of the way in which the life of political subjects is captured within the political sphere in the book *Homo Sacer*. He argues that the qualitative distinction made by Aristotle in his treatise on the formation of the state between biological life (*zoē*) and specific ways of life (*bios*), including political life, effectively excluded natural life from the polis and relegated it entirely to the domestic sphere, as "merely reproductive life."[6] Of this account of the origins of the city-state, Agamben claims that natural life thereby constituted the "inclusive exclusion" that provides the foundation for the political sphere, in that it is excluded from politics, while included by virtue of that exclusion. The first implication to take from this is that biopolitics is intimately linked to the structure of sovereignty, to such an extent that Western politics has always been biopolitical. Indeed, Agamben postulates that the original task of the sovereign was the production of the biopolitical body. The central task of *Homo Sacer*, then, is to elaborate on the logic of the production of the biopolitical body, one key aspect of which is the doctrine of the sacredness of life.

Without going into detail, the upshot of Agamben's discussion of sovereignty and sacrality is that life exposed to death is the originary political element and the figure of *homo sacer* expresses the originary political relation, that is, the relation of abandonment. The idea is that in being abandoned by (and to) the law, *homo sacer* is exposed absolutely to violence; *homo sacer* is simultaneously "free, open to all" and the object without protection of violence. In this role, *homo sacer* appears as the privileged figure of the object of biopolitics, that is, of bare life. According to Agamben, bare life emerges from within the distinction that he derives from Aristotle: it is neither *bios* nor *zoē* but rather the politicized form of natural life, where this politicization occurs through the exposure to violence and death in abandonment. He thus defines "bare life" as "life that is irremediably exposed to death,"[7] and states "not simple natural life, but life exposed to death (bare life or sacred life) is the originary political element."[8] It is important to keep sight of this point that bare life is defined explicitly as life determined by its own negation, that is, life exposed to death. For forgetting it effectively neutralizes the logic of the inclusive exclusion, and moreover, undermines the motivation for realizing a form-of-life beyond the biopolitical machine that is Western politics. I will return to this point in a moment; but first I will consider the major task that Agamben has set for himself—the formulation of an understanding of life that is not condemned to repeat the disasters of biopolitical violence.

The formulation of a new concept of life has been an important thread of Agamben's work for some time, but it is given particular impetus and significance in the *Homo Sacer* project. Agamben makes clear his belief in the political necessity of such a conception of life in *Means without End*, where he writes:

> The "happy life" on which political philosophy should be founded thus cannot be either the naked life that sovereignty posits as a presupposition so as to turn it into its own subject or the impenetrable extraneity of science and of modern biopolitics that everybody today tries in vain to sacralize. This "happy life" should be, rather, an absolutely profane "sufficient life" that has reached the perfection of its own power and of its own communicability—a life over which sovereignty and right no longer have any hold.[9]

In this formulation, Agamben augurs a politico-philosophical redefinition of a life that is no longer founded on the biopolitical separation of natural life and political life, and is instead a life of absolute immanence.

In the essay "Absolute Immanence," Agamben notes that both Foucault and Gilles Deleuze discuss the concept of life in their last essays published prior to their deaths—entitled "Life: Experience and Science"[10] and "Immanence: A Life . . ."[11] respectively. This coincidence, he suggests, bequeaths to future philosophy the concept of life as a central subject, inquiries into which must start from the conjunction of Foucault and Deleuze's essays. While Foucault's essay, which is on the philosophy of life developed by Canguilhem, aims at "a different way of approaching the notion of life" through error, Deleuze seeks "a life that does not consist only in its confrontation with death and an immanence that does not once again produce transcendence."[12] Insofar as these essays provide a "corrective and a stumbling block" for each other, they clear the ground for a genealogy that will, according to Agamben, "demonstrate that 'life' is not a medical and scientific notion but a philosophical, political and theological concept."[13] Such an inquiry would reveal the archaism and irrelevance of the various qualifications of life: animal life and organic life, biological life and contemplative life, and so on, and give way to a new conception of life that recognizes beatitude—blessedness or happiness—as the "movement of absolute immanence."[14] It is toward such a conception of life that Agamben's own philosophy aims: in proposing a typology of modern philosophy in terms of the thinking of transcendence (Kant, Husserl, Levinas, and Derrida via Heidegger) and immanence (Spinoza, Nietzsche, Deleuze, and Foucault via Heidegger), Agamben evidently positions himself as the philosophical heir of Deleuze and Foucault.

This is confirmed in his interpretation of Deleuze's notion of an absolutely immanent nonindividuated life, which is the focus of Agamben's interest in "Absolute Immanence." Deleuze develops this idea through reference to Charles Dickens's novel, *Our Mutual Friend*, in which the character Riderhood wavers on the point of living and dying and compels unprecedented fascination and sympathy in witnesses to his predicament. Deleuze uses this story to develop a conception of a nonsubjective or "impersonal" life, which is composed of "virtualities, events, singularities,"[15] and which may be manifest in but is not reducible to an individual. Commenting further on the Dickens story, Agamben emphasizes the way that this "separable" life exists in the indeterminacy between states of being such as life and death, which he describes as a "happy netherworld" that is "neither in this world nor in the next, but between the two."[16] He goes on to cast the Deleuzian notion of a life of absolute immanence within the conceptual framework of biopolitics proposed in *Homo Sacer*, suggesting that impersonal life risks coinciding with the "bare biological life" of biopolitics. In Agamben's interpretation, Deleuze escapes this apparent declension by virtue of two related factors: first, the insistence on the "absolute immanence" of impersonal life, such that "*a life* . . . is pure potentiality that preserves without acting,"[17] and second, the connection between potentiality and beatitude, whereby the former is immediately blessed in lacking nothing. This means, "*beatitudo* is the movement of absolute immanence."[18] The value, then, of reading the essays by Foucault and Deleuze together is that this complicates both, such that "the element that marks subjection to biopower" must be discerned "in the very paradigm of possible beatitude."[19]

While Agamben indicates his general philosophical orientation to the concept of life in this essay, his more recent work translated as *The Highest Poverty* gives the clearest indication of the potential that he sees in the idea of form-of-life in regards to biopolitics. Agamben casts his study of monasticism as an attempt to construct a form-of-life, understood as "a life that is linked so closely to its form that it proves to be inseparable from it."[20] This requires an investigation of the relationship between life and rule, not to resolve them into a perfect unity, but to bring forth a "third thing," that is, form-of-life, from their "reciprocal tension."[21] This "third thing" comes closest to its realization in the Franciscan emphasis on poverty and the use without rights, especially of ownership, that this entails. However, while Franciscanism can be seen as an "attempt to realize a human life and practice absolutely outside the determinations of the law,"[22] it nevertheless fails to fully realize form-of-life. This is because ultimately the Franciscans were unable to develop a theory of use that was independent of juridical concepts, which could be put into relation with the monks' form of life, situated outside the law. Consequently, the "highest poverty" that was to define Franciscan form-of-life remained harnessed negatively to law. It is not necessary to consider this account in more detail to see that what was at issue in monasticism in Agamben's view is the transformation of a form of life—*bios*—into form-of-life.

If *bios* is thus positioned as the site for political transformation beyond biopolitics, what becomes of *zoē* in Agamben's account? To respond to this question, we must return to bare life. Interestingly, while Agamben appears to define bare life as "life exposed to death," he does not maintain a strict distinction between bare life and natural or biological life and often conflates or conjoins them in variable ways. This is evident, for instance, in his use of the phrase "bare natural life" in his discussion of Hannah Arendt in *Homo Sacer*, throughout which he also attempts to correlate "bare natural life" and birth, or what he terms "the pure fact of birth."[23] While this conceptual confusion has occasionally been noted,[24] its theoretical implications have largely been ignored in the subsequent enthusiasm for the notion of bare life. These are, however, worth considering since, as Jacques Derrida[25] points out, the apparent distinction between *bios* and *zoē* that yields the notion of bare life is itself less clear or strict than Agamben asserts, and further, that bare life *cannot* be clearly distinguished from natural life.

Throughout his various discussions of biopolitics and the concept of life, there are times when Agamben wishes to grasp at something like the "simple" fact of life, of something's being alive rather than dead or animate rather than inanimate. Presumably, the name for this fact of living that he wishes to mobilize is the term *zoē*, derived from Aristotle. Around this use of *zoē*, however, arises a panoply of terms that are supposed to be equivalent but that are not necessarily so: "simple natural life," "biological life as such," "simple living bodies," to name but a few. One point at issue in attempting to comprehend these ideas is the notion of the "simple"—that is, that to which nothing is added, that is single and indivisible.[26] But this confuses rather than clarifies matters since Agamben also notes that "in the syntagm 'bare life,' 'bare' corresponds to the Greek *haplōs*, the term by which first philosophy defines pure Being."[27] This allows Agamben to set up an association between politics and metaphysics, insofar as both "find their foundation and sense in [bare life and pure Being] and them alone."[28] But it also reveals the depth of confusion between natural life and bare life, for now bare life appears only as simple life, that is, as *zoē*. This suggests that the weight of differentiation should in fact fall on the "natural." Unfortunately, Agamben provides no rendition of what "natural" might mean, leaving it to be defined negatively as the nonpolitical, or even noncultural. But if natural life is defined negatively, then it again subsides into bare life.

One implication of this apparent confusion is that natural life cannot be thought apart from its politicization. In other words, life is entirely encompassed within the political, such that biopolitics is just politics, and politics has no limit. This in turn would mean that "biological life as such" provides no opposition to the reach of politics, and this is confirmed in Agamben's closing reflections in *Homo Sacer*, when he rejects Foucault's gesture toward a "new economy of bodies and pleasures" in opposition to the biopolitical management of life. Thus, he writes, "the 'body' is always already a biopolitical body and bare life, and nothing in it or the economy of its pleasure seems to allow us to find solid ground on which to oppose the demands of sovereign power."[29] Given this rejection of the body, and by dint of that, of biological life, as a grounds for opposition to the totalizing impulse of biopolitics, it is no surprise that the positive conception of life that Agamben goes on to propose is founded on a concept that has little obvious biological content, namely, that of beatitude.

Provocative as his affirmation of a life of beatitude is, though, to what extent does it help to articulate the "bio" of modern biopolitics? Perhaps it is misguided to demand this kind of clarification from Agamben since it is at a kind of (non-Derridean) deconstruction of the distinctions between nature and culture, or biology and politics, that his thought aims. Nevertheless, one can ask how successful such a deconstruction can be when one side of the opposition is left almost entirely obscure. The problem is that Agamben's sweeping claim that life is neither a biological nor medical concept forestalls engagement with the specificity of the ways in which biology and medicine have and do contribute to contemporary understandings of life. Furthermore, this leaves him unable to articulate the limits and failures of the machinations of contemporary biopolitics. However, biopolitics is neither all-encompassing, nor entirely efficacious, and a critique of biopolitics could be well served by identifying the points at which the contemporary technologies and rationalities of power reach their limits. In this regard, one might seek to discover "a certain fragility . . . in the very bedrock of existence . . . in those aspects of it that are most familiar, most solid and most intimately related to our bodies and to our everyday behaviour."[30]

Rose and Rabinow: Biopower Today

The contemporary approach to biopower that remains most true to the spirit of Foucault is probably that jointly and separately developed by Paul

Rabinow and Nikolas Rose. Both scholars of various aspects of contemporary medicine and life sciences, Rabinow and Rose propose an approach to biopower that is empirically fastidious, conceptually restrained and analytically focused on the diagnosis of the "near future," in the sense given to this term by Deleuze, that is, "what we are in the process of becoming."[31] They argue that the concept of biopower names a "plane of actuality" that must minimally include

> a form of truth discourse about living beings and an array of authorities considered competent to speak that truth; strategies for intervention upon collective existence in the name of life and health; and modes of subjectification, in which individuals can be brought to work on themselves, under certain forms of authority, in relation to truth discourses, by means of practices of the self, in the name of individual or collective life or health.[32]

For Rose and Rabinow, the accounts of biopower developed by theoreticians such as Agamben fall into an abyss of overarching theories that "describe everything but analyze nothing."[33] Additionally, they take issue with two substantive points in these theories, namely, the construal of modern power in terms of sovereignty and the related emphasis on death.

Of the first of these, they reject Agamben's "totalization" of sovereign power to everyday life. They argue that while it may illuminate the absolutist politics of Nazism and Stalinism, a more "nuanced" understanding of power, including sovereign power, is required to "analyse contemporary rationalities and technologies of biopolitics."[34] In rejecting the centrality of sovereignty to the operations of biopolitics, Rabinow and Rose also challenge the association that Agamben makes between biopolitics and *thanatopolitics*, whereby the supposed politics of life is in actuality a politics of death. For them, biopower is not about "making die" so much as it is about "making live"; they write, "central to the configuration of contemporary biopower are all those endeavours that have life, not death, as their *telos*."[35] The consequence of these two points is that Rabinow and Rose oppose any association of contemporary biopower with the Holocaust, thus indicating their distance from Agamben in particular, for whom Hitler's Germany and concentration camps more generally are the *nomos* of modern biopolitics. This opposition extends to disclaiming that contemporary developments in biopower across the themes of race, reproduction, and genomics can be understood as forms of eugenics.

Setting aside the question of whether the conclusions that Rabinow and Rose draw about *thanatopolitics* and eugenics are right, the significance of their approach is in large part methodological, insofar as they insist on empirical veracity in analyzing the contemporary. The strength of this approach lies in it allowing for recognition of the variability in the mechanisms and strategies that may be operationalized in biopower, and Rose and Rabinow both provide exemplary studies of biosciences that are sensitive to the biological realities of life today. Nevertheless, this approach also has its weaknesses, particularly in terms of elaborating the conception of life that may be informing it. Despite the analytic focus on "making live," at its strictest, the empiricism urged by Rabinow and Rose would appear to forestall any independent attempt to conceptualize life. Consider, for example, Rose's influential and widely read text, *The Politics of Life Itself.*[36] While illuminating in its discussions of advances in biomedicine, at no point does Rose give any account of "life itself," preferring instead to "explore the philosophy of life that is embodied in the ways of thinking and acting espoused by the participants in [the] politics of life itself."[37] Thus, what is under investigation here is not *life*, but *what is said about life.* Within this, the term "life" tends to operate as a signifier without referent, almost infinitely encompassing and divisible, with the consequence that "life itself" is whatever is said about it (as long as the speaker is sufficiently authoritative), and the operations by which life is managed and directed are seen as almost inevitably efficacious.[38]

To summarize, the failure of their approach appears to be the opposite of Agamben's: while he is unable to engage with the specificity of the conception of life proposed by bioscience and medicine, the latter are constrained to repeat it. But the *effect* of these shortcomings is the same: neither works to elucidate the *limits and failures* of biopower. The danger of this is that it elides the ways in which the phenomena of life might exceed and escape the ways in which people think about them, as well as the practices that strive to contain and improve them. Life becomes nothing more than an epiphenomenon of the state, entirely enclosed within the terms of the discourses about it.

Esposito: Immunology and the Normal

If Agamben and Rabinow and Rose can be seen as two poles of biopolitical analysis today, then Esposito's increasingly influential work can usefully be characterized as operating between these poles, sharing some features with each of them while rejecting others. Thus, with Rabinow and Rose, he shares a concern with empirical veracity, while also going beyond this to theorize the unifying rationale of biopolitics, in a manner not unlike

Agamben, though the terms of his analysis are different. Further, in regards to the role of Nazism and the Holocaust in biopolitics, Esposito takes a position between Rabinow and Rose's rejection of an alliance between Nazi eugenics and contemporary technologies of biopolitics, and Agamben's casting of the Holocaust and the camp as the "nomos of the modern." He argues that biopolitics "breaks off into two antithetical but not unrelated forms: Nazism, the biopolitics of the state, and liberalism, the biopolitics of the individual."[39] In this way, Esposito may manage to avoid some of the problems identified in the approaches of Rabinow and Rose, and Agamben. And indeed, his work is explicitly directed toward developing a conception of life that allows for a dimension that is not entirely captured within biopolitics. Thus, in his analysis of the "immunitary paradigm" of modern biopolitics, he has sought to push the logic of immunization beyond itself, to generate a positive conception of life.

The central concept in Esposito's analysis of modern biopolitics is that of immunization. He argues that this concept provides the nexus between the two poles of biopolitics—that is, politics and life, not least because the term "immunity" itself has both a biological and political valence. It refers to both a "natural or induced refractoriness on the part of a living organism when faced with a given disease" and a "temporary or definitive exemption on the part of the subject with regard to concrete obligations or responsibilities that under normal circumstances would bind one to others."[40] Beyond this semantic value, the notion of immunization brings to the fore the way in which biopolitics consists in the protection of life through the contradiction of it; the protection of life requires a dose of the evil that threatens it, precisely to generate the protection required against that evil. In this sense, immunity is a kind of negative protection: "it can prolong life, but only by continuously giving it a taste of death."[41] For Esposito, the critical question is whether life can be preserved in a way that does not pursue this negative protection. To this, he rejects "an immediately affirmative response . . . that situates the development of life in a horizon that is radically external" to the paradigm of immunization. Instead, he argues that the point is to "[deepen] the internal contradiction" of the logic of immunization.[42]

Speaking more directly to the characteristics of biopolitics, Esposito undertakes a similar task in *Bios*, where he follows the logic prescribed by the *thanatopolitics* of Nazism to push it to yield a positive biopolitics. In *Bios*, he shows that the German Nazi regime relied on the expertise of biomedicine to justify and carry out its murderous plans in the camps and institutions such as T-4. The Nazi operations, he argues, were effectively a "biocracy" in which the legitimacy of the biomedical sciences gave strength to the political powers, and in return, the regime provided the bodies required for biomedical experimentation. From this characterization of the negative biopolitical core of Nazism, Esposito seeks an affirmative biopolitics uncontaminated by the *thanatopolitics* that emerges in modern politics. He outlines three axes along which what he calls the "immunitary dispositif" of biopolitics must be overturned. These are the *double enclosure of the body*, the *preemptive suppression of birth*, and the *normativization of life*.[43] I focus on the third of these here, for it is specifically in rethinking the conjunction of life and norms that Esposito sees the possibility for an affirmative politics of life.

In the closing pages of *Bios*, Esposito argues, contra Agamben, that the Nazi regime was characterized by an absolute normativization of life, such that this regime did not derive its power from the subjective decision in the shadow of the suspension of law but rather in the derivation of a normative framework from the very "vital necessities of the German people." The relation between law and life at stake in this, he argues, entails a double presupposition whereby the juridical norm presupposes the facticity of life, and life presupposes "the caesura of the norm as its preventative definition." Thus, he concludes that Nazism created a "norm of life," not however, in the sense that it "adapted its own norms to the demands of life," but in the sense that it "closed the entire extension of life within the borders of a norm that was destined to reverse it into its opposite," that is, into death.[44] The problem for Esposito at this point is to suggest a way forward to a genuine politics of life or an affirmative biopolitics that breaks this deadly knot in which life and norm are entwined and mutually presupposed. He argues that attempts to distinguish more clearly between life and norm, such as in transcendental normativism and juris-naturalism, are unsatisfactory responses, however, since neither the absolutization of the norm nor the primacy of nature can be considered external to Nazism. Instead, then, Esposito looks for resources in philosophical traditions that have emphasized the radical immanence of life and norm, and which in that way undermine the double presupposition that ties them together in Nazism.

Of these resources, he suggests that the theorization of vital norms developed by Canguilhem may be especially valuable since it allows for the "maximum deconstruction of the immunitary paradigm and the opening to a different biopolitical lexicon."[45] To reach this conclusion, Esposito references the radical vitalization of the norm that Canguilhem proposes in his work on the concepts of the normal and the pathological in the history of medicine. He argues that life is internally and necessarily normative, since even at the simplest level "living means preference and exclusion."[46] Living necessarily involves polarities of valuation, such that an organism cannot be understood as indifferent to the environment in which it finds itself. Esposito goes on to emphasize that this means that disease and health are both normative states in that both indicate new forms of life for the organism and, moreover, reveal the normal functioning of the body. Conditions of disease or biological abnormality are not simply deviations from a fixed prototype of the normal: they are instead normative forms of a qualitatively different order. Similarly, to be "normal" is not to coincide with a preestablished norm but rather to be able to harness and maintain one's own normative power: to be normal is to be able to create new norms. In view of this radicalized immanence of life and norm, Esposito writes that "if Nazism stripped away every form of life, nailing it to its nude material existence, Canguilhem reconsigns every life to its form, making of it something unique and unrepeatable."[47]

For Esposito, the productive power of Canguilhem's thinking is that the immanence of norms in life undermines the separation and mutual presupposition of the facticity of life and normative transcendentalism. Moreover, this analysis rejects an objectivist approach to life and emphasizes instead the vital potential in life, in terms of the capacity to generate norms. However, it is exactly this productive power of the immanent normativity of life that points the way toward identifying several shortcomings in Esposito's approach. For this normative capacity, the power to create norms that inheres in life, is itself conditioned by the environment or milieu in which an organism finds itself. One point that Canguilhem is wholly committed to but that Esposito tends to skip over is that an organism by itself is never normal—rather, what can be considered "normal" is the relationship between the organism and its environment. Canguilhem writes:

> Taken separately, the living being and his environment are not normal: it is their relationship that makes them such. For any given form of life the environment is normal to the extent that it allows it fertility and a corresponding variety of forms such that, should changes in the environment occur, life will be able to find the solution to the problem of adaptation . . . in one of these forms.[48]

Thus, life is inherently normative, in the sense that it aims at the restoration of functional or "normal" relations between an individual organism and its environment. This correction to foreground the relationship between the organism and its environment may seem like a relatively minor interpretive point; but I want to suggest that it actually has important implications, two of which I will mention here.

The first point goes to the fact that the environment in which human beings are located is necessarily social and, as such, cross-cut with the force of social norms. As Canguilhem suggests, human norms are "determined as an organism's possibilities for action in a social situation rather than as an organism's functions envisaged as a mechanism coupled with the physical environment. The form and functions of the human body are the expression not only of conditions imposed upon life by the environment but also of socially adopted modes of living in the environment."[49] This means that the "normal" is always an effect of a complex comingling and expression of vital norms in the midst of socially defined ways of living. Human life is never simply biological; and nor, for that matter, is it ever simply social or political. That Esposito leaves aside the necessary embeddedness of an organism in its environment means that he also risks obfuscating the ways that social norms cut across the vital norms of the living human being. In doing so, his analysis runs surprisingly close to the arguments of liberal eugenicists and transhumanists, who valorize the possible plurality of bodily norms that technologies of enhancement are supposed to engender, without consideration of the ways in which those possibilities are delimited in advance by social norms that are lived in often less than conscious ways.[50] In this, he risks a version of the libertarian fantasy of escape from the fundamental conditions of existence of humanity.

The second point derives from this, for while the existence of human beings is conditioned by social norms, it cannot be assumed that vital and social (or legal) norms are equivalent in the manner that Esposito treats them. Rather, what needs to be taken into account is the disjuncture between vital and social norms, and consequently, what requires explanation is the means by which they intermingle. In other words, vital and social norms may be empirically inseparable, but they are nevertheless analytically distinct. In the postscript to *The Normal and the Pathological,* Canguilhem argues that while physiological norms are immanent to the organism, social norms have no equivalent immanence. In a living organism, norms are "presented without being represented, acting without deliberation or calculation," such that there is "no divergence, no delay between rule and regulation." In contrast, rules in a social organization must be "represented, learned, remembered, applied."[51] In light of this insistence on the exteriority of social norms, we would do well to qualify Esposito's thesis on the "vitalization of the norm." While Canguilhem's work develops a philosophy of life that emphasizes the productive power of the living in terms of the capacity to create norms, he also resists a complete vitalization of the norm, insisting on a more differentiated approach to norms and normalization. This is important because while the exteriority—perhaps even transcendence—of social norms is indicated by the capacity to question those norms, it also opens them to such questioning and, ultimately, to transformation.

In this regard, Esposito also has little to say about another aspect of the productive power of the living that Canguilhem insists on. This is the notion that life is characterized by an internal errancy, or capacity for error. Interestingly, it is this capacity for error that Foucault focuses on in his essay "Life: Experience and Science," suggesting that "Canguilhem has proposed a philosophy of error, of the concept of the living, as a different way of approaching the notion of life."[52] While Esposito follows Agamben in privileging Deleuze's essay on life and immanence, I want instead to return to Foucault's treatment of the capacity for error in "Life: Experience and Science" and its implications for conceptualizing life in biopower.

Foucault: Life and Error

In the final chapter of the first volume of *History of Sexuality*, Foucault makes his now infamous argument that during the eighteenth century, Western politics underwent a fundamental transformation from the principles and practices of sovereignty to a new regime of biopower, in which biological life itself became the object and target of political power. Biopower incorporates both disciplinary techniques geared toward mastering the forces of the individual body and a biopolitics centered around the regulation and management of the life of a new political subject, the population.[53] This new regime of political power operates according to the maxim of "fostering life or disallowing it" and signals for Foucault the "threshold" of our modernity. It entails new forms of government and social regulation, such that power no longer operates through violence imposed on subjects, but through apparatuses that regularize, administer, and foster the life of subjects through the "normalization of life processes."[54] In this, the field of biopower is marked out by the "the body [of the individual] as one pole and the population as the other," in a continual circuit of mutual presupposition and reference.[55]

Foucault's account of biopower thus gives a central role to normalization as a form of social and political regulation, suggesting at one point that "a normalizing society is the historical outcome of a technology of power centered on life."[56] As a technique of biopower, normalization is irreducible to the institutions and force of the law, and arises from the socio-political authority of statistics.[57] Interestingly, Foucault claims that normalization works in opposing ways in discipline and a biopolitics of population.[58] In the former, infractions of the norm are produced as a consequence of the prior application of the norm, insofar as the phenomenal particularity of an individual is itself identified and calibrated through the application of a norm. Normalization produces individuals as the necessary mode and counterpart of the operation of norms, that is, as a material artifact of power.[59] In a biopolitics of population, however, norms are mobilized in exactly the opposite way, insofar as "the normal comes first and the norm is deduced from it." The biopolitics of populations, and the apparatuses of security that Foucault identifies as crucial to it, involves "a plotting of the normal and the abnormal, of different curves of normality, and the operation of normalization consists in establishing an interplay between these different distributions of normality and in acting to bring the most unfavourable into line with the most favourable."[60] Given these differences in the operation of normalization, detailed studies of the mobilization of norms in regard to specific instances of the management of life processes today are required to understand the operation of biopower.

A precept of such studies would be that social and vital norms are simultaneously inseparable and irreducible; they do not determine each other, but neither can one be determined in the absence of the other. This condition of living in two worlds at one and the same time points to the ambivalence in the concept of life, where its meaning is often determined in its accompanying qualification, as, for instance, biological or social. As I noted earlier, Agamben sees such qualifications as themselves part of the operation of biopolitics, and because of this he resists any engagement with biological conceptions of life. However, his revivification of the distinction between *zoē* and *bios* can be seen as a kind of anachronism since it is not obvious that the term *zoē* really does accord with biological life today. In Foucault's view, biopower is intimately related to the appearance of the biological in the sphere of politics; but biology, as a "discipline" or regime of truth, is a historically specific phenomenon, and its categories and concepts cannot simply be read back into Aristotle or vice versa. For a start, the contemporary "molecularization" of life has to a large extent overtaken the organic biology of the nineteenth century, to say nothing of the preceding discourses of life in natural history.[61] Further, it is undoubtedly true that discourses and practices of molecular life, which often tie individual and population identities to genetics through an integration with capital, are aligned with biopolitical strategies. However, this should not in itself render all engagement with biology suspect, for it may also be that less hegemonic conceptions of life emergent within contemporary theoretical biology provide ways of thinking beyond "recombinant biopolitics."[62]

Finally, and not unrelated to this, in a little analyzed moment in his discussions of biopower in *History of Sexuality*, Foucault offers the caveat that one should not imagine that life has been totally administered and controlled by governmental techniques; rather, he states, life constantly escapes or exceeds the techniques that govern and administer it.[63] But what is meant by the term "life" such that what it refers to is able to escape the political techniques that seek to control it? What enables this moment of escape, and in what form is it realized? At this point Foucault's discussion of the idea of an inherent potential for error in life developed in the thought of Canguilhem becomes important. In his short essay on Canguilhem mentioned earlier, Foucault argues that at the center of the problems that preoccupy Canguilhem resides "a chance occurrence . . . like a disturbance in the informative system, something like a 'mistake,'" in short, "error"; Foucault states, "life—and this is its radical feature—is that which is capable of error."[64] As such, the error that is borne within life as its necessary potentiality provides the radical contingency around which the history of life and the development of human beings are twined for Canguilhem, which enabled him to identify and draw out the relation of life and knowledge. Foucault writes,

> if one grants that the concept is the reply that life itself has given to that chance process, one must agree that error is the root of what produces human thought and its history. The opposition of the true and the false, the values that are attributed to the one and the other, the power effects that different societies and different institutions link to that division—all this may be nothing but the most belated response to that possibility of error inherent in life.[65]

Thus, it is through the notion of error that life is placed in a relation of contiguity and contingency with truth and structures within which it is told. "Error," or the inherent capacity of life to "err" both establishes the relation of life to truth and undermines that relation by disentangling humanity from the structures of truth and power that respond to the potential for error. Hence, "with man, life has led to a living being that is never completely in the right place, that is destined to "err" and to be "wrong."[66]

If this is so, the potential for error in life directs us to an important point about the operation of biopower, specifically, that the biopolitical state is simply the latest response to the possibility of error. Biopower is less a matter of controlling life than it is a matter of managing error—or rather, to the extent that it is the former, it is so by virtue of the latter. This would then mean that the biopolitical state is systematically *reactive*, wherein the errancy internal to life constantly provokes the biopolitical state, forcing it to respond to the contingencies of the living and the phenomena of life. Nevertheless, the mystery of biopower is to make it appear *as if* the state controlled and mastered life. There is a cliché that it is not so much the masters who walk their dogs as the dogs who walk their masters. To elaborate, it is not simply that humans tame dogs as pets—but rather that somehow dogs have managed to tame humans to such an extent that the latter will spend thousands of dollars and hours in keeping their pets alive and well. Similarly, we might consider the ways that life has demanded that the state care for it, has demanded—often quite successfully—that the state foster it by providing the conditions for its flourishing in manifold ways. This is not to say that the biopolitical state does not also involve itself in the production of death—it evidently does; but when it does, it typically does so for the sake of the living. This suggests that biopower cannot be understood in terms of oppositions such as those of "making live" *or* "causing death," "care" *or* "violence"; instead, it establishes a mutually reinforcing relation between care and violence, between life and death, wherein each presupposes the other. Today, a biopolitical state *cannot not react* to the provocations of life, even if that reaction entails disallowing life. Thus, errancy allows for a different construal of the relation between life and politics, where this relation is not simply one of privation and negation, but of plural positivity. Life produces politics.

Notes

1. Paolo Virno, *A Grammar of the Multitude*, trans. Isabella Bertoletti, James Cascaito, and Andrea Casson (New York: Zone, 2004); Paul Patton, "Agamben and Foucault on Biopower and Biopolitics," in *Giorgio Agamben: Sovereignty and Life*, ed. Steven DeCaroli and Matthew Calarco (Stanford, CA: Stanford University Press, 2007).
2. Eugene Thacker, *After Life* (Chicago: University of Chicago Press, 2010), ix.
3. Giorgio Agamben, *Homo Sacer: Sovereign Power and Bare Life*, trans. Daniel Heller-Roazen (Stanford, CA: Stanford University Press, 1998).
4. Eric Santner, *On Creaturely Life: Rilke, Benjamin, Sebald* (Chicago: University of Chicago Press, 2006).
5. Melinda Cooper, *Life as Surplus: Biotechnology and Capitalism in the Neoliberal Era* (Seattle: University of Washington Press, 2008).
6. Agamben, *Homo Sacer*, 2.

7. Ibid.
8. Ibid., 88.
9. Giorgio Agamben, *Means without End: Notes on Politics*, trans. Cesare Casarino and Vincenzo Binetti (Minneapolis: University of Minnesota Press, 2000), 114–15.
10. Michel Foucault, "Life: Experience and Science," in *Aesthetics, Method and Epistemology: Essential Works of Foucault 1954–1984*, vol. 2, ed. James Faubion (London: Allen Lane, 1998). This essay was initially published in 1978 as the introduction of the English translation of Georges Canguilhem, *Le normal et le pathologique* (Paris: Presses Universitaires de France, 1966), reprinted in Georges Canguilhem, *The Normal and the Pathological*, trans. Carolyn Fawcett (New York: Zone, 1991). Another version of it was published in *Revue de métaphysique et le morale*, appearing in 1985, shortly after Foucault's death.
11. Gilles Deleuze, "L'immanence: Une Vie . . . ," *Philosophie* 47, no. 1 (1995), republished as "Immanence: A Life," in *Pure Immanence: Essays on a Life*, ed. John Rajchman (New York: Urzone, 2001).
12. Giorgio Agamben, "Absolute Immanence," in *Potentialities: Collected Essays in Philosophy*, ed. Daniel Heller-Roazen (Stanford, CA: Stanford University Press, 1999), 238.
13. Ibid., 239.
14. Ibid., 238.
15. Deleuze, "Immanence: A Life," 31.
16. Agamben, "Absolute Immanence," 229.
17. Ibid., 234.
18. Ibid., 238.
19. Ibid.
20. Giorgio Agamben, *The Highest Poverty: Monastic Rules and Form-of-Life* (Stanford, CA: Stanford University Press, 2013), xi.
21. Ibid., xii.
22. Ibid., 110.
23. E.g., Agamben, *Homo Sacer*, 127–28.
24. E.g., Ernesto Laclau, "Bare Life or Social Indeterminacy," in *Giorgio Agamben: Sovereignty and Life*, ed. Matthew Calarco and Steven DeCaroli (Stanford: Stanford University Press, 2007), 11–22; Paul Patton, "Agamben and Foucault on Biopower and Biopolitics," in Calarco and DeCaroli, *Giorgio Agamben*, 203–18; Andrew Norris, "Giorgio Agamben and the Politics of the Living Dead," *Diacritics* 30, no. 4 (2000).
25. Jacques Derrida, *The Beast and the Sovereign*, ed. Michel Lisse, Marie-Louise Mallet, and Ginette Michaud, trans. Geoffrey Bennington (Chicago: University of Chicago Press), 1:434. Note that Derrida understands bare life as synonymous with *zoē*.
26. Given this, we might do well to question the very idea of a "simple" living body, especially when used in relation to human beings.
27. Agamben, *Homo Sacer*, 182.
28. Ibid.
29. Ibid., 187.
30. Michel Foucault, "Two Lectures," in *Power/Knowledge: Selected Interviews and Other Writings 1972–1977*, ed. Colin Gordon, trans. Colin Gordon, Leo Marshall, John Mepham, Kate Soper (New York: Pantheon, 1980), 80.
31. Gilles Deleuze, "What is a *Dispositif*?," in *Michel Foucault: Philosopher*, trans. Timothy J. Armstrong (Hertfordshire: Harvester Wheatsheaf, 1992), 164.
32. Paul Rabinow and Nikolas Rose, "Biopower Today," *BioSocieties* 1 (2006):197, 203–4.

33. Ibid., 199.
34. Ibid., 202.
35. Ibid., 203.
36. Nikolas Rose, *The Politics of Life Itself: Biomedicine, Power and Subjectivity in the Twenty-First Century* (Princeton, NJ: Princeton University Press, 2007).
37. Ibid., 49.
38. It is important to recognize one limit on what life can mean or be for Rabinow and Rose, and that is that a discourse about life must be "in the true" to have authoritative force. Thus, they argue that biological claims about race were no longer "in the true" in either political or biological discourse in the late twentieth century. They are, however, reentering the domain of biological truth through a "molecular gaze." Rabinow and Rose, "Biopower Today," 205–6.
39. Roberto Esposito, "Totalitarianism or Biopolitics? Concerning a Philosophical Interpretation of the Twentieth Century," *Critical Inquiry* 34, no. 4 (2008): 642.
40. Roberto Esposito, "The Immunization Paradigm," *Diacritics* 36, no. 2 (2006): 24.
41. Roberto Esposito, *Immunitas: The Negation and Protection of Life* (Cambridge: Polity Press, 2011), 9.
42. Ibid., 16.
43. Roberto Esposito, *Bios: Biopolitics and Philosophy*, trans. Timothy Campbell (Minneapolis: University of Minnesota Press) 138–45.
44. Ibid., 184.
45. Ibid., 191.
46. Canguilhem, *Normal and the Pathological*, 136.
47. Esposito, *Bios*, 189.
48. Canguilhem, *Normal and the Pathological*, 143–144.
49. Ibid., 269.
50. For further discussion, see Catherine Mills, *Futures of Reproduction: Bioethics and Biopolitics* (Dordrecht: Springer, 2011), especially chap. 2, "Normal Life: Liberal Eugenics, Value Pluralism and Normalisation."
51. Ibid., 250.
52. Foucault, "Life: Experience and Science," 477.
53. Michel Foucault, *The History of Sexuality*, vol. 1, *An Introduction*, trans. Richard Hurley (London: Penguin, 1981), 135–45. See also Foucault, *"Society Must Be Defended": Lectures at the College De France, 1975–76*, eds. Mauro Bertani and Alessandro Fontana, trans. David Macey (New York: Picador, 2003), and Michel Foucault, *Security, Territory, Population: Lectures at the College De France, 1977–78*, trans. Graham Burchell (New York: Palgrave MacMillan, 2007). As this suggests, Foucault occasionally makes a useful distinction between "biopolitics" and "biopower," wherein the former term refers to the constitution and incorporation of the population as a new subject of governance, and the latter is a broader term that encompasses both biopolitics and discipline. I use "biopower" in the discussion of Foucault to specifically indicate a technology of power that incorporates both discipline and a biopolitics of population.
54. Georges Canguilhem, "On *Histoire de la folie* as an Event," in *Foucault and His Interlocutors*, ed. Arnold I. Davidson (Chicago: University of Chicago Press, 1997), 32.
55. Foucault, *"Society Must Be Defended,"* 253.
56. Foucault, *History of Sexuality*, 1:144.
57. On the history of statistics, see Ian Hacking, *The Taming of Chance* (Cambridge: Cambridge University Press, 1990). For a compelling account of the importance

of statistics for Foucault, see Mary Beth Mader, *Sleights of Reason: Norm, Bisexuality, Development* (Albany: State University of New York Press, 2011).

58. See Michel Foucault, *Discipline and Punish: The Birth of the Prison*, trans. Alan Sheridan (London: Penguin, 1979), 177–83, and Foucault, *Security, Territory, Population*, 57–63.
59. Foucault, *Discipline and Punish*, 184.
60. Foucault, *Security, Territory, Population*, 63.
61. See Rose, *Politics of Life Itself.*
62. Michael Dillon, and Julian Reid, "Global Liberal Governance: Biopolitics, Security and War," *Millennium Journal of International Studies* 30, no. 1 (2001).
63. Foucault, *History of Sexuality*, 1:143.
64. Foucault, "Life: Experience and Science," 476. While this is more Canguilhem's understanding of life than Foucault's, it is possible to see the former's identification of a potential for error within life as at least a point of inspiration for the latter.
65. Ibid.
66. Ibid.

FIVE

Power and Biopower in Foucault

PAUL PATTON

Michel Foucault is widely credited with having invented new concepts of biopower and biopolitics, despite the fact that they do not play a major role in his work.[1] These terms appeared in lectures in 1976 and 1977, and in volume 1 of *The History of Sexuality*, first published in December 1976.[2] Thereafter they were superseded by the concepts associated with his analysis of the different ways in which governmental power has been exercised by modern states. In fact, they are confused and confusing terms that never achieved the status of determinate concepts. After retracing their brief career in Foucault's work, this chapter will identify some of the confusions involved in his use of these terms and suggest that there are good reasons why the concepts of biopower and biopolitics disappeared from his subsequent work.

"Biopolitics" appears for the first time in a 1974 lecture devoted to the emergence of public health.[3] It appears again, along with "biopower," in the final lecture of his 1975–76 course at the Collège de France. Here "biopolitics" referred to the manner in which, from the end of the eighteenth century, the power of government began to be exercised over the biological dimension of human existence, that is, over "man in so far as he is a living being."[4] From this point onward, he argues, modern political societies were governed by "a power that has taken control of both the body and life or that has, if you like taken control of life in general—with the body as one pole and the population as the other."[5]

The terms "biopolitics" and "biopower" both appeared in part 5 of *The History of Sexuality*, volume 1. Here Foucault draws an epochal contrast between a type of society in which power was exercised primarily as a means of subtraction, though mechanisms that appropriated a portion of the labor, wealth, or time of individual subjects, and a society in which power sought to exercise a more positive control over human life, endeavoring "to administer, optimize and multiply it, subjecting it to precise controls and comprehensive regulations."[6] The emergence from the seventeenth century onward of disciplinary techniques for the subjugation of bodies followed by mechanisms for the control of populations "marked the beginning of an era of 'biopower.' "[7]

Foucault's course the following year, *Security, Territory, Population,* began with the statement that he would undertake a study of the "mechanisms through which the basic biological features of the human species became the object of a political strategy."[8] Accordingly, the first three lectures dealt with mechanisms for the government of urban populations in relation to the circulation of people and materials, the avoidance of grain shortage, and the control of disease. They sought to demonstrate how these "security mechanisms" were substantially different from early modern techniques for the exercise of sovereign power over subjects and also from disciplinary mechanisms of power: the object, aims, and mode of operation were not the same in each case. However, by the fourth lecture, the focus of this course shifted from the object of government, namely certain conditions of the biological existence of human beings, to the means of exercising sovereign power or techniques of government. Thereafter the analysis of specific forms of "governmentality" occupied center stage for the remainder of the lectures in 1977–78 and 1978–79.

Even though the 1978–79 course was entitled *The Birth of Biopolitics,* it was devoted to the analysis of liberal and then neoliberal forms of government, in particular the indirect forms of action on the actions of individuals exercised through market mechanisms.[9] At the end of the first lecture, Foucault sought to justify the shift of focus by suggesting that the distinctly liberal art of government formed the historical framework within which "something like biopolitics could be formed" and that only when we know what liberalism was will we be able to grasp "what biopolitics is."[10] In the fourth lecture of this course, he outlined the problems that he hoped to discuss within this framework of liberal governmentality: the problem of law and order, the opposition between state and civil society and then, finally, "the problem of biopolitics and the problem of life."[11] He never reached this third problem, and he never made further references to biopolitics apart from his assurance at the beginning of lecture eight that he "really did intend to talk about biopolitics."[12] Thereafter the terminology of biopower and biopolitics largely disappeared from Foucault's work.

The Scope of Biopolitics

A first ambiguity concerns the scope of the terms "biopolitics" and "biopower." When these are first introduced in the final lecture of Foucault's 1975–76 course, they refer to a new technology of power that is explicitly contrasted with the disciplinary technology that emerged during the course of the seventeenth century:

> After the anatomo-politics of the human body established in the course of the eighteenth century, we have, at the end of that century, the emergence of something that is no longer an anatomo-politics of the human body, but what I would call a "biopolitics" of the human race.[13]

Discipline involved the training of individual bodies to increase their productive forces while also maximizing control over their activities. It sought to organize the spatial distribution and sequential activity of individuals organized into functional multiplicities of various kinds (military, industrial, educational, and so on), all the while ensuring control by means of constant surveillance. By contrast, biopolitical technology operated by different means, on a different scale. It sought to regulate birth and death rates, the prevalence and spread of illness, levels of nourishment, and the physical environments in which people lived. The period from the latter half of the eighteenth century onward saw the development of state policies in relation to rates of birth and the ratio of births to deaths. It also saw the development of policies in relation to the rate of deaths in so far as this was affected by diseases endemic within a given population. During this period, the specific branch of medicine concerned with public health and hygiene took shape. Other fields of intervention associated with biopolitics included the range of phenomena associated with old age, accidents, infirmities, and other biological disabilities. These lead to the introduction of insurance and other means of collective saving, along with measures designed to preserve both lives and the quality of lives. Finally, biopower involved various forms of control over the relations between human beings and their environment, including features of the natural environment such as swamps and the conditions of urban life. These are the initial domains of biopower: "biopolitics will derive its knowledge from, and define its power's field of intervention in terms of, the birth rate, the mortality rate, various biological disabilities, and the effects of the environment."[14]

A distinguishing feature of this biopolitics was the fact that it operated on populations rather than individual bodies: "Biopolitics deals with the population, with the population as political problem, as a problem that is at once scientific and political, as a biological problem and as power's problem."[15] A further distinguishing feature was the nature of the phenomena addressed by this new biopower: these are "aleatory and unpredictable" at the individual level but become predictable and measurable at the level of the collective.[16] They are not punctual but serial phenomena the effects of which are experienced over time. Finally, since biopolitics does not intervene at the level of individuals but at the level of populations, where the aim is to establish and maintain an equilibrium, it employs very different mechanisms to those employed in the exercise of disciplinary power. Various kinds of statistical measures and forecasts are deployed so that rates of mortality rate can be lowered, life expectancy increased, and birthrates improved: "In a word, *security mechanisms* have to be installed around the random element inherent in a population of living beings so as to optimize a state of life."[17]

Even though the concepts of biopower and biopolitics are first introduced in contrast to disciplinary power, in the lecture of March 17, 1976, and in volume 1 of *The History of Sexuality*, Foucault drew attention to the complementarity as well as the differences between these two distinct technologies. The fact that they do not exist on the same level or apply to the same objects meant that they could operate in concert: "they are not mutually exclusive and can be articulated with each other."[18] Certain phenomena, such as health and sexuality, could be addressed by both technologies at the same time, at the level of individual bodies and at the level of populations. This led him to propose a broad sense of biopower that encompassed both disciplinary technologies and mechanisms of security:

> To say that power took possession of life in the nineteenth century . . . is to say that it has, thanks to the play of technologies of discipline on the one hand and technologies of regulation on the other, succeeded in covering the whole surface that lies between the organic and the biological, between body and population.[19]

It is in this broader sense of the term that he can speak of "a power that has taken control of both the body and life or that has, if you like, taken control of life in general."[20] It is this broad sense of the term that is invoked in the epochal claim in volume 1 of *The History of Sexuality* that, from the end of the eighteenth century onward, Western society entered into an "era of biopower."[21] The ambiguity concerning the scope of the terms "biopower" and "biopolitics" emerges clearly when we consider the forms of power with which they are contrasted. On the one hand, they are contrasted with disciplinary power and the associated "anatomo-politics" of the human body. On the other hand, they are contrasted with the old, subtractive power of sovereignty:

> Beneath that great absolute power, beneath the dramatic and sombre absolute power that was the power of sovereignty, and which consisted in the power to take life, we now have the emergence, with this technology of biopower, of this technology of power over "the" population as such, over men insofar as they are living beings. It is continuous, scientific, and it is the power to make live. Sovereignty took life and let live. And now we have the emergence of a power that I would call the power of regularization, and it, in contrast, consists in making live and letting die.[22]

There are good reasons to be skeptical of such epochal claims. A first reason is the fact that Foucault made a series of such claims about the nature of power in modern societies. His lectures in 1975–76 began with a statement of an intellectual crisis and his intention to reconsider the trajectory followed up to this point. He sought to reexamine the concept of power and the manner in which he had conceived of it hitherto, primarily in terms of struggle between contending forces. He sought answers to both a conceptual question—what is power and what concepts are appropriate for its analysis?—and to a diagnostic question: "What is this power whose irruption, force, impact, and absurdity have become palpably obvious over the last forty years, as a result of both the collapse of Nazism and the retreat of Stalinism?"[23] Foucault did not answer the conceptual question in the course of his 1975–76 lectures. However, he offered at least two answers to the diagnostic question: in the second lecture he suggested that Western societies are ruled by a complex combination of disciplinary power and a democratized version of sovereign power according to which the sovereign right of individuals was delegated to the state: "A right of sovereignty and a mechanics of discipline. It is, I think, between these two limits that power is exercised."[24] In the final lecture he suggested that the regime of power in modern societies is one of biopower.

The first lecture of the 1977–78 course, *Security, Territory, Population,* outlines a more detailed account of the mechanisms of security through which power is exercised over the conditions of life. In contrast to both juridical and disciplinary mechanisms, security mechanisms are applied to populations rather than to individuals or groups. They deal with probable rather than actual events and seek to respond to these on the basis of calculations of cost and in terms of a norm of acceptable outcomes rather than a binary division between the permitted and prohibited. At this point, Foucault entertains a further answer to the diagnostic question by asking whether it might not be the case that "the general economy of power in our societies is becoming a domain of security?"[25]

After the shift of focus from mechanisms of security to the forms and rationalities of government in the 1977–78 lectures, Foucault offered further epochal claims about the nature of power. He insisted on the unique character of pastoral power as it developed within European Christianity and its difference from political power up until the end of the eighteenth century, but also on its role in the development of modern forms of state government: pastoral power was "one of the decisive moments in the history of power in Western societies."[26] In "The Subject and Power," he described contemporary state power as a unique combination of totalizing and individualizing power, in part because of the degree to which it has integrated and transformed the techniques of Christian pastoral power. The contemporary state, he suggests, can be seen as "a modern matrix of individualization, or a new form of pastoral power."[27] Finally, in the second lecture of his 1978–79 course, *The Birth of Biopolitics*, Foucault discusses the idea of "frugal government" as this term was used in the eighteenth century to mean government that does not govern too much. He equates this with liberal government and comments that "actually at this moment we are entering what could be called the epoch of frugal government."[28] Foucault's apparent tendency to advance epochal claims about whatever type of power he is investigating is a reason not to attach too much importance to the claim that we live in an era of biopower.

A second reason for skepticism with regard to Foucault's large-scale epochal concept of biopower concerns the lack of any detailed specification of this concept. Consider the manner in which he typically proceeds to elaborate concepts, whether of discourse, power, or ethics: in formulating concepts of practices in these domains, he tends to rely on the four elements of the Aristotelian concept of causality: material, formal, efficient, and final cause. Applied to ethical practice, this schema allows a characterization of distinct practices in terms of their object, mode of subjectification, ethical work, and desired effect.[29] Applied to practices of punishment, it allows us to distinguish, for example, the different technologies of power proposed to deal with criminals at the end of the eighteenth century.[30] Foucault follows a similar procedure in contrasting biopower understood in the narrow sense of the term with disciplinary power: they do not have the same objects (populations as opposed to bodies and their forces), mechanisms (mechanisms of security as opposed to techniques of training), modes of operation (regulation of collective phenomena as opposed to control of individuals and groups), or goals (statistical norms as opposed to docile bodies). In the lectures devoted to mechanisms of security at the beginning of his 1997–98 course, Foucault sought to specify differences between these two technologies of power in terms of their different relationships to space, to events, and to norms or normalization.[31] His aim was to show that biopower involved a different substance, mechanism, mode of operation, and objectives to disciplinary power. To the extent that he succeeded in doing so, the result would have been a well-formed concept of biopower and the biopolitics associated with it, in contrast to disciplinary power and its associated "anatomo-politics" of the body. However, he does not develop the epochal concept of biopower to the same level of specification. In relation to this broad usage of the term, there is no determinate concept of biopower.

Biopower as Right versus Biopolitical Technology

A second area of confusion surrounding the concept of biopower involves the difference between the study of power at the level of its exercise and at the level of its theorization and legitimation. Throughout his analyses of power, Foucault contrasted his own approach to the study of power with the "juridical" theory that provided the historical and conceptual framework of political philosophical approaches to power. Whereas the latter addressed the nature and limits of the legitimate exercise of power, he insisted on a descriptive analysis oriented toward the question of *how* power is exercised: "How is it exercised?" and "What happens when individuals exert (as we say) power over others?"[32] He repeatedly distanced himself from the representations of power in theories of sovereignty or the state in favor of a historical analysis of techniques of power. He abandoned questions of legitimation in favor of a descriptive approach to the technologies and forms of rationality implicit in the exercise of political power.

In accordance with this focus on the question of how power is exercised, he introduced the term "biopolitics" with reference to the specific technology of power that emerged toward the end of the eighteenth century and that served to bring the biological conditions of human life under state control. The transformation that interested him in relation to the contrast drawn at the end of volume 1 of *The History of Sexuality* between a "right of death" and a "power over life" involved the forms of exercise of state power rather than the forms of its representation in political theory. This transformation occurs at the level of the "mechanisms, techniques and technologies of power."[33] This new technology operates on different objects, at a different level, on a different scale, and with different instruments than those that he had earlier associated with disciplinary power.

Foucault's descriptions of the regulatory mechanisms of biopower make it clear that these involve techniques for the exercise of power rather than forms of its representation. Despite this, he confuses the issue by presenting the concept of biopower initially not with reference to techniques of power but with reference to the representation of state power in the classical theory of sovereignty. At the outset of his lecture of March 17, 1976, he describes sovereign power in terms of its forms of juridical and political representation rather than in terms of its effective exercise. He describes the basic attribute of classical sovereign power as "the right of life and death" and suggests that this is a strange right "even at the theoretical level."[34] It is strange because, to the extent that the sovereign really does have the right to decide whether subjects live or die, the subject is suspended between life and death. Qua subject, he or she has no right to live or die independently of the will of the sovereign:

> In terms of his relationship with the sovereign, the subject is, by rights, neither dead nor alive. From the point of view of life and death, the subject is neutral, and it is thanks to the sovereign that the subject has the right to be alive or, possibly, the right to be dead.[35]

The material life or death of the subject is doubled by the right to live or the obligation to be dead in accordance with the sovereign will: "the lives and deaths of subjects become rights only as a result of the will of the sovereign."[36] The body of the subject is therefore doubled in precisely the same way that, according to Ernst Kantorowicz the body of the sovereign is doubled by virtue of occupying the position of supreme political power.[37]

However, while in theory the sovereign has power over the life and death of the subject, in practice this power or right "is always exercised in an unbalanced way: the balance is always tipped in favour of death."[38] In other words, as Foucault had already argued in *Discipline and Punish*, the exercise of sovereign power in the premodern period was primarily negative. This was a power that took life, or threatened to do so, just as it extracted tribute or deducted taxes on the basis of a creation of wealth that lay largely outside its understanding or control. In terms of the contrast that frames all of Foucault's analyses of power between representation and technique, or between the theory and practice of power, the asymmetry that he points to here in relation to classical sovereign power applies at the level of the exercise of power rather than at the level of representation. It is at this level that the transformation in sovereign power takes place: different objects, different scale of application, and different mechanisms for the regulation of the conditions of life in a given population.

Foucault confuses the exercise and the representation of power in his lecture of March 17, 1976, when he suggests that "the very essence of the right of life and death is actually the right to kill" and that the right of the classical sovereign is "essentially the right of the sword."[39] Since he is here referring to the forms of exercise of sovereign power rather than to the theory, there is no contradiction between suggesting that there is symmetry at the heart of the sovereign's right over life and death and dissymmetry at the level of its exercise. If it is true that the dissymmetry arises only at the level of the exercise of sovereign right, then it need not affect the essence of sovereign right as such. This is one way to understand the force of Giorgio Agamben's contribution to the debate over the character of sovereign power in *Homo Sacer*.[40] He insists on the exclusive inclusion of the lives of subjects within the sphere of sovereign right in the manner outlined by Foucault: from this point of view, "the subject is, by rights, neither dead nor alive."[41] However, by the same token, the thesis that the biological life of subjects (*zoē* as opposed to *bios*) was always included within the scope of sovereign power is already anticipated in Foucault's analysis of classical sovereignty. Although he does not discuss the figure of the *homo sacer* who is at once excluded from the protection of both the political and the divine order, Foucault's suggestion that the subject of sovereign power is neutral with respect to his right to live or die already implies the possibility of the withdrawal of the status of subject. In turn, this implies the possibility of exclusion from the sphere of sovereign power, thereby exposing the exiled figure to the loss of protection of his life as much as to the freedom from the threat of death at the hands of the sovereign. In this sense, his analysis of classical sovereign right and its doubling of the body of the subject is entirely consistent with Agamben's suggestion that bare life and the possibility of the ban or exclusion has long constituted the essential structure of sovereign power.[42]

There is further confusion between the exercise and the representation of state power in the lecture of March 17, 1976, when Foucault suggests that "one of the greatest transformations" in political right during the nineteenth century involved the permeation of the old right to take life or let live by the new right to make live and let die. This confusion takes a terminological turn when, in the midst of characterizing this new right he refers to it as "the *power* to 'make' live and 'let' die."[43] It is clear that he is talking here about a transformation in the mode of exercise of sovereign power rather than its form of representation. Whereas in practice the old right of sovereign power amounted to "the right to take life or let live," the new right that emerged amounted to "the right to make live and to let die."[44] He goes on to suggest that this nineteenth-century transformation in political right can be traced "in the theory of right."[45] However, his example is not drawn from nineteenth century political philosophy, as we might expect, but from the "jurists" of the seventeenth and eighteenth centuries. These theorists provided a new account of the sovereign right of life and death, in which the lives of subjects were excluded from the contract that established sovereign power. To the extent that individuals only enter into the contract on the basis of the desire to protect their own lives, Foucault asks, does it make sense to suppose that they would grant a right over their life to the sovereign? "Isn't life the foundation of the sovereign's right, and can the sovereign actually demand that his subjects grant him the right to exercise the power of life and death over them, or in other words, simply the power to kill them? Mustn't life remain outside the contract to the extent that it was the first, initial and foundational reason for the contract itself?"[46]

Foucault does not provide examples from the seventeenth- and eighteenth-century theorists of sovereign right. However, we can readily fill in the details of this argument by reference to the classical social contract theorists. For example, Thomas Hobbes and John Locke both believed that individuals had rights in the prepolitical state of nature, although they gave different accounts of the basis of such natural rights. For both, the right to life was foremost among these natural rights. In the case of Locke, the reason that compels men to leave the state of nature and enter into political society is the mutual preservation of their "Lives, Liberties and Estates," which is what he means by their "Property."[47] It is because the natural right to the enjoyment of this property is threatened in the state of nature that rational individuals will choose to enter into political society under sovereign rule. Hobbes and Locke give different accounts of the reasons for insecurity in the prepolitical state and different accounts of the reasoning behind the decision to contract to establish a sovereign power. In both cases, however, the preservation of individual life is the fundamental reason for the establishment of government. In both cases, too, the natural right to life sets limits to the extent of the rights of the sovereign. Locke defended a natural right of rebellion when the sovereign failed in its obligation to protect the property, including the lives, of subjects. In the case of Hobbes, for whom the sovereign is accountable to subjects only via the performance of the ends for which the commonwealth was established in the first place, "the obligation of subjects to the Sovereign is understood to last as long, and no longer, than the power lasteth by which he is able to protect them."[48]

In effect, Foucault argues that the eighteenth- and nineteenth-century transformation in the forms of exercise of sovereign right, whereby it became biopower, was preceded by a transformation in the theory of sovereign right, whereby the lives of subjects were no longer subject to the will of the sovereign but natural, inalienable, and prior to the contract that established sovereign right. In other words, sovereign power was biopolitical in theory long before it acquired the technological capacity to preserve, maintain or enhance the lives of its subjects. Whatever its merits in relation to the history of early modern political thought, this argument raises further questions about the subsequent development of theories of political right. Was there in the course of the nineteenth century a further transformation in the light of this new biopolitical technology? If there was, where do we find the juridical or political representation of government that corresponds to the late modern society in which power is exercised over the biological conditions of human life?

It should be remembered that the concept of biopower is introduced in the final lecture of Foucault's 1975–76 course devoted to the themes of war and race and the role that these have played and continue to play in the history of political thought. Foucault refers toward the end of this lecture to the previous discussion of biopower as a "long digression," entered into to provide an explanation of how it came about that racism became a basic mechanism of modern state power. His argument is that the discourse of race and racism enabled a form of power whose primary objective is to promote life to have recourse to the old sovereign right to kill. Racism enables the sovereign right of death to be exercised within a regime otherwise founded on biopower. It is not that racism only came into existence at this point—"It had already been in existence for a very long time"—but rather that it was "the emergence of this biopower that inscribes it in the mechanisms of the state" to the point that modern states "can scarcely function without becoming involved with racism at some point."[49]

Racism does this by effecting a division of the biological continuum of human life addressed by the techniques of biopower. It divides the species into distinct races and allows for a biological conception of the relationship between them, for example, when the relationship between the different races is thought of along the lines of competition between species. Other races can be represented as a threat to the very survival of the chosen race, thereby legitimizing the state's right to eliminate those other races.

Foucault suggests that the conjunction between the biopolitical state and racism allows us to make sense of a range of phenomena during the nineteenth century. Colonization and the role of social Darwinism was one particularly privileged moment in this history: "Racism first develops with colonization, or in other words with colonizing genocide."[50] The evolution of modern warfare into the attempted destruction of an enemy race was another such moment. Foucault argues that, in general terms, racism came to play a crucial role in the economy of biopower: "The specificity of modern racism, or what gives it its specificity, is not bound up with mentalities, ideologies, or the lies of power. It is bound up with the technique of power, with the technology of power."[51] The paradigm of a biopolitical power that sought legitimation in racial terms is, of course, Nazism. No state relied more heavily on disciplinary power, nor "was there any other State in which the biological was so tightly, so insistently regulated."[52] Foucault presents the Nazi state as exemplary in both its generalization of biopower and its generalization of the sovereign right to kill: a racist, murderous, and suicidal state. However, his point is not that this was an exception but rather that the relationship between biopower and racism was widespread throughout nineteenth- and twentieth-century political thought. Racism was by no means the prerogative of Fascist representations of biopolitical states since it could be found even in the different socialist traditions.

Conclusion

Foucault's "digression" regarding biopower and biopolitics comes at the end of a course that began by raising a series of questions about power and the appropriate means to conceptualize power. In the course of these lectures, he does not really answer the questions posed at the outset about the utility of the "war-repression" model on which much of his own work had relied. Commentators have pointed out, however, that the effects of biopower and the means by which they were achieved are not readily explained by reference to binary relationships of domination or violent struggle between contending forces.[53] In reality, Foucault's efforts to specify the technology of biopower in the narrow sense, through his analysis of mechanisms of security, led to a quite different conception of power, even if this was not spelt out until his 1982 essay "The Subject and Power." Power is here defined in a manner that draws on the lectures in 1978 and 1979 devoted to forms of governmentality. The conception of power as struggle between contending forces is superseded by a conception of power understood as action on the actions of others: "In effect, what defines a relationship of power is that it is a mode of action that does not act directly and immediately on others. Instead, it acts upon their actions: an action upon an action, on possible or actual future or present actions."[54] Action on the actions of others can take a variety of forms, of course. The organization and maintenance of markets is among the most influential ways in which, in the course of the twentieth century, governments have sought to regulate the behavior of those whom they govern. One of Foucault's examples of mechanisms of security developed in the eighteenth century to deal with the problem of grain shortage effectively involved reliance on a free market in grain: food security was to be addressed by establishing conditions under which market incentives would cause individuals to act in ways that address any problem of shortage.[55] As a mechanism of government, this implies a form of power that works on economic subjects or subjects of interest who are also free to act in a variety of ways. Rather than biopower, it is this indirect economic power that becomes the focus of Foucault's theorization of power. Rather than pursue the emergence of the different technologies that made it possible to regulate the conditions of life, in the years that followed his 1975–76 course he became increasingly concerned with the different "regimes of truth" within which the exercise of sovereign power was conceived and rationalized. Depending on the particular form of governmentality in question, the immediate target of state power was not so much life or the conditions of life but the well-being of individuals, civil society, or the proper functioning of market interactions. In Foucault's lectures, biopolitics gave way to governmentality and biopower to the government of free economic agents.

Notes

Parts of this chapter were previously published in Paul Patton, "Life, Legitimation and Government," Constellations *18, no. 1 (2011): 35–45.*

1. Thomas Lemke, *Biopolitics: An Advanced Introduction* (New York: New York University Press, 2011), 33. See also Catherine Mills, "Biopolitics and the Concept of Life,"chapter 4 in this volume.
2. Daniel Defert, "Chronology," in *A Companion to Foucault,* ed. Christopher Falzon, Timothy O'Leary, and Jana Sawicki (Chichester: Wiley-Blackwell, 2013), 62.
3. Foucault writes in "The Birth of Social Medicine" that "society's control over individuals was accomplished not only through consciousness or ideology but also in the body and with the body. For capitalist society, it was biopolitics, the biological, the somatic, the corporal, that mattered more than anything else. The body is a biopolitical reality; medicine is a biopolitical strategy." "The Birth of Social Medicine," in *Essential Works of Foucault 1954–1984*, vol. 3, *Power*, ed. James D. Faubion, trans. Robert Hurley et al. (New York: New Press, 2000), 137.

4. Michel Foucault, *"Society Must Be Defended": Lectures at the Collège de France 1975–1976*, ed. Mauro Bertani and Alessandro Fontana, trans. David Macey (New York: Picador, 2003), 239.
5. Ibid., 253.
6. Michel Foucault, *The History of Sexuality*, vol. 1, *An Introduction*, trans. Robert Hurley (London: Allen Lane/Penguin, 1979), 137.
7. Ibid., 140.
8. Michel Foucault, *Security, Territory, Population: Lectures at the Collège de France 1977–1978*, ed. Michel Senellart, trans. Graham Burchell (Houndmills: Palgrave Macmillan, 2007), 1.
9. Michel Foucault, *The Birth of Biopolitics: Lectures at the Collège de France 1978–1979*, ed. Michel Senellart, trans. Graham Burchell (Houndmills: Palgrave Macmillan, 2008).
10. Ibid., 21–22.
11. Ibid., 78.
12. Ibid., 185.
13. Foucault, *"Society Must Be Defended,"* 243.
14. Ibid., 245.
15. Ibid.
16. Ibid., 246.
17. Ibid. (emphasis added).
18. Ibid., 250.
19. Ibid., 253.
20. Ibid.
21. Foucault, *History of Sexuality*, 140.
22. Foucault, *"Society Must Be Defended,"* 247.
23. Ibid., 12–13. For further analysis of his answers to these questions, see Paul Patton, "From Resistance to Government: Foucault's Lectures 1976–1979," in *A Companion to Foucault*, ed. Christopher Falzon, Timothy O'Leary, and Jana Sawicki, 172–88 (Oxford: Blackwell, 2013).
24. Foucault, *"Society Must Be Defended,"* 37.
25. Foucault, *Security, Territory, Population*, 10–11. At the end of the fourth lecture (February 1, 1978), following the reorientation of this course toward a history of governmentality, he suggests that we still live in the era of a governmentality that acts on the population, that relies on a form of economic knowledge or political economy, and that "would correspond to a society controlled by apparatuses of security" (ibid., 110).
26. Ibid., 185.
27. *Essential Works of Foucault 1954–1984*, vol. 3, *Power*, 334
28. Foucault, *Birth of Biopolitics*, 28.
29. Foucault, *The Use of Pleasure: The History of Sexuality*, vol. 2 (New York: Pantheon, 1985), 26–28.
30. Mitchell Dean draws attention to the parallels between Foucault's conceptualization of ethical practice and his characterization of practices of punishment in *Discipline and Punish*: see his *Critical and Effective Histories: Foucault's Methods and Historical Sociology* (London: Routledge, 1994), 160–61 and 196*ff*.
31. Foucault, *Security, Territory, Population*, 1–86.
32. *Essential Works of Foucault 1954–1984*, vol. 3, *Power*, 337.
33. Foucault, *"Society Must Be Defended,"* 241.

34. Ibid., 240.
35. Ibid.
36. Ibid.
37. Ernst Kantorowicz, *The King's Two Bodies: A Study in Medieval Political Theology* (Princeton, NJ: Princeton University Press, 1957). Foucault refers to Kantorowicz's study in *Discipline and Punish: The Birth of the Prison*, trans. Alan Sheridan (London: Penguin, 1977), 28–29.
38. Foucault, *"Society Must Be Defended,"* 240.
39. Ibid.
40. Giorgio Agamben, *Homo Sacer: Sovereign Power and Bare Life*, trans. D. Heller-Roazen (Stanford, CA: Stanford University Press, 1998).
41. Foucault, *"Society Must Be Defended,"* 240.
42. Agamben, *Homo Sacer*, 111. For further comments on Agamben's reading of Foucault, see Paul Patton, "Agamben and Foucault on Biopower and Biopolitics," in *Giorgio Agamben: Sovereignty and Life*, ed. Matthew Calarco and Steven DeCaroli, 203–18 (Stanford, CA: Stanford University Press, 2007).
43. Foucault, *"Society Must Be Defended,"* 241 (emphasis added).
44. Ibid.
45. Ibid.
46. Ibid.
47. John Locke, *Two Treatises of Government*, ed. Peter Laslett (Cambridge: Cambridge University Press, 1988), 350. Later, discussing the extent of sovereign power, Locke argues that since no one has an absolute arbitrary power over himself in the state of nature and so cannot "destroy his own life," so he cannot give such a power to the sovereign. The power of government is limited to the public good and therefore involves no right to "destroy, enslave or designedly to impoverish the Subjects" (ibid. 357).
48. Thomas Hobbes, *Leviathan*, ed. Richard Tuck (Cambridge: Cambridge University Press, 1991), 153.
49. Foucault, *"Society Must Be Defended,"* 254.
50. Ibid., 257.
51. Ibid., 258.
52. Ibid., 259.
53. See Alessandro Fontana and Mauro Bertani, "Situating the Lectures" in Foucault, *"Society Must Be Defended,"* 283; Guillaume Le Blanc and Jean Terrel, "Foucault au Collège de France: un itinéraire," in *Foucault au Collège de France: un itinéraire*, ed. Guillaume Le Blanc and Jean Terrel (Bordeaux: Presses Universitaires de Bordeaux, 2003), 16.
54. *Essential Works of Foucault 1954–1984*, vol. 3, *Power*, 340.
55. Foucault, *Security, Territory, Population*, 35–44.

Bibliography

Agamben, Giorgio. *Homo Sacer: Sovereign Power and Bare Life.* Translated by D. Heller-Roazen. Stanford, CA: Stanford University Press, 1998.

Calarco, Matthew, and Steven DeCaroli. *Giorgio Agamben: Sovereignty and Life.* Edited by Matthew Calarco and Steven DeCaroli. Stanford, CA: Stanford University Press, 2007.

Dean, Mitchell. *Critical and Effective Histories: Foucault's Methods and Historical Sociology.* London: Routledge, 1994.

Falzon, Christopher, Timothy O'Leary, and Jana Sawicki, eds. *A Companion to Foucault.* Chichester: Wiley-Blackwell, 2013.

Foucault, Michel. *The Birth of Biopolitics: Lectures at the Collège de France 1978–1979.* Edited by Michel Senellart. Translated by Graham Burchell. Houndmills: Palgrave Macmillan, 2008.

———. *Discipline and Punish.* Translated by Alan Sheridan. London: Allen Lane/Penguin, 1977.

———. *Essential Works of Foucault 1954–1984.* Vol. 3, *Power.* Edited by James D. Faubion. Translated by Robert Hurley et al. New York: New Press, 2000.

———. *The History of Sexuality.* Vol. 1, *An Introduction.* Translated by Robert Hurley. London: Allen Lane/Penguin, 1978.

———. *The History of Sexuality.* Vol. 2, *The Use of Pleasure.* Translated by Robert Hurley. New York: Pantheon, 1985.

———. *Security, Territory, Population: Lectures at the Collège de France 1977–1978.* Edited by Michel Senellart. Translated by Graham Burchell. Houndmills: Palgrave Macmillan, 2007.

———. *"Society Must Be Defended": Lectures at the Collège de France 1975–1976.* Edited by Mauro Bertani and Alessandro Fontana. Translated by David Macey. New York: Picador, 2003.

Hobbes, Thomas. *Leviathan.* Edited by Richard Tuck. Cambridge: Cambridge University Press, 1991.

Kantorowicz, Ernst H. *The King's Two Bodies: A Study in Medieval Political Theology.* Princeton, NJ: Princeton University Press, 1957.

Le Blanc, Guillaume, and Jean Terrel. "Foucault au Collège de France: un itinéraire." In *Foucault au Collège de France: un itinéraire,* edited by Guillaume Le Blanc and Jean Terrel, 7–26. Bordeaux: Presses Universitaires de Bordeaux, 2003.

Lemke, Thomas. *Biopolitics: An Advanced Introduction.* New York: New York University Press, 2011.

Locke, John. *Two Treatises of Government.* Edited by Peter Laslett. Cambridge: Cambridge University Press, 1988.

Patton, Paul. "Agamben and Foucault on Biopower and Biopolitics." In *Giorgio Agamben: Sovereignty and Life,* edited by Matthew Calarco and Steven DeCaroli, 203–18. Stanford, CA: Stanford University Press, 2007

———. "From Resistance to Government: Foucault's Lectures 1976–1979." In *A Companion to Foucault,* edited by Christopher Falzon, Timothy O'Leary, and Jana Sawicki, 172–88. Oxford: Blackwell, 2013.

PART TWO

The Question of Life

SIX

Foucault, Cuvier, and the Science of Life

MARY BETH MADER

Over the course of his writing career, Michel Foucault treated the concept of life, including the conditions for its emergence as an object of study in the science of life, and its importance as the object of a new modern form of power that he termed "biopower." His discussions of these points do not evidently cohere, since there are multiple variants of the concept of life as it develops in biological thought, and since Foucault does not always clearly indicate the specific senses of the notion of life that are critical to his conception of biopower. Further, despite its current fertility for philosophical and political inquiries into how contemporary versions of power relate to the notion of life, it is not clear which aspects of the varying versions of the concept of life analyzed by Foucault do or ought to ground this conception of biopower or to impress us as analytically and politically important today.

This chapter exposes some of the conceptual bases for a warranted present interest in Foucault's analyses of conceptions of life drawn from the history of biological thought and of early biological inquiry. Its chief thesis is that through his account of the relations between classical natural history and Cuvier's introduction of a functionalist notion of life, Foucault exposes specific versions of visibility, possibility, and abstraction that can contribute to our contemporary philosophical thinking about life and its sciences. Since he does not always do so directly, the paper seeks to present and to highlight what are at certain points quick claims or implicit suggestions about these specific kinds of visibility, possibility, and abstraction. The argument of the chapter is developed by attention to Foucault's thought in *Les Mots et les choses* (1966)[1] and in "La situation de Cuvier dans l'histoire de la biologie," his presentation at a 1969 colloquium devoted to examination of the thesis on Cuvier that Foucault advanced in the

1966 work.[2] The chapter is mainly limited to drawing out the specificity of the kinds of visibility, possibility, and abstraction—and their necessary relations—treated by Foucault.

In 1969, the Institut d'Histoire des Sciences et des Techniques of the University of Paris organized a colloquium for the bicentennial of the birth of the French naturalist and zoologist Georges Cuvier (1769–1832).[3] The colloquium presented an occasion to discuss the work of one of the Institute's research groups that had undertaken to critically examine a thesis Foucault had propounded in his 1966 publication, *Les Mots et les choses.* Colloquium participants included Francis Courtès, Camille Limoges, François Dagognet, Michel Foucault, Yvette Conry, Bernard Balan, and Georges Canguilhem. The presentations and discussion were published the following year in the *Revue d'histoire des sciences et de leurs applications*. The main thesis examined in the colloquium was Foucault's position that contrary to construals of Cuvier's contributions to biological thought as essentially fixist—and therefore as obstacles to the development of evolutionary biology—Cuvier's functionalist taxonomies in fact conceptually prepared the way for Charles Darwin's liberation of animal forms from atemporal, agenealogical understandings of speciation. The present chapter aims to shed light on the importance of Foucault's thesis and its defense by means of an analysis of the specific notions of visibility, possibility, and abstraction that can be found in it. We begin with a brief synopsis of Cuvier's thought on life.

Cuvier's Thought on Life

In his 1805 *Leçons d'anatomie comparée,*[4] Georges Cuvier opens his reflections on the conception of life with a consideration of the living body of a healthy, young woman, noting that all her features of color, warmth, gracious movement, and charm can disappear in sudden death. Her death is construed as the exit of fluids, evaporations, of both the attraction and dissipation of humidity. Death is a final separation of elements—air, water, heat—from the body, and their sympathetic return to those same surrounding elements outside of the body. But, Cuvier wonders, if bodies are always surrounded by these elements, how could death as the separation and exit of the body's internal elements of air, water, and heat be avoided? Cuvier reasons that prior to death the force of this elemental "elective attraction," which would otherwise act to withdraw the elements from the body, must be counteracted by another, stronger cohesive force. This more powerful force that would act to bind or retain the internal elements so as to prevent their exit from the body must be the cardinal feature of the "general idea of life."[5] This dynamic concept of life is then modified, however, to include the view that it must at least consist in a continuous, mutual circulation of elemental "molecules" between outside and inside a living body.[6] This conclusion is reached from the observational premise that elements are constantly entering and leaving living beings and that they do so without living beings either growing or living unceasingly. That living beings stabilize at relatively fixed sizes and that they die imply not merely the existence of a binding, retentive internal *force vitale*. They imply the partial release of some elements from the living being during its lifetime and the eventual overcoming of the binding force of life by the power of external elements. Living bodies, then, are centers into which dead substances enter to combine and function, taking part in a generalized circulation of elements that constitutes life. For elements within a body to return to the surrounding world amounts to a separation from the vital forces that occupy not the individual parts of an organism or an organized body, but only the body as a whole. That body parts severed from the organism "die" demonstrates the holistic and interrelated nature of the property of life; that property should be ascribed to the living body in its entirety, Cuvier notes, citing Immanuel Kant's distinction between living and inert bodies.[7]

One of Cuvier's principal contributions to the history of biological thought, including according to Foucault's reconstruction and contestation of portions of that history, has been to determine the notion of a function and its classificatory use for the ordering of natural beings. The critical point about Cuvier's use of the notion of a function is its classificatory one. Representations of the relations between natural beings are radically reconceived on Cuvier's schema by being ordered on the basis of common combinations of functions, and hence of systems of organs and types of bodily organization.

We can present in summary form some of the principal theses of Cuvier's understanding of life. First, all functions concern essentially the circulation and transformation of fluids.[8] Second, functions may be exercised in different individuals by different organs and systems of organs. Cuvier emphasizes that a single function, such as respiration, is carried out in different classes of animal by organs whose structures have no common feature. (In fish and mollusks, respiration—of water—is accomplished by gills; in all animals, respiration of air is carried out by lungs. Lungs and gills purportedly have no common structural feature.[9]) So, it is important to note that both structure and organ are dissociated from function, on his view. Cuvier specifies: "these differences in organs of the same sort are precisely the object of comparative anatomy."[10] Third, and most important, Cuvier groups species into higher taxa on the basis of their shared concomitant functions, or resemblances between systems of organs. It should be noted that Cuvier uses already existing terms for species of organisms, but his functional taxonomy means that organisms will be grouped together in novel ways. Fourth, the combinations of functions, as indicated earlier, yield biological incompatibilities:

> These combinations, which appear possible when considered in an abstract way, do not all exist in nature; for in the state of life, organs are not simply proximate to each other but they act upon each other, and cooperate for a common end. In this way, modifications of one organ exert an influence upon the modifications of all of the other organs. Certain of these modifications that cannot exist together mutually exclude each other, while others call for each other, so to speak. And this is so not only for the organs that are immediately related, but even for those that appear at first glance to be most distant and independent from each other.[11]

So, for example, the functions of respiration and circulation mutually imply each other since respiration serves chiefly to bring the surrounding element, whether air or water, into the circulating blood.[12] Predator animals that can digest only flesh, Cuvier holds, must also have faculties for tracking, seizing, subduing, and tearing. Thus, the organs for locomotion, perception, and sensation are in a relation of mutual dependence with the function of digestion—on pain of death of the organism.[13] Fifth, Cuvier claims that the mutual dependence of functions, as well as the dependence of the systems of organs they require, is governed by laws whose necessity is "equal to that of metaphysical or mathematical" laws.[14] Sixth, an epistemological consequence of this necessity is that knowing one of the parts of an animal permits the deduction of most of the other parts from this single telltale one, and on occasion the inference to organs with other functions. So, since teeth suited for tearing flesh should never be found on hooved animals, coming across a hooved animal will permit the exclusion of the hypothesis that it bears teeth suited for tearing flesh.[15]

This conception of life implies a point that was of particular interest to Foucault: Cuvier explicitly takes his classificatory prioritization of the function to imply the falsity of the traditional scale or chain of being, in which all beings—living and inert, organic and inorganic—were ranged in a hierarchy of being and perfection, from lowly minerals to plants, animals, human beings, angels, and, at times, God. This is an expression of the foundational homogeneity and continuity guaranteed by the creative unity of divine power, a kind of natural theodicy, that governed thought about the natural world into the classical period. Cuvier admits a hierarchy of importance for organs, in relation to their execution of functions which themselves must be ranked according to their centrality and necessity for the maintenance of the life of the organism, within "the limits prescribed by the necessary limits of existence."[16] The functions of digestion and sensation are more important for an individual organism, in most cases, than is the function of generation. Further, given this sort of hierarchy of function and organ, which varies by species of organism, organisms share combinations of functions and organs, and thus could be ordered on the basis of the simplicity or complexity of their organic instantiations of their functional requirements. Cuvier admits that functions and organs can be determined to be more or less simple, as well as more or less perfect, according to a traditional, classical ordering criterion for the natural world. However, he denies the continuity and graduated nature of the relations of organisms represented in the scale or chain of being.

In one of Cuvier's arguments against the chain of being, he glimpses one of the consequences of his view that the proper starting point for the classification of natural beings is the level of functions or the organs that satisfy them: it is difficult to know which function or type of organ ultimately should serve as the fundamental taxonomic starting point. Cuvier clearly struggles with this, at points elevating digestion or respiration over other functions for this foundational taxonomic role.[17] But he sees, at points, that if functions are not ranked for the purpose of taxonomic ordering, the result will be an unruly multiplication of competing taxonomies, depending on which function is taken as the classificatory starting point. The result of his occasional insistence that functions must be ranked is the conclusion that there is no necessity for a continuity of forms of living beings. One cannot suppose that natural beings are in a continuous relation for the reason that there simply are not necessarily natural beings that are *functional* intermediates. On this basis, Foucault will argue that the continuity of extended being found in natural history meets its end in the modern ontology of the function.[18]

Visibility

"Natural history is nothing more than the nomination of the visible."[19]

On Foucault's account in *Les Mots et les choses*, Carl Linneaus (1707–1778), Swedish botanist, zoologist, and taxonomist, is the prime representative of classical taxonomy and its tabular, homogeneous form of representation. For Foucault, the classical age conceived of the relations of natural beings to each other without the modern, synthetic notion of life. Instead, it conceived of natural beings as related in a chain of being in which all beings were ultimately related through their fundamental sameness as extended beings. It is Cuvier who, on Foucault's interpretation, was responsible for the conceptual weakening of the Linnaean conception of the origins, identities, and relations of natural beings. But how exactly are natural beings purportedly related to each other without and prior to the concept of life? What does the advent of the concept of life mean for the relations of natural beings to each other? Notably, in Foucault's recounting of the emergence of the science of life from the domain of natural history, "natural beings" are not synonymous with "*living* forms." Natural beings are linked in an ontological relation of possibility while living forms are related by (the) conditions for life. On Foucault's view, the differentiation of natural beings in the classical age rests on a continuity of *being* instead of a continuity of life. The spatial juxtaposition of natural beings on the table of the classical age represents the fact that their external relation to each other crosses no ontological gap; further, natural beings are not beings whose being depends on a relation to external milieus, to abstract vital functions, or to competing natural threats. The fact that they are disposed in one single space displays the ontological homogeneity, and its continuity, its intrinsic scaling possibilities, that presume that natural beings are ontologically of a piece.

The continuity of the space of the spatial representation in the natural history of the classical age aptly represents the external relations of natural beings to each other and the continuity of nature. For this spatial continuity also represents the fact that these external relations all take place within a single dimension—the dimension of the visible. For this reason Foucault describes the classical form of representation as flat or as superficial. Visible features are related to visible features and animal forms are related by continuously differentiating, gradating divergence from each other as visible, and not, notably, from some invisible, insensible, abstract, and functional template in relation to which a set of living forms would vary.

One important aspect of this account concerns the specificity of the classical representative project. In particular, its one-dimensionality stems from the functionally metonymic relation of the visible to itself. Visible things, in their visibility, can represent the visibility of other visible things. This means that the visible nature of other visible things can be and *is* represented in the classical table of natural beings. It means that Foucault interprets this classical exposition of natural beings to be displaying the fact that in this period natural beings are conceived to be essentially and exhaustively available to sight, to be completely presentable to sight, and then to be entirely *re*presented to sight as well. Thus, it is not merely the fact that natural beings were considered to be representable that is important here. It is also that their representability is thought to be complete, that they are considered to be representable without remainder and that their representability is itself susceptible to display. In addition to being each individually visible, they share this same visibility, as well.

By contrast, visibility in the modern epoch will itself undergo a change since it is no longer chiefly autologically related to itself but put into communication with aspects that are not visible and hence not representable to sight. Importantly, this means that what *is* counted as visible changes in its nature since it is now related to the invisible. What kind of visibility advenes, at this point, if and when the visible is pressed to convey complex and abstract relations between biological functions and their corresponding organic structures, neither of which is in itself locatable in any individual organism or even organisms? The critical factor here is that the biological function is an entirely abstract and general notion that can find itself instantiated in any number of different types of organisms. Individual organisms and species of organism are instantiations of general problematic templates; they are solutions to species' or organisms' problems of survival. But notably, since a function is abstract and applies to the conditions of existence of many differing organisms and types of organisms, in principle there can be no visible representation of it or of all of its instantiations. For example, the function of *respiration in general* cannot be represented by a representation of any given breathing organism or indeed by a representation of all breathing organisms. It cannot be represented by a representation of any given breathing animal because such a representation would not show the other instantiated versions of respiration in general. These versions differ from each other precisely because biological functions by definition, for Cuvier, are integrally interrelated or correlative, and because organic structure itself follows from achieved functional solutions to the problems of individual organisms and species in relation to their milieus and to other living forms.

Clearly, this representative failure contrasts with the metonymic representability of natural beings in the classical *épistémè*. For in that period, there was no limit in principle to the capacity of one visible natural being to represent, as a scaled version of other natural beings, or as a representative *type* of classified natural being, as an instance of a class of beings, others of its type. But to share a type was to share a *visible* type or a type in which the salient relations of parts and subparts were all equally visible and representable, as well as to share a classificatory place in the table. It is not that there were not important relations between these characters for the classical thinker of the natural world, but those characters and relations all occupied the same "plane" since they were all equally visible and thus representable. By contrast, in the modern era, no visible element can stand in for or substitute either for any other visible element or for an invisible element.[20]

Possibility

For Foucault, one of the chief aspects of the ontological difference between a function and the living individuals in which it is instantiated is that the multiple realizability of the function implies a new sort possibility. In the classical era, the tabular representation of the relations of natural beings to each other, itself an ontological sector of the greater expanse of being itself, employed spatial juxtaposition to represent the graduated nature of the relations of these natural beings to each other. Foucault takes Linnaeus's classificatory schema to exemplify the pre-Cuvier genre of taxonomy that exploited the spatial and temporal homogeneities of the "non-temporal rectangle" found in classical "herbariums, collections, gardens," as well as in the better-known grid or table.[21] In this Linnaean schema, the nomination of natural beings reposed on the four variables that can be applied to their elements: number, form, disposition in space, and size.[22] Representations of plant and animal forms arranged on the classificatory basis of these visible characteristics grounded the display of the relations of natural beings on varieties of their extended features. They thus employed a philosophical category—the primary quality of extension—as the chief medium of taxonomic classification. By its very extended nature, the spatial extension of the Linnaean table captures the extended features of the natural beings represented in it.

Of course, to most ready readers of the tabular form of representation, notice of this spatial correspondence must initially appear a trivial or vacuous observation. For such an observation to have a hope of carrying any philosophical, historical, or conceptual significance, it must consistently be considered in a comparative light. Thus, in this case, we can seek to show the implication of the Cuvierist form of classification precisely by contrast to the Linnaean representative schema. One way that we can do this is with respect to their modal differences, that is, with respect to the differing kinds of possibility that they imply or employ.

In the Linnaean tabular schema, representations of natural beings are set out in relation to each other in such a way that undepicted natural forms are literally envisionable or foreseeable.[23] Such undepicted forms can be interposed imaginatively between existing representations; viewers of the well-ordered garden, table, or herbarium can envision unrepresented plant or animal forms that would be intermediate between the represented forms. All that need be done to insert into the space of representation possible plant and animal forms not already depicted there is to envision modifications of the four Linnaean variables. Possibility in the classical age, then, is a matter of the extrapolation of extended features and is essentially the envisioning of the modulation of depthless, extended, visible, countable, and measurable features.[24] This species of natural possibility is fundamentally anticipatable through the application of means for the transformation of extended quantities. Thus, all change in the four variables is essentially geometrical change, and the foundational kind of homogeneity of natural beings themselves, as well as in relation to each other, is fully expressible in terms that are ultimately geometrical.

A consequence of grounding thought about natural beings on the representation of their extended features and, more importantly, on an undergirding continuity of being, was that naturalists created tables of natural beings that included spaces for purely conjectural creatures. French philosopher and historian of science Henri Daudin (1881–1947) provides an instructive example of this. In *Les Classes Zoologiques et l'idée de série animale: Cuvier et Lamarck,* Daudin discusses the tabular results advanced by Bernard-Germain-Etienne Delaville, Comte de Lapécède (1756–1825), a French zoologist from a generation older than Cuvier but very much influenced by him.[25] Daudin writes that Lapécède's method of classification produced, in addition to "really given forms," categorical voids ready to be completed with beings that remained to be discovered. About the divisions devised in Lapécède's method, Daudin explains that "a good number . . . remain *purely virtual,* since the combinations of characters that it predicts are presented, in fact by no known animal."[26] It is important to note, however, that this empty tabular space is not a kind of discontinuity; rather, a virtual place in a table, prepared for eventual occupation by a possible being, differs radically from a break between organismic series unsupported by any suppositional continuous ground to be filled in.

Lapécède's tables in *Histoire Naturelle* (1804) show this conjectural space clearly. For instance, he draws up a "table of combinations of the different numbers of front foot toes and rear foot toes in oviparous quadrupeds."[27] The table is a sheer combinatoric grid of all of the permutations for combining various numbers of front and rear toes with their matching types of quadruped species. The possible combinations of numbers of toes are purely arithmetic pairings of numbers: 5 front + 5 rear; 5 front + 4 rear; 5 front + 3 rear; 5 front + 2 rear; 5 front + 1 rear; 4 front + 5 rear; 4 front + 4 rear, and so on. Lapécède notes that there is a combination for which a corresponding reptile is missing: "An oviparous quadruped didactyl, that is, one that would have two toes on each foot, would still be needed to fill the empty space that we find in a series of quadrupeds arranged according to the number of toes on each of their four feet. We must believe that this still unknown species exists, and that it will be discovered, as were the tetradactyl and the monodactyl."[28] The kind of possibility conceivable here is literally disposed in representational space and limited by the combinatorics of visible characteristics.

A different sort of possibility, based on a different sort of *im*possibility, is discernable with the advent of the classificatory schema introduced by Cuvier and based on the notion of the function. This revelation can be characterized as the exposure not of undepicted though possible forms, but as the indication of impossible living forms. Due to the foundational role of life functions in Cuvier's classification of living beings, what show up instead of possible natural beings that are geometrical extrapolations of extended forms are vitally impossible beings. That is, the obstacle to their existence is not some purely geometrical impossibility—the impossibility of a sort of natural square circle—but the reality that certain conceivable combinations of functions would in fact exclude the life of the organism that is supposed to be organized by and on the basis of those functions. It is the notion of "biological incompatibility" and, hence, biological impossibility that replaces the natural possibility of the classical age.

As we have seen, this natural possibility is simply the set of conceptual consequences of the homogeneity of spatial extension. But the new sort of possibility that emerges with Cuvier's comparative anatomy is the set of organismic possibilities secondarily delimited by primary incompatibilities located on the functional level. These incompatibilities are of two sorts: first, only certain combinations of instantiations of life functions (chiefly, respiration, digestion, circulation, sensation, generation, and locomotion) can coexist. It is the impossibility of the coexistence of instantiated types of life function that determines the possibility of the life of a type of organism as a whole. That is, the limits of functional concomitance restrict the constitutional possibilities on the organismic and species levels. As mentioned above, only certain solutions to the functional problems of ingestion can coexist with certain solutions to the problems of locomotion. Purportedly, if an animal form must move to live, this is because, unlike plants, it cannot absorb and incorporate external elements into itself through its roots.[29] Thus, the life functions of locomotion and digestion (or absorption) are linked, such that a deracinated organism must move since roots both fix an organism in place and are a particular instantiation of the general function of absorption. Second, allegedly, certain sets of functions, along with their organs, cannot coexist; so, claws cannot be found on ruminant animals, and herbivores will never have teeth suitable for tearing flesh. These two forms of functional incompatibility imply that there exist limits to the combinations of functions and to the combinations of organs that execute those functions; these limits are at once limits to the conceivability of combinations *and* to their sheer existence. These functional limits to organismic existence, as developed by Cuvier, imply that the extrapolative insertion of imagined intermediate forms into the classical taxonomy, which characterized some classificatory thought that followed the Linnean taxonomic practice, as well as hybrid schemes such as that of Lapécède, could well illegitimately interpolate impossible forms by failing to attend to the biological impossibilities knowable on the basis of functional incompatibilities. In addition, these limits do not yield conceptual spaces that are *a priori* determinable as open to being realized by an actual organism. Instead, they create micro-series of organs and macro-series of functions, which two types of series are discontinuous relative to each other, and which offer no intrinsic reason to suppose that their discontinuities will eventually be closed by union with other functional or organismic series.[30]

Abstraction

One might be tempted to think that the obstacle allegedly posed to representation by the abstract function, on Foucault's account, would be removed through application of a standard type-token distinction.[31] On this view, the alleged difficulty would contain nothing of particular significance for biological thought, but would merely be an instance of a common distinction that obtains between a categorial type of general scope and its instantiation(s) in one or more tokens of a particular nature. In this way, we would interpret the supposed problem of the representability of the abstract function as simply the "obstacle" that any general categorial term or type would encounter: the type itself as a type is not representable, but its tokens as tokens and, more important, as tokens of their type, are representable, we might thus propose. Foucault's error would be to hold that the advent of the abstract function in natural history presents a revealing problem of representation when it is merely a case of the well-known difference between a type (in this case, say, the abstract function of respiration) and its token (for Cuvier, this particular form of respiration by means of gills). One could further note that this kind of abstraction, namely, the abstraction of function, would not be any *more* abstract than any instance of a type-token distinction. On this view, the abstraction of the species or genus identified on the basis of types of functional organization would be no more abstract in relation to instances of living forms than would be the abstraction of species or genus identified as a type of natural being in classical natural history, relative to instances of those natural beings.

This question of the particular nature of the abstraction at issue allows us to approach Foucault's famous periodizing claims about the break in forms of knowledge between the classical and modern ages—the "new epistemological arrangement"[32]—from the angle of what might be termed a genealogy of abstractive practices. More precisely, it permits us to suggest that the distinction between a classical kind of abstractive practice, in which naming species and higher taxa on the basis of visible primary qualities and their variations must *amount to* classifying them, is, with respect to the history of thought, strikingly different from a modern kind of abstractive practice in which the sortal criteria are *never visible either as types or as tokens*. To use the language of type and token, anachronistically, we could say that while visible tokens of natural beings and their extensively varying parts (leaves, jaws, shells, limbs, stalks, etc.) serve as the matter for classifying individual natural beings by means of naming in the classical schema, the modern approach relies rather on nonextended, *permanently* imperceptible processes whose imperceptibility is not merely the consequence of their being generic categories instead of particular instances. More precisely, the specific imperceptibility at issue in the case of the abstract function is not due to its typal, or generic, nature, but rather is due to the facts that (1) functions are (teleological) processes, (2) processes are not visibly apprehensible as modifications of extended matter, (3) physiological processes take place both superficially and in a body's interior, and (4) they also imply interactions with the organism's surrounding milieu.

So a function is already in itself "abstract" simply relative to the classificatory, criterial variables of visibility, extension, and surface.

However, to understand the sort of abstraction in question here, we must consider the specific imperceptibility of the function in conjunction with the fact that it comes to serve as a classifying character. To grasp this point, we would do well to heed Foucault's characterization of the transformation that the notion of a function undergoes via Cuvier's work as the "transition of the function into *evident* invisibility."[33] We find an important example of this transition that the notion of a function purportedly undergoes in Foucault's account of the challenge that Cuvier's thought represents to a dominant conception of reproduction in the classical age. Foucault holds that the gradual discovery, through anatomical dissection, of the material reality of reproductive processes is the basis for a radical change in the conceptions of life, the individual organism, and the means of species classification. The classical age, on this view, essentially considers reproduction to be a form of growth and of the continuity of life. Foucault writes that "life is defined by its possibility for extension. Life is that which continues something and that which itself continues."[34] Thus, in reproduction considered as this temporal extension of life, the individual organism is continued by addition, through its corporeal growth, until it is then continued by multiplication through procreation. Strictly speaking, then, there is no separate life function of "reproduction" in the classical age; there is only growth by the multiplication that is life's continuity. On this account, the view of reproduction as a function of an individual organism emerges gradually in the modern development of biology. Although the function is "invisible," it becomes "evident" as a distinct feature attributable to an organism. Prior to this, the "growth" account of reproduction dominated the classical age, which simply saw procreation as "the apex of the individual," as "the point at which its growth became proliferation."[35] In the modern period, by contrast, reproduction becomes "evident" as a separate, determinate function, instead of being simply a consequence of the continuity of life, and of individual growth; but, of course, as a function, it was not visible.

The important conclusion to draw here is that the particular *sort* of imperceptibility of the abstract function is the imperceptibility of something that is *also* evident. The reason that the type-token distinction fails properly to apply here is that nothing about it implies this particular combination of the *historical* emergence of an evident reality—here, reproduction as an organismic or life function—that is also empirically imperceptible. Like life, the "reproductive" function was not evident in the classical age, while in the modern age it is evident and invisible. Of course, it is invisible because, as a process, it is not an extended reality graspable by the faculty of vision, and because, as a physiological process that coordinates many sub-elements of the organism, and thus is a mode of organic organization, it must be sought in the depths of the living form. But the abstract function is "evident" thanks to its manifest capacity to ground the new taxonomic practices introduced by Cuvier; these practices precisely split off taxonomy from nomenclature, or classification (division) from naming, so as to introduce classifiability into the very constitution of the organism. Instead of the organism being classifiable because it *is* in the same way that naming beings (words) *are*, it is newly classifiable because its being is now life, and because life forms are to be grouped by life functions, or processes aimed at the continuation of life. The being of the modern organism, or living form, is to be classifiable because it is living, whereas the being of the classical natural being was to be spatially, visibly of a piece with all beings, living or not. The classical natural being is classifiable because of its shared being with all beings, including words; the modern living form is classifiable because life, as the concatenating telos of its functions, is now intrinsic to the form.

Foucault traces, with excruciatingly fine differentiations, the varieties of abstraction, and of what I have termed abstractive practices, in the history of Western thought. To describe the transition from natural history to biology via the abstract function as either an unspecified sort of abstraction or as an exploitation of the type-token distinction would appear to be misguided. For Foucault is arguing that there are historical-epistemological limits to the kinds of abstraction we can employ. What counts as abstraction will vary historically, and exposing the history of this variation is precisely one of the aims of Foucault's work. Here, then, we can spell out some of the difficulties in identifying abstraction in this case. For Foucault claims that words and things in the classical age occupy a common ontological matrix. To name is to classify; nomenclature and taxonomy are isomorphic. In the case of natural history and biology, it is the emergence of biology that breaks apart this consonance. So, in what sense would words be abstractions relative to what they designate in the classical age? There are, of course, many senses of abstraction. We might mean that an abstraction is of a markedly different ontological sort relative to that from or of which it is abstracted. On this sense, even words of Foucault's classical sort, which classify by naming, are importantly *not* abstractions relative to that which they classify. They can designate, but they do not do so across an ontological or epistemological divide. Both words as designators and things as designated are beings. Designating beings are beings as are designated beings. Moreover, the source of their designating power reposes precisely on the fact of this shared being. If a kind of abstraction rightly describes the classical relation of designation, it is not this sense of abstraction.

How should we understand the abstraction at issue in the abstract function in modernity? A canonical version of abstraction as the conceptual neglect of sensible qualities accompanied by the preservation of intelligible ones will not aptly describe the specific nature of the abstraction at issue in the modern abstract function. But neither will the type-token distinction satisfy, here, if we attend to the particularity of the case at hand. The abstractness of the abstract function of reproduction has an ontological seat in the individual, whereas prior to this period, even if procreation by growth was itself a perfectly generic concept in that it potentially applied to numerous particular tokens, it was in fact not a direct feature of individuals; it was not even an effect of individuals, but was rather a sign of the power of life exercised through individuals. In other words, to arrive at the notion of the abstract function of reproduction, one did not abstract away, say, from particular versions of growth and procreation (in various species, say), to derive a purified generic concept of "reproduction." Reproduction is an abstract function of an individual, whereas procreation, or growth in the form of generating offspring, is the continuation of life through individuals. It is, strictly speaking, not a function of an organism. Reproduction has an ontological seat in the causal individual, while procreation has an epistemological place in the signifying individual. With procreation, offspring are *signs* of life's power of continuity, not *effects* of the causal reproductive power of an individual organism.

The relatively empty, indeterminate concepts of generality and particularity used in the type-token distinction are themselves likely products of postclassical, modern, and postmodern thought. For the varieties of abstraction found in classical and modern thought on the natural world are either, in the classical case, prior to a strong ontological distinction between designator and designated, or, in the modern case, still radically embedded in causally active individual organisms. Even modern abstraction, then, crafted out of a simultaneous invisibility and evidence, with all of the internal tension that this conjunction implies, does not achieve the sort of truly blank, intrinsic abstraction of the type-token distinction. The foregoing exposition is intended to shed light on the sort of historico-epistemic distinctions to which Foucault seeks to draw our myopic attention and which ground his claims that the successive *épistémès* he identifies have meant that human beings have experienced markedly different forms of order across time. Contemporary philosophical approaches to understanding the natural world would do well to develop their own means for attending to the various remnants, reuses, and mixtures of past versions of visibility, possibility, and abstraction that our present practices continue, in order not to be determined in our current conceptions by undiscerned vestigial presuppositions that are in fact never explicitly posited.

Notes

1. Michel Foucault, *Les Mots et les choses* (Paris: Gallimard, 1966), trans. Alan Sheridan as *The Order of Things: An Archaeology of the Human Sciences* (New York: Random House, 1970).
2. Michel Foucault, "La situation de Cuvier dans l'histoire de la biologie," *Revue d'histoire des sciences et de leurs applications* 23, no. 1 (1970): 63–92, http://www.persee.fr/web/revues/home/prescript/article/rhs_0048-7996_1970_num_23_1_3117; and in Michel Foucault, *Dits et Ecrits*, vol. 1, *1954–1988*, ed. Daniel Defert and François Ewald (Paris: Editions Gallimard, 2001), 898–934.
3. Institut d'Histoire des Sciences, Université de Paris, May 30–31, 1969.
4. Georges Cuvier, *Leçons d'anatomie comparée*, vol. 1 (Paris: Baudouin, 1805), 2 (repr. Bruxelles: Impression Anastaltique, Culture et Civilisation, 1969).
5. Ibid., 3.
6. Ibid., 5.
7. Ibid., 6.
8. Ibid., 33.
9. Ibid., 41.
10. Ibid., 35.
11. Ibid., 46.
12. Ibid.
13. Ibid., 55 ("et on sent que s'il ne trouvoit pas dans ses sens et dans ses organes du mouvement les moyens de distinguer et de se procurer ces sortes d'alimens, il ne pourroit subsister").
14. Ibid., 47.
15. Ibid., 56.
16. Ibid., 58.
17. Foucault outlines this struggle in *Order of Things*, 260.
18. In "La situation de Cuvier dans l'histoire de la biologie," Foucault specifies that "Cuvier criticizes the chain of beings and not continuity" (81). In *The Order of Things*, Foucault stresses the kinds of discontinuities that Cuvier's conception of life introduces into the classical schema's ontology of natural beings, but he also plainly argues that these discontinuities themselves are accompanied by another, new kind of continuity between the organism—even if closed on itself according to this new conception of life—and its surroundings, which act as external conditions for the organism's existence.
19. Foucault, *Order of Things*, 132.
20. However, Foucault does suggest that some biologists of the modern period seek to use visible elements to represent the invisible; in such cases, the character can be a sign of the invisible organic structure. Indeed, I would add that Cuvier himself at

times affirmed the proposition that some internal characters were linked with external characters and were thus "indicating characters" (*"caractères indicteurs"*).

21. Foucault, *Order of Things*, 131.
22. Ibid., 134. In "La situation de Cuvier dans l'histoire de la biologie," Foucault adds that continuity in the classical age "is not simply spatial, but temporal" (90).
23. Foucault, *Order of Things*, 130.
24. Ibid.
25. Henri Daudin, *Les Classes Zoologiques et l'idée de série animale: Cuvier et Lamarck*, vol. 2 (Paris: Librairie Félix Alcan, 1926). More precisely, instead of being a canonical Linnaean thinker, Lapécède is better considered a hybrid thinker who joined some of Cuvier's functional analysis with the more classical tabular sort of zoological thought.
26. Ibid., 39 (my emphasis).
27. Bernard-Germain-Etienne Delaville, Comte de Lapécède, *Histoire Naturelle*, vol. 1, new ed. (Paris: Jouvet, 1881), 273.
28. Ibid.
29. Cuvier, *Leçons d'anatomie comparée*, 1:12–13.
30. For more on this point, see Foucault, *Order of Things*, 270–71.
31. My thanks to Stephan Blatti for insightful discussions on this point.
32. Foucault, *Order of Things*, 257.
33. Ibid., 264.
34. Foucault, "La situation de Cuvier dans l'histoire de la biologie," 90 (my translation).
35. Ibid., 91 (my translation).

Bibliography

Cuvier, Georges. *Leçons d'anatomie compare*. Vol. 1. Paris: Baudouin, 1805 (Première leçon, Article premier). Reprinted edition: Bruxelles: Impression Anastaltique, Culture et Civilisation, 1969.

Daudin, Henri. *Les Classes Zoologiques et l'idée de série animale: Cuvier et Lamarck*. 2 vols. Paris: Librairie Félix Alcan, 1926.

Foucault, Michel. *Les Mots et les choses*. Paris: Gallimard, 1966.

———. *The Order of Things: An Archaeology of the Human Sciences*. Translated by Alan Sheridan. New York: Random House, 1970.

———. "La situation de Cuvier dans l'histoire de la biologie." *Revue d'histoire des sciences et de leurs applications* 23, no. 1 (1970): 63–92. In *Dits et Ecrits*, vol. 1, *1954–1988*, edited by Daniel Defert and François Ewald, 898–934. Paris: Editions Gallimard, 2001.

Lapécède, Bernard-Germain-Etienne Delaville, Comte de. *Histoire Naturelle*. Vol. 1, new ed. Paris: Jouvet, 1881.

Revue d'histoire des sciences et de leurs applications 23, no. 1 (1970). http://www.persee.fr/web/revues/home/prescript/article/rhs_0048-7996_1970_num_23_1_3117.

SEVEN

The Archaeology of Biopower: From Plant to Animal Life in *The Order of Things*

JEFFREY T. NEALON

If cross-disciplinary movements in the North American university function like financial instruments (which, of course, they do and they don't), the strongest "buy" orders of recent years would have to come from the burgeoning discourse surrounding biopower and the related body of work dedicated to animal studies. One supposes that this collection is itself evidence of continuing strong scholarly interest in biopower. And animal studies, for its part, is making a strong pitch to be the "next big thing" in the academy, or so the *New York Times* has announced.[1]

These two emerging fields of practice are intimately related, of course: if biopolitical studies began by pointing out that questions pertaining to human "life" have become *the* political topics of the modern era (revolving around practices of identity, health, and sexuality), animal studies steps in to show how that notion of human-centered biopower is itself based on an originary exclusion and abjection of its other, animal life. In classic deconstructive form (not surprising that Jacques Derrida is one of the most often cited figures), animal studies shows how the privileged term of "biopower" (human life itself) is made possible and remains hegemonic through its illegitimate forgetting of animal life: the hidden suffering and slaughter of animals on the factory farm is literally what makes the on-the-go meals available for the homo economicus of biopower, today's busy lifestyle consumer.

Not surprisingly, Michel Foucault's work figures quite prominently in these emerging fields of study: Foucault, of course, coins the word "biopower" in volume 1 of *The History of Sexuality*. And in his lecture courses touching on the concept (*Society Must Be Defended* and *The Birth of Biopolitics*), Foucault discusses the ways in which biopower might differ from the form of power he famously calls "discipline" (which aims at modifying individual behaviors and is always mediated through institutions). As Foucault explains in his 1975–76 lecture course *Society Must Be Defended,* biopower comprises

> a new technology of power, but this time it is not disciplinary. This technology of power does not exclude the former, does not exclude disciplinary technology, but it does dovetail into it, integrate it, modify, it to some extent, and above all, use it by sort of infiltrating it, embedding itself in existing disciplinary techniques. This new technique does not simply do away with the disciplinary technique, because it exists at a different level, on a different scale, and because it has a different bearing area, and makes use of very different instruments. Unlike discipline, which is addressed to bodies, the new non-disciplinary power is applied not to man-as-body but to the living man, to man-as-living-being.[2]

As Foucault insists, this new form of biopolitical power does not simply replace discipline but extends and intensifies the reach and scope of power's effects by freeing them from the disciplinary focus on the "exercise" and the "institution." Biopower, one might say, radically expands the scale of power's sway: by moving beyond discipline's "retail" emphasis on training individual bodies at linked institutional sites (e.g., family, school, army, factory, hospital), biopower enables an additional kind of "wholesale" saturation of power effects, smearing them across the entire social field. What Foucault calls this "different scale" and much larger "bearing area" for the practices of power make it possible for biopower to produce more continuous effects because one's whole life (one's identity, sexuality, diet, health) is saturated by power's effects, rather than power relying on particular training functions carried out in the discontinuous domain of X or Y institution (dealing with health in the clinic, diet at the supermarket and the farm, sexuality in the family and at the nightclub, and so on). Hence biopower works primarily to extend and intensify the reach of power's effects: not everyone has a shared disciplinary or institutional identity (as a soldier, mother, nurse, student, or politician); but everyone does have an investment in biopolitical categories like "sexuality," "health," or "quality of life"—our own, as well as our community's. If discipline forged an enabling link between subjective aptitude and docility,[3] biopower forges an analogous enabling link between the individual's life and the life of the socius: the only thing that we as biopolitical subjects have in common, one might say, is that we are all individuals, charged with the task of creating and maintaining our lives. And that power-saturated task is performed not solely at scattered institutional sites but virtually everywhere, all the time.

The ethical challenge presented by animal studies arrives hot on the heels of a triumphant human biopolitics: most centrally, of course, there would seem to be serious concerns about the ethics of sacrificing animal life solely for human benefit (under a regime dominated by an intense concern for "life," why do we live and they die?), not to mention the sustainability consequences (both for individual health and the life of the ecosphere) that are being wrought by the huge corporate animal farms required to feed a growing global hunger for animal flesh.[4]

In any case, it seems clear that Foucault's texts of the 1970s and 1980s constitute linchpin sites for both biopolitical analysis and the related fields that cluster under the rubric of animal studies.[5] Though we should note right from the beginning that Foucault seems to be of more use within contemporary biopolitical theory and practice than in the field of animal studies, where Foucault's work on biopower serves less as a common touchstone to build on and more as a kind of negative jumping-off point—similar to the way that Theodor Adorno's supposed dismissal of popular culture served for many years as an enabling negative horizon for academic work in cultural studies.[6] In short, Foucault is routinely chastised for not paying enough (or really any) attention to the question of animality within his discussions of life and/as biopower. Most infamously, Foucault's work has been charged with "species chauvinism" (Donna Haraway's accusation in *When Species Meet*). Haraway in fact reveals that her book was spawned by the realization that Foucault's critical project didn't go far enough: "I had read Michel Foucault, and I knew all about biopower and the proliferative powers of biological discourses. . . . I had read *Birth of the Clinic* and *The History of Sexuality*, and I had written about the technobiopolitics of cyborgs. I felt I could not be surprised by anything. But I was wrong. Foucault's own species chauvinism had fooled me into forgetting that dogs too might live in the domains of technobiopower."[7] Perhaps, as I suggest below, if she had spent more time with *The Order of Things* and *History of Madness*, things might have been different.[8]

Foucault is given credit in animal studies for calling attention to the central question of life within modern political existence; but he is just as quickly disciplined for confining his analysis to humans, thereby doubling down on the nefarious ethical exclusion of animal life from the discussion. Nicole Shukin writes in *Animal Capital*, for example: "The pivotal insight enabled by Foucault—that biopower augurs 'nothing less than the entry of life into history, that is, the entry of phenomena of the life of the human species into the order of knowledge and power'—bumps against its own internal limit at the species line. The biopolitical analyses he has inspired, in turn, are constrained by their reluctance to pursue power's effects beyond the production of the human social and/or species life and into the zoo-politics of animal capital."[9] There are undoubtedly a whole series of Foucaultian ways that one could respond to this kind of claim. I suppose the most obvious is that if Foucault is a booster for the human species—offering sunny thoughts like "man is an invention of recent date. And one perhaps nearing its end" (*OT* 387)—I would shudder to think what *critics* of the species might think. But rather than pursue this kind of defense (as Foucault himself insisted, polemical back-and-forth argumentation is unlikely to lead us beyond our present consensus),[10] I would like to take a different tack. One way to go would be to suggest avenues whereby Foucault's work on human biopower could be harnessed for thinking about animal life. As Stephen Thierman writes, insofar as "Foucault did not write" extensively about nonhuman animals, perhaps "it is left to those of us who think his methods and conceptual tools can be fruitfully employed to explore our relationships with other animals to fill in the blanks."[11] And there is already plenty of work being done in this emergent area of research, analyses inspired by Foucault that interrogate the question of life across the human-animal divide.[12]

However productive that path might be, I'd like to take a somewhat different direction in this chapter by trying to think about new directions in Foucaultian biopower, not so much by extending his analyses into animal formations and institutions that he *did not* study (contemporary corporate farming practices, genetic manipulation, the companion animal phenomenon, etc.) but by looking at neglected formulations concerning animality and the emergence of biopower in Foucault's own work. I would begin simply by noting that Foucault hardly ignored animals altogether, especially in his early archaeological work. *History of Madness* (1961), for example, contains a substantial bestiary of reflection on the myriad historical ways that "the animal realm . . . serves to reveal the dark rage and sterile folly that lurks in the heart of mankind" (*HM* 19). Foucault in fact insists that much of the discourse on madness, in the classical age and beyond, "took its face from the mask of the beast. . . . Madness . . . was for the classical age a direct relation between man and his animality, without reference to a beyond and without appeal" (*HM* 147–48). Human madness, as Foucault demonstrates, was for a very long time understood and treated as a kind of animality.

Even more centrally, an extended interrogation of animality figures in what we might call the *first* birth of biopower within Foucault's corpus, in 1966's *The Order of Things.* To begin reexamining the Foucaultian emergence of life as a central concern for power, one needs merely to cite these controversial lines from *The Order of Things*: "if biology was unknown [in eighteenth-century Europe], there was a very simple reason for it: life itself did not exist. All that existed was living beings, which were viewed through a grid of knowledge constituted by natural history" (*OT* 127–28). Of course, this provocation—life did not exist until the nineteenth century, specifically until 1802 when Lamark was the first to use the word "biology"—functions as a bit of a dry run for Foucault's later, seemingly just as "outrageous" idea that homosexuality was invented in 1870, or his declarations that "author" and indeed even "man" itself are products of recent invention.[13] This type of sentence-level provocation is one of Foucault's characteristic means for dramatizing, in a very stark way, the crucial importance of social and historical emergence. Alongside the Foucaultian pleasure evident in the perversity performed by sentences like "up to the end of the 18th century, in fact, life does not exist" (*OT* 160), there is a relatively consistent philosophical point being advanced: the historical emergence of a new way of handling topic X (here, "life") gives rise to different problematics, different practices, and thereby different objects. A new form of practice literally remakes the (supposedly preexisting) object of the discourse: biology *creates*, rather than discovers, this object of study called life. This emphasis on discursive emergence constitutes Foucaultian Archaeology 101 and keeps us focused on the most basic terrain for all of Foucault's work: the question of how today is different from yesterday.

In schematic terms, Foucault's prey in *The Order of Things* is the epistemic shift from a classical regime of representation (natural history, where classification and nomination of visible "living things" are the key practices) to a regime of modern knowledge-transcendentals like life, labor, and language (where the "object" of knowledge is no longer readily available to classification, but rather disappears into the shadowy halflight of discursive practice). In the birth of biology, the question of life unhinges itself from a practice of representation (the discourse is freed from what Foucault calls the "pure tabulation of things" [*OT* 131] in natural history's grids) and attaches itself instead to a mode of speculation about this murky thing called life—now understood not as a visible manifestation of similitude, but as a darkly hidden secret that connects living things. This movement from surface to depth signals the decline of natural history and the birth of biology, the emergence of the science of life. And in this movement across spheres, the modern human sciences and their era of transcendentals begin to replace the representational episteme—just as representation, in its turn, had replaced the early modern regime of fabulation or magic.

And it is here, in the interstices of this grand narrative about knowledge in the West, where animals make their appearance in *The Order of Things:* "To the Renaissance, the strangeness of animals was a spectacle: it was featured in fairs, in tournaments, in fictitious or real combats, in reconstitutions of legends in which the bestiary displayed its ageless fables. The natural history room and the garden, as created in the Classical period, replace the circular procession of the 'show' with the arrangement of a 'table,' . . . a new way of connecting things both the eye and to discourse" (*OT* 131). As the historical a priori of representation emerges and later mutates into the era of the human sciences, the "being" of animals changes as well: in the era of representational natural history, "the plant and the animal are not seen so much in their organic unity as by the visible patterning of their organs. They are hoofs and paws, fruits and flowers, before being respiratory systems or internal liquids. Natural history traverses an area of the visible . . . without any internal relation of subordination or organization" (*OT* 137). For a representational regime, it is the visible surface of living things, rather than some buried "organic unity," that is the bearing area of discursive power. Natural history constitutes a series of practices whereby living things—plants and animals—find their proper classification through organization by common visible traits, rather than hidden animating principles.

But on Foucault's account, it is with the rise of the transcendentals, in the era of the human sciences, that animals begin to take priority over plants as the privileged form or figure of life itself. In an era of natural history where knowledge was characterized by "the apparent simplicity of a description of the visible," "the area common to words and things constituted a much more accommodating, a much less 'black' grid for plants than for animals" (*OT* 137). Most animals, simply put, have more hidden, interior space than plants, and thereby present a greater volume of "black" or blank space to the gaze of the classifying naturalist. As Foucault writes about this era of representation, "Because it was possible to know and to say only within a taxonomic area of visibility, the knowledge of plants was bound to prove more extensive than that of animals" (*OT* 137), precisely because plants can be pulled up out of the ground and thereby rendered fully visible, from the tip of the roots to the outermost edges of the flower or leaf.

At the dawn of the nineteenth century, however, Foucault traces a mutation of the dominant epistemic procedures—from a representational discourse that maps external similitude and resemblance, to the emergence of a speculative discourse that takes as its object hidden internal processes. In short, we see emerge a discourse that "opposed historical knowledge of the visible to philosophical knowledge of the invisible" (*OT* 138): knowledge's privileged practices abandon the surface of objects to plumb their hidden depths instead. And first and foremost among those transcendental "invisibles" was a little thing we like to call "life": "The naturalist is the man concerned with the structure of the visible world and its denomination according to characters. Not with life" (*OT* 161), Foucault insists, because life is not representable. Life is in fact a kind of unplumbable depth, animating the organism from a hidden origin somewhere within. This birth of biology—which is to say, the emergence of "life" itself as a bearing area for discursive power and a depth to be explored—constitutes the first birth of biopower, this one in Foucault's work of the mid-1960s.

So, why is this archaeology of biopower important, in terms of using Foucault in the present and maybe leading us toward new directions in the future? Well, I suppose there are myriad answers to that question, but I would like to suggest one particular line of useful inquiry here, one that also involves a way to respond to the animal studies critique of Foucaultian biopower with which we began—that Foucault essentially ignores the question of animal life, and thereby extends the unearned privileges of human biopower rather than questioning them.

In short, Foucault's work on biopower 1.0 shows that animal life is not in fact jettisoned or abjected at the dawn of humanist biopower in the nineteenth century, but instead that *animality is fully incorporated into biopower as the template for life itself.* As Foucault puts it,

> the animal, whose great threat or radical strangeness had been left suspended and as it were disarmed at the end of the Renaissance, discovers fantastic new powers in the 19th century. In the interval, Classical nature had given precedence to vegetable values . . . with all its forms on display, from stem to seed, from root to fruit; with all its secrets made generously visible, the vegetable kingdom formed a pure and transparent object for thought as tabulation. But when the characters and structures are arranged in vertical steps toward life—that sovereign vanishing point, indefinitely distant but constituent—then it is the animal that becomes the privileged form, with its hidden structures, its buried organs, so many invisible functions. . . . If living beings are a classification, the plant is best able to express its limpid essence; but if they are a manifestation of life, the animal is better equipped to make its enigma perceptible. (*OT* 277)

In short, Foucault argues that with the emergence of the human sciences at the birth of biopower, the animal is not excluded or forgotten, but quite the opposite: animality comprises the dominant apparatus for investigating both what life is and what life does. The living is no longer primarily vegetable (sessile and awaiting mere categorization), but understood as evolving, appetite-driven, secret, discontinuous, mendacious, inscrutable, always on the prowl, looking for an opening to break free. As Foucault puts it, "Transferring its most secret essence from the vegetable to the animal kingdom, life has left the tabulated space of order and become wild once more" (*OT* 277).

And this, of course, is not just a development within the narrow confines of biology. Foucault could in fact cue here the advent of philosophical modernity itself. One might speculate that Hegel's 1807 *Phenomenology of Spirit* (where human life itself is refashioned as nothing other than unfathomable discontinuity and animal appetite—in short, desire) shows the way for later nineteenth-century thought, which in turn opens a path directly to our day: from Darwin's evolution of life, through Freud's life of the unconscious and Nietzsche's life of self-overcoming, all the way to Schumpeter's neoliberal life of creative destruction. All of these formations depend completely on the bedrock connection of life to an animating, hidden, "wild" animality of desire: both prior to and beyond the human, yet somehow still constituting that very humanity as its secret essence.

In his *History of Madness*, Foucault charts a similar archaeological shift in the role and status of animals within European discourse on the mad. As he notes, from the beginning "it was probably essential for Western culture to link its perception of madness to imaginary forms of the relation between men and animals" (*HM* 151), and for a very long time in the West animals played the role of that mad, menacing opposite to man's reason: "animals were more often thought of as being part of what might be termed a counter-nature, a negativity that menaced the order of things and constantly threatened the wisdom of nature with its wild frenzy" (*HM* 151). "In the classical age," Foucault concludes, "madness was still thought of as the counter-natural violence of the animal world" (*HM* 151). However, further into his *History of Madness* (foreshadowing the analysis in *The Order of Things*), Foucault will note a decisive series of breaks in the Western discourse of madness and animality, at precisely the Enlightenment moment when madness begins to get incorporated into the definition of reason (as a potential malady or even a secret romantic source of intellectual power, as opposed to the classical formation wherein madness functions as a negative counterpoint to the discourse of reason): "From the moment when philosophy became anthropology, and men decided to find their place in the plenitude of the natural order, the animal lost that power of negativity, and assumed the positive form of an evolution between the determinism of nature and the reason of man" (*HM* 151). As in *The Order of Things*, Foucault shows us in *History of Madness* that animality is not jettisoned at the birth of anthropological biopower (at least in part because such an abjected binary otherness was the role of animality in a prior, classical era); rather, "when philosophy became anthropology," when thought became focused on human life, the animal became incorporated into reason. Animality, then, becomes less a binary all-or-nothing function in relation to madness, but operates on a sliding scale. With the anthropological turn to life, the madman is recast: not as the subhuman animal other, but as our less-fortunate sibling—beset not by the classical era's understanding of animality as the absence of rationality, but by too much of a good thing: biopower's embrace of life as animal desire.

Likewise, in *History of Madness* Foucault also calls our attention to the Aristotlean definitions of man (as rational and political animal) that will play a pivotal role in his thinking about biopower in the 1970s and 1980s: "Did the fact that, after Aristotle, men had spent two thousand years thinking of themselves as reasonable animals necessarily imply that they accepted the possibility that reason and animality were of a common order? Or that the definition of man as a 'rational animal' provided a blueprint for understanding man's place in natural positivity? Independently of whatever Aristotle meant with that definition, it might be the case that for the Western world 'rational animal' meant the manner in which the freedom of reason took off from a space of unchained reason and marked itself off from it, ultimately forming its opposite" (*HM* 151). But as philosophy becomes anthropology at the birth of biopower in the early nineteenth century in Europe, Foucault argues that the classical age of the animal's binary alterity is also eclipsed: "At that point, the meaning of the term 'rational animal' underwent a radical change. . . . From then on, madness had to follow the determinism of a humanity perceived as natural in its own animality" (*HM* 151).

In the *History of Madness*, Foucault suggests we look to "the French poet Lautréaumont" for "proof" of the "wild frenzy" (*HM* 151) that is life-as-animality; but consider, just as a passing example nearer to the present, Frank O'Hara's 1950 poem "Animals":

> Have you forgotten what we were like then
> when we were still first rate
> and the day came fat with an apple in its mouth

it's no use worrying about Time
but we did have a few tricks up our sleeves
and turned some sharp corners

the whole pasture looked like our meal
we didn't need speedometers
we could manage cocktails out of ice and water

I wouldn't want to be faster
or greener than now if you were with me O you
were the best of all my days[14]

Here in O'Hara's poem, as Foucault suggests in the larger biopolitical realm of modernity, animals function less as our excluded "other" than as very intense markers for our hidden, better or former—perhaps more authentic—selves. Our unconscious drives, O'Hara's poem suggests, are animal in nature, and it is those unbridled desires that make us "first rate," most truly who we are. As O'Hara's poem puts it, "the best of all my days" were animal in nature—when I did not worry about time, cars, or stocking the bar, and "the whole pasture looked like our meal." Here, animality is not demented or predatory (as Foucault suggests the classical age understood the mad, irrational "otherness" of animals), but takes the form of a biopolitical or "more gentle form of animality, which did not destroy its human truth in violence, but allowed instead one of nature's secrets to emerge: the rediscovery of the familiar but forgotten resemblance [of the madman] with tame animals and children" (*HM* 435). This I take to resonate with O'Hara's conclusion, where it is the precocious, desiring animal in us all who is addressed in the final lines: "O you / were the best of all my days."

In short, the archaeology of biopower that Foucault performs in *The Order of Things* and *History of Madness* shows decisively that the jettisoned, negative, or forgotten other of our biopolitical conception of life is most assuredly *not* the animal, insofar as animality is the subtending paradigm for our era of humanist, neoliberal biopower—where it's all appetite and appropriation all the time. Tweaked and intensified a bit, Foucault's archaeology of animality may compel us to ask whether contemporary animal studies, far from constituting a critique of an all-too-humanist biopower (exposing the imperialism of human life over animal life), tends to function in fact as an intense extension of that very biopower. A Foucaultian provocation might suggest that animals are important within contemporary academic discourse, or at least they are more important than plants or other forms of life, not because animals function ethically as "wholly other" to humans, but primarily because they are "like us" in an originary way: they experience intense feelings; they are born and they die; they like to go walkies in the park at sunset. Indeed, animals are wild, just like we are in our best and freest moments. As such, though, animals function less as our ethical "absolute other" than as our hidden, better selves. Animals are more our life "companions" (to steal another phrase from Haraway) than our "others" (those figures excluded, forgotten, wholly unlike us, but still we depend on them absolutely). Following Foucault's reading, one might suggest that role of abjected other as been played throughout the biopolitical era not by the animal, but by the plant—which was indeed forgotten as the privileged form of life at the dawn of biopower. In this context, it is probably worth recalling that the biomass of plant life on Earth's terra firma does remain approximately one thousand times greater than the combined zoomass of all humans and other animals.[15]

And perhaps one related upshot of this genealogy, in terms of new directions in biopolitical ethics, might be an imperative to look at the strange and consistent elision of plants within the voluminous work on life within contemporary theory and philosophy. There are myriad tantalizing places in Heidegger, Agamben, and Derrida (all of them rereading Aristotle, especially from *De Anima* Book II) where the proximity and difficulty posed by plant life is highlighted, only to be dropped quickly and consistently by all these thinkers, to stay on the trail of the human/animal distinction.[16] Indeed, if animal studies scholars can charge Foucault with "speciesism," then in turn Foucault's archaeology of life (from the privileged plant form in an earlier era to the biopower's fetishizing of the animal) might suggest that animal studies, in its foundational abjection of plant life, is guilty of "kingdomism"—ignoring not just a species, but an entire kingdom, which one would assume is a much greater crime on the "kingdom-phylum-class" sliding scale of differentiation. Likewise, continuing research into plant behavior and intelligence confirms that plants are not, as was thought for centuries following Plato and Aristotle, sessile and insentient:[17] recent research has uncovered that plants evidence active, purposeful, future-oriented movement, and exhibit both competitive and defensive behavior. Plants, it seems, also have a certain kind of language—they share information concerning soil conditions and the presence of predators.[18]

Given this series of revelations, it becomes even harder to draw the lines among human, animal, and vegetable life. Indeed, recent botanical research even confirms Aristotle's 2,400-year-old quip about feeling plants' pain: "A plant which is fixed in the ground does not like to be separated from it."[19] And there are several recent scholarly works that bring up the question of vegetative life—for example, Matthew Hall's *Plants as Persons* and Michael Marder's *Plant-Thinking*. Both of these books are notable, for my purposes here, not so much because they attempt to question the (animal) life-as-hidden secret model that Foucault diagnoses for us, but because they work very hard to extend that "hidden life" paradigm (and the anthropomorphic biopolitical logic that rules over it) to plants as well. In short, both Hall and Marder argue that plants are the new animals—our new privileged figures for "the other," and the new subtending connection to this open-ended, multifarious thing called life.[20] Likewise, there exists a robust emergent literature in both environmental and machine-robot ethics,[21] which suggests if nothing else that the question, "What counts as an ethically compelling form of 'life'?" is and will remain an open and hazardous one.

In any case, as Derrida persuasively insists in his work on animals, the primary stake of interrogating animality is not asking what we humans ethically need to "grant" to animals (personhood, thinking, recognition, a voice, a face, and so on) or to treat animals as identical to humans, but in asking whether humans can somehow separate themselves from (or elevate themselves above) their conception of "subhuman" others; the Derridean project is undertaken not in the name of granting human privileges like rationality, communication, and agency to animals, but wondering whether, in the end, humans have any less fettered access to those things than animals do. And the conundrum would be ethically similiar, it seems to me, for whatever we might have to say in the future about plants or other forms of biopolitical life: the project is less offering some of our human privilege to plants or machines or the earth itself (in short, anthropomorphizing them), but paying close attention to the power effects rendered by the myriad practices by which we do in fact differentiate ourselves from other forms of life, and what violences those practices inevitably inflict.

Foucault, of course, parts ethical company from Derrida (and, I would suggest, from the founding principles of much of animal studies) around the binary pathos of "totalization or non-totalization," which constitutes nearly the whole field of ethics in a deconstructive context: if totalization or the violent desire for completion can be disrupted, if an originary *différance* of undecidability can be mobilized and demonstrated, then some positive deconstructive work has been accomplished. However, such a supposedly ethical gesture toward the unfathomable or untotalizable other, as Foucault will insist throughout his work, poses no essential question (ethical or otherwise) to the human sciences because those contemporary sciences do not require or even desire totalization. As Foucault demonstrates in his work on the emergence of life in Europe, the Western human sciences need constantly to refashion an unfathomable depth, and inexhaustible other, so they can continue to do their work. The insistence on the primacy of some nontotalizable "other" does not cripple the human sciences, but rather constitutes an essential component of their work: as Foucault concisely puts it, "an unveiling of the non-conscious is constitutive of all the sciences of man" (*OT* 364). (Economics, for example, does not know what value is any more than theology knows what God is or biology knows what life is—that is why you have a robust discourse to study it.) So the trading-places game of ethical alterity—the nonhuman other is best figured as the unconscious, the animal, the plant, the earth, the robot, and so forth—tends primarily to extend and deepen the constitutive work of the human sciences (the production of undecidability, which in turn produces more commentary), rather than to disrupt that work in some essential way.

In fact, barely a page of Foucault's methodological treatise *The Archaeology of Knowledge*[22] goes by without some kind of stinging critique of any and all discourses of absent origin, hidden depth, or undecidability. As Foucault clearly writes, what he seeks in archaeology is "not a condition of possibility but a law of coexistence": discourse is constituted by "something more than a series of traces" (*AK* 107) of lost origins, and he insists that the statements comprising and transforming the archive are "not defined by their truth—that is, not gauged by the presence of a secret content" (*AK* 120). In his most pointed criticism of Derrida in the *Archaeology*, Foucault questions the manner in which discourse "can be purified in the problematic of trace, which, prior to all speech, is the opening of inscription, the gap of deferred time [*écart du temps différeré*]: it is always the historico-transcendental theme that is reinvested" (*AK* 121). And, mirroring language he uses to criticize Derrida's reading of *History of Madness*,[23] Foucault lays waste to any discourse, scientific or philosophical, "which finds, beneath events, another, more serious, more sober, more secret, more fundamental history, closer to the origin, more firmly linked to its ultimate horizon (and consequently more in control of all its determinations)" (*AK* 121). In the end, for Foucault one might say that deconstruction is merely a marker and bearer of the "nontotalizing" *symptoms* of biopower and the human sciences, rather than forwarding any kind of corrective to these formations.[24]

I think one can see more clearly Foucault's difficulties with the whole transcendentalist discourse obsessed with impossible totalization and hidden connection by looking briefly in conclusion at Giorgio Agamben's reading of Foucault. Agamben famously outlines his project in *Homo Sacer: Sovereign Power and Bare Life*[25] like this: "the Foucaultian thesis [on biopower] will have to be corrected or, at least completed" (9). Agamben gives Foucault credit for calling our attention to something like the move we have just highlighted in Foucault's work, the biopolitical subsumption of animal life (Agamben's notion of "zoe") within the realm of social power ("bios"): "In the last years of his life . . . Michel Foucault began to direct his inquiries with increasing insistence toward the study of what he defined as biopolitics, that is, the growing inclusion of man's natural life in the mechanisms and calculations of power" (119). And certainly Agamben's work on the "exception" follows a logic somewhat similar to Foucault's on the "norm": alterity is incorporated precisely by being designated as other; the abnormal exception reinforces the power of the norm. However, Agamben's completion of Foucault functions not through a series of minor corrections, but through a wholesale rejection of Foucault's privileged biopolitical sites of historical analysis—the human sciences, sexuality, market economics, or the fashioning of the self—in favor of Agamben's narrow focus on Nazi concentration camps as the "exemplary places of modern biopolitics" (119). "Today," he writes, "it is not the city but rather the [concentration] camp that is the fundamental biopolitical paradigm of the West" (181).

In short, Agamben rejects the entirety of Foucault's historicism (his axiomatic sense that power works differently today than it did in the past), reducing the complex drama of power and resistance "today" to the brutalities of a sovereign power that works primarily to totalize, to eradicate difference (a sovereign form whose dominant practices, on Foucault's account, would have been rendered antique in the West by the eighteenth century). This, then, is Agamben's "correction" of Foucault: Agamben rejects the idea that power has become more subtle and effective, suffused through our everyday lives (even in our sexuality and our everyday consumer existence); he argues instead that power remains sovereign, brutal, literally animalizing its others so they can be eradicated. We in the first-world West live not in a panopticon or in an endless marketplace, but in a concentration camp. Here Agamben quotes Foucault on the ways in which "the importance of life for . . . problems of political power increases": "a kind of animalization of man through the most sophisticated political techniques results. Both the development of the possibilities of the human and social sciences, and the simultaneous possibility of protecting life and of the holocaust make their historical appearance."[26]

But if the "progress" of the human sciences for Foucault has an obvious relation to the "animalization" of "man," I have tried to show in what sense this functions in *The Order of Things*: biopower "animalizes" *not* through an chiasmic reversal or "reduction" of man to its supposed other, "bare" "animal" life (through what Agamben calls biopolitical tattooing). Rather, biopower "animalizes" in Foucault through its subtending incorporation of an understanding of "life" primarily as desire (from the representational era's picture of life as a kind of sessile plant, to the roaming, restless animal who becomes the incorporated figure for biopower). Insofar as the human animal is almost wholly understood in terms of its desires under the regime of biopower, it certainly follows for Foucault that those desires can be worked on by sophisticated political techniques—cue what Adorno famously called "the culture industries." Indeed, compared to the relatively crude techniques of sovereign power (torture and public execution), biopower needs to build on disciplinary exercises by inventing evermore supple techniques. The sophistication of biopolitical techniques, in other words, is not for Foucault primarily to be found in their brutish totalized or totalitarian *intentions* (power's reduction of us all to beasts, leashed and branded by power), but in those techniques' lightness, their increasing saturation within the socius and their heightened economic *effectiveness*, what I have elsewhere highlighted as Foucaultian power's "intensity."[27]

In other words, when Foucault insists that there is an "animalization of man" involved in biopower's birth and functioning, he means it quite literally: we have incorporated the beast into the contemporary biopolitical definition of "man" as endless, unthematizable animal desire, with the practices of sexuality and neoliberal capitalism its two most intense markers in the present. Perhaps this explains why Foucault good-naturedly tells Deleuze, "I cannot bear the word *desire*,"[28] because for Foucault the theme of desire quite literally continues to play out a well-worn transcendentalist drama born at the dawn of the nineteenth century: animal desires brutally repressed by the constraints of civilization.

For Agamben, on the other hand, bestialization constitutes less a contemporary set of practices or a historical phenomenon and remains primarily a transhistorical metaphor or simile for the human condition, as are (despite Agamben's protests to the contrary) his emphasis on the concentration camp or sovereign power. For Agamben, twenty-first-century Western society is *like* a concentration camp or *like* an absolute monarchy; we are treated *like* animals when we have to surrender our DNA or fingerprints. But if animal studies has taught us nothing else, its emphasis on the material practices of the food industry should make us suspicious of metaphors suggesting that humans are treated as animals by advanced capitalism: recall that well more than 50 billion chickens are slaughtered each year globally (more than 9 billion in the United States alone in 2010), the vast majority living their short, genetically engineered lives in cages that are too small for them to stand up or turn around, many subsisting on feed made from the ground-up corpses of their diseased brethren, too unfit for sale. As utterly terrible as global poverty is to endure for half the humans on this planet, most people are not in fact treated like corporate farm animals—slaughtered by the billions in what is indeed a sovereign manner, with little or no afterthought.

Indeed, as Foucault puts it in his formula for biopower, today's global poor are made to live or left to die—an important difference from the sovereign, concentration-camp edicts of making die or letting live (which most animals presently endure: the feed animals are made to die, while the dwindling populations of "wild" animals are merely left to live). Of course, ethnic cleansing, torture, and mass slaughter of humans (Foucault's sovereign "making die") still does happen in our world, but those sovereign practices are hardly the primary modality of power's functioning today (small solace, however, if you happened to be subject to these horrific practices). Or at least this is Foucault's bedrock argument about power's mutations in the West, pivoting over the last four hundred years from the sovereign execution of the regicide (described in the horrific opening pages of *Discipline & Punish*), to a focus on the biopolitcal intricacies of your identity, sexuality, or consumer desires.

And this diagnostic difference between sovereignty and biopower is crucial at least partially because sovereign power, while notoriously difficult (if not impossible) to resist, tends to be relatively easy to spot, diagnose, and denounce: in short, someone else is always wielding "sovereign power." On the other hand, the biopolitics of "making live and letting die" is a regime in which all of us are implicated: who gets health care and who doesn't? Are the rich countries willing to pay more taxes or endure weaker corporate profits so that millions of poor people in remote regions can live longer? What's to be done about global warming or the destruction of the flora in the Amazon and elsewhere throughout the globe? Are such policy matters reducible to a model of sovereign "decision"? Who would make such a decision?

In short and in conclusion, Agamben performs an intense doubling-down on the move that Foucault diagnoses for us as the founding of biopower—the withdrawal of "life" from a regime of everyday practices and its disappearance into a realm of hidden enigma. In Agamben, life becomes something we need endlessly to worry over, but in the end, there is really not much to be done—as adding more cultural fixes to the questions of life will bring only more brutality, more victims: bios becomes zoe endlessly. This is to say, at least partially, that for Foucault, the primary *problem* with the question of "life" in the biopolitical era is that it has stubbornly remained a kind of shadowy transhistorical theoretico-moral question that is everywhere and nowhere, thereby concerning everyone and no one. Of course, this state of affairs (where there is a lot to worry over, a lot to talk about, but very little to do)—makes much of the contemporary discourse on "life" tailor-made for the Febreeze capitalism of our biopolitical era. As Foucault consistently reminds us, the relations among various forms of power and life are neither good nor bad (they are not primarily theoretical or moral relations), but they are dangerous.[29]

Notes

1. See James Gorman, "Animal Studies Cross Campus to the Lecture Hall," *New York Times*, January 2, 2012, http://www.nytimes.com/2012/01/03/science/animal-studies-move-from-the-lab-to-the-lecture-hall.html.
2. *"Society Must Be Defended": Lectures at the College de France, 1975–76*, trans. David Macey (New York: Picador, 2003), 242.
3. As Foucault concisely puts it, the disciplinary body becomes "more obedient as it becomes more useful." *Discipline & Punish*, trans. Alan Sheridan (New York: Vintage, 1979), 138.
4. As the best-known examples of a vast and growing popular literature concerning sustainability and the global meat industry, see Robert Kenner's documentary film *Food Inc.*, and Jonathan Safran Foer's *Eating Animals* (Boston: Back Bay Books, 2010).
5. See Wolfe's extensive engagement with Foucault throughout *Before the Law: Humans and Other Animals in a Biopolitical Frame* (Chicago: University of Chicago Press, 2013).
6. See, for example, Simon During's headnote to Adorno and Horkheimer's "The Culture Industry: Enlightenment as Mass Deception" in *The Cultural Studies Reader* (London: Routledge, 2007): "Adorno and Horkheimer neglect what was to become central to cultural studies: the ways in which the cultural industry, while in the service of organized capital, also provides for all kinds of individual and collective creativity and decoding" (32).
7. Donna Haraway, *When Species Meet* (Minneapolis: University of Minnesota Press, 2007), 59–60.
8. Michel Foucault, *History of Madness*, trans. Jonathan Murphy (New York: Routledge, 2006) (hereafter cited in the text as *HM*); Michel Foucault, *The Order of Things: An Archaeology of the Human Sciences*, trans. Alan Sheridan (New York: Pantheon, 1970) (hereafter cited in the text as *OT*).
9. Nicole Shukin, *Animal Capital: Rendering Life in Biopolitical Times* (Minneapolis: University of Minnesota Press, 2009), 11.
10. On Foucault's dislike for polemic argumentation, see his interview, "Polemics, Politics, and Problematizations" in *Foucault Live* (New York: Semiotexte, 1996).

11. Stephen Thierman, "Apparatuses of Animality: Foucault Goes to a Slaughterhouse," *Foucault Studies* 9 (September 2010): 90.
12. See, for example, Chloe Taylor's overview in "Foucault and Critical Animal Studies," *Philosophy Compass* 8, no. 6 (2013): 539–51.
13. Regarding 1870 and homosexuality, see Foucault's March 19, 1975, lecture in *Abnormal: Lectures at the College de France, 1974–75*, trans. Graham Burchell (New York: Picador, 2004), and *History of Sexuality*, vol. 1, trans. Robert Hurley (New York: Vintage, 1980), 43.
14. Frank O'Hara, "Animals," 1950, Frank O'Hara.org, http://www.frankohara.org/writing.html#animals.
15. "Biomass (ecology)," *Wikipedia*, last modified March 6, 2015, http://en.wikipedia.org/wiki/Biomass_%28ecology%29.
16. See, for example, Heidegger's work on "world" in his 1929 lecture course, *Fundamental Concepts of Metaphysics: World, Finitude, Solitude*, trans. William McNeil and Nicholas Walker (Bloomington: Indiana University Press, 2001), where he infamously suggests (reworking Aristotle on *psukhe*) that humans are world building while animals are poor in world, and the stone is worldless. The figure of the plant, of course, haunts this tripartite distinction, sometimes remaining with the stone and earth (as completely worldless) and sometimes being associated with the animal (as merely poor in world). Clearly the plant is alive and thereby differs from the stone, but Heidegger never really nails down the plant's status in terms of "world." For his part, Derrida recognizes this liminal state of the plant in his extensive readings of Heidegger's lecture course (especially in the two volumes of his *Beast and the Sovereign*, trans. Geoffrey Bennington [Chicago: University of Chicago Press, 2011]), and points out the odd place of plant life several times—only to abandon it altogether to keep his focus on animals. Likewise, Agamben follows this Aristotle-Heidegger pairing in *The Open: Man and Animal*, trans. Kevin Attell (Stanford, CA: Stanford University Press, 2003), even suggesting in section 4, with reference to Foucault no less, that "vegetative life" is in fact the proper name for Agamben's concept of "bare life" (15)! He then very quickly backs off that association, and in the short space between paragraphs we're back to the central pathos of the human-animal distinction, where man's "distance and proximity to the animal have been measured and recognized first of all in the closest and most intimate space. . . . We must learn to think instead of man as what results from the incongruity of these two elements" (16): man and animal, bios and zoe. Exeunt plants, pursued by an animal.
17. See Aristotle, *Parts of Animals* (656a1–10): "Plants, again, inasmuch as they are without locomotion, present no great variety in their heterogeneous parts. For, where the functions are but few, few also are the organs required to effect them." Compare Plato on the plant: "Wherefore it lives and does not differ from a living being, but is fixed and rooted in the same spot, having no power of self-motion. . . . The superior powers had created all these natures to be food for us." (*Timaeus* 77a-c). (For online editions of both Aristotle and Plato, see the Internet Classic Archive housed at MIT: classics.mit.edu/). As St. Thomas Aquinas sums up the age-old prejudice, even "brute animals are more noble than plants" (*Summa Theolgica*, pt. 3, issue 2—question 44, art. 3, http://www.newadvent.org/summa/4044.htm).
18. For an excellent overview of research in the area, see Anthony Trewavas, "What Is Plant Behaviour?," *Plant, Cell and Environment* 32 (2009): 606–16.
19. Aristotle, *On Plants*: 820a10, in *The Complete Works of Aristotle: The Revised Oxford*

Translation, vol. 2, ed. Jonathan Barnes (Princeton, NJ: Princeton University Press, 1984). On recent research concerning plant perception and its relations to so-called ethical vegetarianism, see Natalie Angier, "Sorry, Vegans, Brussels Sprouts Like to Live, Too," *New York Times*, December 21, 2009, http://www.nytimes.com/2009/12/22/science/22angi.html; Carol Kaesuk Yoon, "No Face, but Plants Like Life Too," *New York Times*, March 14, 2011, http://www.nytimes.com/2011/03/15/science/15food.html.

20. See Marder's *Plant-Thinking: A Philosophy of Vegetal Life* (New York: Columbia University Press, 2013), and Hall's *Plants as Persons* (Albany: SUNY Press, 2010). For a fine book on vegetable life that does not fall prey to this logic, see Doyle's *Darwin's Pharmacy: Sex, Plants, and the Noosphere* (Seattle: University of Washington Press, 2013).
21. On environmental ethics, see, for example, the essays collected in *Rethinking Nature*, ed. Bruce V. Foltz and Robert Frodeman (Albany: SUNY Press, 2004), as well as Timothy Morton's *Ecology without Nature* (Cambridge: Harvard University Press, 2009); on machine ethics, see Wendell Wallach and Colin Allen's *Moral Machines* (New York: Oxford University Press, 2010) and David Gunkel's *The Machine Question* (Cambridge: MIT Press, 2012).
22. *The Archaeology of Knowledge*, trans. A. M. Sheridan Smith (New York: Pantheon, 1972) (hereafter cited in the text as *AK*).
23. Recall the context in which Foucault infamously tags deconstruction as a "historically determined little pedagogy": such a generous ventriquilizing reanimation of "undecidable" objects and texts "inversely gives to the master's voice that unlimited sovereignty that allows it indefinitely to re-say the text" (*HM* 573).
24. For better or for worse, this is a characteristic move in Foucault's work—Derrida remains uninteresting to him because Derrida is merely the bearer of a biopolitical symptom created many years before, in the same way that Marx (and his reliance on Ricardo's labor theory of value) remains completely continuous with the discourses he's ostensibly critiquing: "At the deepest level of Western knowledge, Marxism introduced no real discontinuity; it found its place without difficulty, as full, quiet, comfortable, and goodness knows, satisfying for a time (its own), within an epistemological arrangement that welcomed it gladly" (*OT* 261).
25. Giorgio Agamben, *Homo Sacer: Sovereign Power and Bare Life*, trans. Daniel Heller-Roazen (Stanford, CA: Stanford University Press, 1998).
26. The quotation is found in Hubert Dreyfus and Paul Rabinow, *Michel Foucault: Beyond Structuralism and Hermeneutics* (University of Chicago Press, 1982), 138; cited in Agamben, *Homo Sacer*, 3—where it is misattributed to *Dits et écrits*. This linkage of animalization to the Holocaust is clearly important to Agamben (in fact, it sets up his entire project), but it is sadly nowhere to be found in Foucault's published works. It exists, as far as I can tell, only as quoted in Dreyfus and Rabinow's *Michel Foucault: Beyond Structuralism and Hermeneutics*, and seems to have been taken from a transcript of a discussion at Stanford after Foucault gave "Omnes et Singulatim" (in English) as the Tanner Lecture in 1979. Presumably, Foucault was asked by an audience member a question about the Holocaust, and this was his improvised response (again, in English); or if this was indeed part of the original lecture, Foucault excised this reference to the Holocaust from all subsequent versions.

    For his exact wording, Agamben is clearly translating—into Italian—the French rendering of Dreyfus and Rabinow's Foucault book (published by Gallimard as *Michel Foucault: Un parcours philosophique* in a translation by Fabienne Durand Bogaert in 1984); Foucault's original English is clumsy (all the verbs at the end of the sentences) and, as I say, not included by Foucault in any publication of the lecture. Specifically in context, Foucault is commenting on how "the importance of life for these problems of political power increases": "a kind of animalization of man through the most sophisticated political techniques results. Both the development of the possibilities of the human and social sciences, and the simultaneous possibility of protecting life and of the holocaust make their historical appearance" (Dreyfus and Rabinow 138).

Foucault's somewhat inelegant English is rendered much more lyrically by Bogaert's French translation: "il en résulte une sorte animalisation de l'homme au travers des techniques politiques les plus sophistiqées. Apparaissent alors dans l'histoire le déploiement des possibilités des sciences humaines et sociales ainsi que la possibilité simultanée de protéger la vie et d'autoriser l'holocauste" (*Michel Foucault: Un parcours philosophique*, 202).

Compare Agamben's Italian rendering (clearly following the specific wording of the French translation, not the English original): "Ne risulta una sorta di animalizzazione dell'uomo attuata attraverso le più sofisticate tecniche politiche. Appaiono allora nella storia sia il diffondersi delle possibilità delle scienze uname e sociali, sia la simultanea possibilità di proteggere la vita e di autorizzarne l'olocausto" (*Homo Sacer: Il potere sovrano e la nuda vita* [Torino: Einaudi, 1995], 5).

27. See my *Foucault Beyond Foucault: Power and Its Intensifications since 1984* (Stanford, CA: Stanford University Press, 2008).
28. Reported by Deleuze in his "Desire and Pleasure," in *Foucault and His Interlocutors*, ed. Arnold Davidson (Chicago: University of Chicago Press, 1996), 189.
29. This chapter is excerpted in part from my *Plant Theory: Biopower and Vegetable Life* (Stanford, CA: Stanford University Press, 2015).

EIGHT

The Biotechnological *Scala Naturae* and Interspecies Cosmopolitanism: *Patricia Piccinini, Jane Alexander, and Guillermo Gómez-Peña*

EDUARDO MENDIETA

> *The four errors.*—Man has been educated by his errors: first, he saw himself only incompletely; second, he endowed himself with fictitious attributes; thirdly, he placed himself in a false rank order in relation to animals and nature; fourthly, he invented ever new tables of goods and for a time took them to be eternal and unconditional—so that now this, now that human drive and condition occupied first place and was ennobled as a result of this valuation. If one discounts the effects of these four errors, one has also discounted humanity, humaneness, and "human dignity."
>
> —Friedrich Nietzsche, *The Gay Science*[1]

When Zarathustra came into the nearest town lying on the edge of the forest, he found many people gathered in the market place, for it had been promised that a tightrope walker would perform. And Zarathustra spoke thus to the people:

> "*I teach you the overman* [Übermensch]. Human being is something that must be overcome. What have you done to overcome him?
>
> All creatures so far created something beyond themselves; and you want to be the ebb of this great flood and would even rather go back to animals than overcome humans?
>
> . . .

Now Zarathustra looked at the people and he was amazed. Then he spoke thus:

> "Mankind is a rope fastened between animal and overman—a rope over an abyss.
>
> A dangerous crossing, a dangerous on-the-way, a dangerous looking back, a dangerous shuddering and standing still.

> What is great about human beings is that they are a bridge and not a purpose: what is lovable about human beings is that they are a *crossing over* and a *going under*."
>
> —Friedrich Nietzsche, *Thus Spoke Zarathustra*[2]

> In man *creature* and *creator* are united: in man there is material, fragment, excess, clay, dirt, nonsense, chaos; but in man there is also creator, form-giver, hammer hardness, spectator divinity, and seventh day: do you understand this contrast? And that *your* pity is for the "creature in man," for what must be formed, broken, forged, torn, burnt, made incandescent, and purified—that which *necessarily* must and *should* suffer? And *our* pity—do you not comprehend for whom our *converse* pity is when it resists your pity as the worst of all pamperings and weaknesses? But it is pity *versus* pity.
>
> —Friedrich Nietzsche, *Beyond Good and Evil*[3]

Friedrich Nietzsche's work is populated by one of the most diverse faunae in the history of philosophy, perhaps only comparable to that of Aristotle.[4] There is a way in which we can read Neitzsche's *Thus Spoke Zarathustra* as an analog of Aristotle's *History of Animals*. But, like Aristotle's work, Nietzsche's is at the same time a variation on the bestiary. The bestiary, which was inaugurated by Aristotle writing on animals, was an important device of moral edification and imagination. It achieved its highest popularity in the late middle ages, when it competed with the bible and breviaries as one of most popular and produced books.[5] Bestiaries imparted moral lessons by means of beautiful illustrations of both real and fictions animals: lions, camels, wolfs, along with griffins, dragons, all kinds of chimeras and fabulous creatures. The bestiary as a pedagogical tool, however, was framed by a moral ecology that rather than being unambiguous was layered. The bestiary was a palimpsest of moral warning. Competing with the bestiary as a dispositif of moral education was the medieval Christian notion of a *scala naturae*, or the great ladder of creation.[6] The *scala naturae* is a hierarchical ontology of nature that places inanimate entities on the lowest rang, followed by plants, animals, humans, angelic beings, and then God. Creation is ordered in accordance with a principle of ascent toward the spiritual. In this great chain of being, humans reside between animals and angels. It is evident that to be human is to be above animals and below angelic beings. Humans are neither animals nor angels. At best they are fallen angels. What is noteworthy is that between the bestiary and a Christian great chain

The Great Chain of Being from *Retorica Christiana* by Didacus Valades, printed in 1579 (woodcut) (b/w photo), Italian School (16th century) / Private Collection / Bridgeman Images.

Scala naturae, cut from Charles Bonnet, *Œvres d'historie naturelle et de philosophie de Charles Bonnet*, 8 vols. (Neuchatel, S. Fauche, 1779–83), vol. 4, pt. 1, p. 1

of creation, we have two semiotic registers, or rather two different grammars that organize the ecology of nature entailing an ecology of ethics: on the one hand, we have a palimpsest that displaces and jumbles the Judeo-Christian proclivity for hierarchies; on the other, we have the inviolable logic of an order of either ascent toward divinity or descent towards the inanimate.

In her fascinating and highly original work *Beasts of the Modern Imagination*, Margot Norris urges us to read Nietzsche as belonging to what she calls a "biocentric" tradition, one that navigates around anthropocentrism or reductive biologism.[7] She articulates Nietzsche's contribution to this tradition in the following way: "Careful to avoid the anthrophormizations of Nature and natural man implicit in Romanticism, Nietzsche elaborates and complicates this model of natural reclamation to keep it consistent with biocentric objectives. . . . Nietzsche rigorously hedges the 'return to Nature' with qualifications. Nature is not a lost Eden or a privileged origin, and a simple return is impossible because 'there has never been a natural humanness' "[8] In other words, Nietzsche complicates the relationship between nature and culture at the same time that he disrupts the wedge driven by Christianity between animals and humans. The Acamporas have taught us more recently that we can read Nietzsche's work as a bestiary that disrupts the logic of domestic versus wild, while also dismantling the *scala naturae* that elevates the human to the fictitious place beyond animality precisely because humans are alleged to be "above" all animals. If man is a bridge between beast and Übermensch, stretched with the flesh of domesticated animals, then to cross over it means to dedomesticate, to undomesticate, to render wild again, the beast in the docile animal humans have made of themselves. To become Übermensch means to go wild, to cease to be human and to become animal, again. As Christa Davis Acampora and Ralph Acampora put it, "In calling for and cultivating an experimental education of the human animal from civilized domesticity to atavistic wilderness, Nietzsche can be seen to develop a feral philosophy of cultural zoology."[9] To become Übermensch is to become feral. Yet, in as much as humans are the docile animals that have become so through the domestication of other animals (i.e., by domesticating itself in the process), there are neither purely wild nor purely domestic animals. We cannot entirely be undomesticated ourselves, in the same way that the dog can never be undomesticated. The Übermensch is not simply an "overman" that is beyond animality. The Übemensch cannot but be an Übertier, an overanimal, a superbeast.[10] You may take the pet out of the dog, but the dog remains a companion species of codomestication. For this reason, and more precisely, to become Übermensch means to become liminally feral. To quote the Acamporas again: "Zoologically speaking, however, it is possible for an animal to be neither fully tame nor entirely wild but rather liminally feral. Ferity, then, is the hermeneutic middle excluded from standard readings of Nietzschean animality; ironically, it is central to Nietzsche's own retranslation of the text he calls *homo natura.*"[11] What is this "hermeneutic middle" between tameness and wildness? I want to suggest that this middle, this being liminally feral, is what I would call interspecies cosmopolitanism.[12] Indeed, Nietzsche's *Thus Spoke Zarathustra,* the *locus classicus* of Nietzsche's bestiary, disrupts the Judeo-Christian-Darwinian *scala naturae* to urge us to turn into good animals, animals that can dwell in the forest of nature, as neither simply a beast of prey nor a lowly beast of burden, a *Sklaventiere,* a slave animal. Nietzsche's bestiary is a manifesto of interspecies cosmopolitanism, an attack on the ontological anthropocentrism that hinders a proper zootic philosophy of companion species coexistence.

While I am not directly concerned with pursuing an exegesis of Nietzsche's bestiary, I am interested in how Nietzsche's trope of the Übermensch allows us to rethink Michel Foucault's own approach to biopolitics in a different light. One of the most succinct definitions of biopolitics one can find in Foucault's corpus could be that biopolitics is the new social order in which "politics places his [modern man] existence as a living being in question."[13] If we think of politics as the realm of the possible, where collectivities enable possibilities, and biology the realm of necessity, where bodies are caught in web of causalities, then biopolitics could be understood as that new horizon in which some possibilities are turned into necessities, and necessities into possibilities. There is a border, a hermeneutical middle, between the necessary and the possible, but one that is not fixed. This border, this in-between, waxes and wanes, shifts and relocates according to the index of our imagination. The encounter between necessity and possibility, the required and the optional, is determined by our imagination, which is why we talk about a political imaginary. Here, I am interested in instigating a biotic imaginary, one in which the political and the biological are always at play, in codetermination and codependence.[14]

It is for this reason that I want to use Nietzsche's critique of the Judeo-Christian *scala naturae* and the rewriting of the medieval bestiary as backgrounds against which to interpret three artists who have engaged in similar projects but from the side of the arts. In the following I discuss these three artists, whose works explore in very striking, innovative, and eerie ways what we have called, following Nietzsche, becoming animal and ceasing to be human. Each one of these artists, tellingly, comes from three distinct "border cultures," or what Haraway has called "frontier practices."[15] Patricia Piccinini is an Australian artist, Jane Alexander is South African, and Guillermo Gómez-Peña is a Chicano border artist. Each artist has explored in his or her work over the last two decades not simply the border between human/animal but also the borders among cultures, as well as between nature and technology. Their collective work illustrates brilliantly how to become an interspecies cosmopolitan agent requires that we transgress these three borders, namely the borders that are created among cultures through political devices and machines, the borders that are created by their deployments of technology, and how these technologies mark borders between the human and the animal. In other words, I want to show that to be cosmopolitan today, in the most generic and pedestrian sense of the term, it is not enough to have a cosmopolitan intent, a certain epistemic attitude toward other cultures. Today, to be cosmopolitan requires we become self-reflexive about how we make ourselves into the type of humans who distinguish themselves from others and animals by means of human technologies, whether they are political, prosthetic, or metaphorical. We have to become reflexive of the ways in which technologies are the means through which we draw boundaries, lines, thresholds between animal and human, culture and nature, human and human.

Patricia Piccinini—Biotechnologies of Care

Piccinini's work has explored most audaciously what it would mean to become transhuman, what is at the other end of the process of splicing

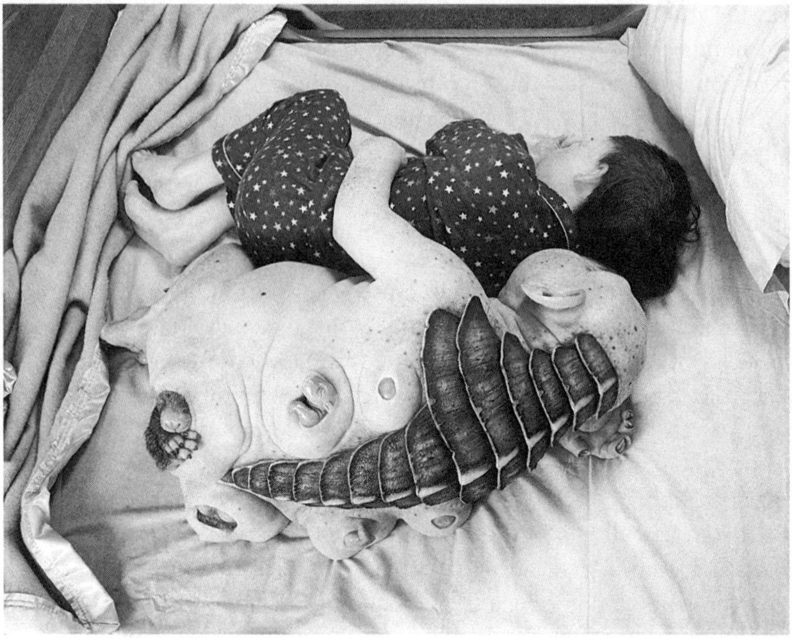

Undivided, 2004, silicone, fibreglass, human hair, flannelette, mixed medium, 74 cm h × 101 × 127 cm. Photo: Graham Baring. Permissions by Patricia Piccinini.

human and animals genes.[16] Her work constitutes a postnatural, posttranshuman, posttransgenic bestiary. In it we find creatures that have been biotechnologically engineered to be care providers: note, for instance, "nature's little helpers."

These images from some of Piccinini's installations give us a sense of her work. The first installation, titled *Undivided,* presents a "surrogate" taking care of a human child. *Surrogate* is the ultimate "nature's helper." She/he has been engineered to have several pouches in her back, where she nestles and nurtures other creatures. She is an interspecies supermom. The same is true of *Big Mother,* who is shown here breastfeeding a human baby. What is striking about *Big Mother* is that from the front she appears asexual. She is simply a muscular mother, who can hold with her arms all kinds of sucklings. However, when we walk around her we notice that her sexual organs are all in her back. *The Long Awaited* iterates the theme of nurturing and caring. Here we are shown a prepubescent child nestling, resting, tendering, with what appears to be a grandmother. It is not clear who is providing whom with care and tenderness.

Since space here is too limited to go into other of her work, let me simply affirm that most of Piccinini's work concerns affectively charged relationships with not simply organic but also nonorganic forms of life.[17] She is not interested only in the ways in which the boundaries between human and animal have been crossed but also those between animate and inanimate.[18] Some of her work has been about anthropomorphizing "machines." In a series of works from 2006 and 2006, *Nest* and *Stags*, Piccinini

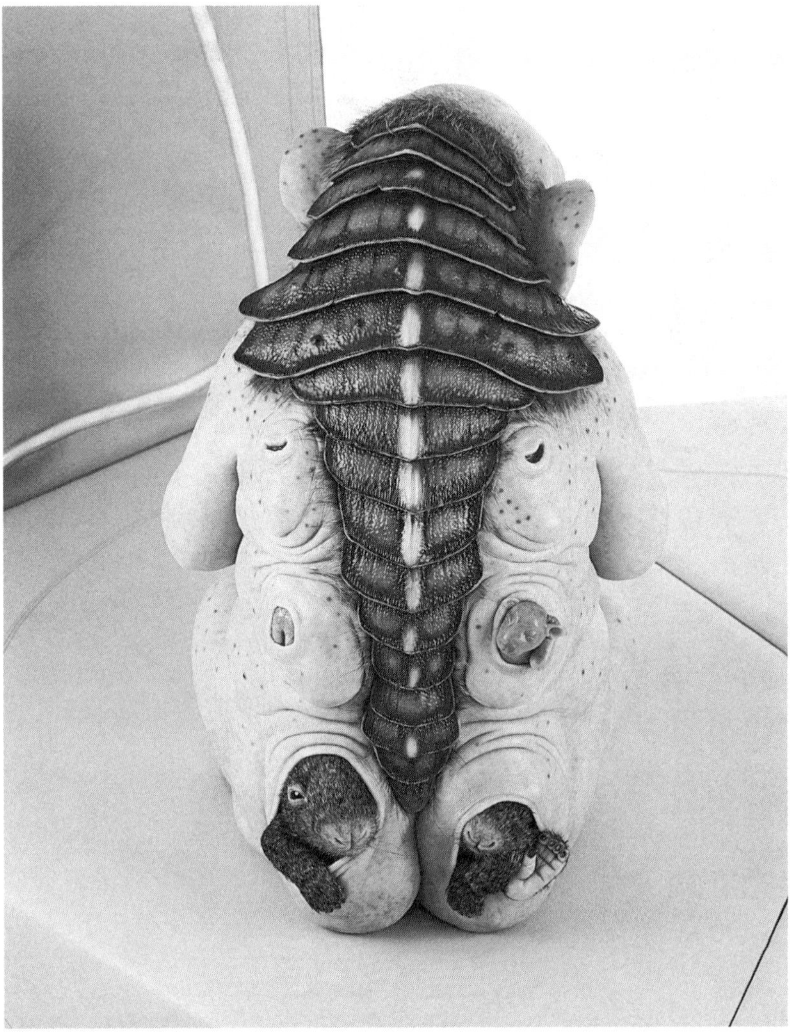

Surrogate, 2005, silicone, fibreglass, leather, plywood, human hair, 103 cm h × 180 × 306 cm. Permissions by Patricia Piccinini.

Big Mother, 2005, silicone, fibreglass, leather, human hair,
175 cm h × 90 cm × 85 cm (including satchels, dimensions variable).
Photo: Graham Baring. Permissions by Patricia Piccinini.

The Long Awaited, 2008, silicone, fibreglass, human hair, plywood, leather, clothing, 92 cm h × 152 × 80 cm. Photo: Graham Baring. Permissions by Patricia Piccinini.

explores how intelligent machines may actually evolve rituals of affect. In these installations we see infantilized motorcycles frolicking and horsing around like cubs. We created machines to facilitate producing ourselves, but in making these machines we also have entered into specific relations of dependency-care. We have to take care not simply for other living creatures, but also those entities that may not be alive, but nonetheless have acquired their own forms for care. Indeed, one of Piccinini's themes is the bidirectionality of care. Being taken care of entails having to take care. We cannot allow someone else to care for us unless we are willing at some level to be ready to care for the one providing care. An ethics of care demands some modicum of reciprocity. This taking care of, however, is a taking care of environments. What is explicit about Piccinini's work is that it references environments. She has deliberately picked creatures that refer to the fragile environments of Australia, specifically the wombat. Surrogate is an Australian marsupial introduced to help with the endangered species from the Australian bush. What is distinctive about Piccinini's trans-species care providers is that they are situated in some affective space beyond terror and disgust. We are eerily fascinated by these creatures that are all too familiar yet also uncanny. We cannot hate. We cannot fear them. We would like to feel revulsion at them, but their tenderness, their maternal evocations simply provokes our nurturing feelings.[19] What art critic Jacqueline

Millner wrote about Piccinini captures very aptly the kind of interspecies ethics of coexistence that I think animates the kind of interspecies cosmopolitanism that we need to espouse: "Unlike Dr Frankenstein who grew to hate his creation and suffered the consequences, Piccinini would urge us to bring an attitude of love to the products of technology, to accept our ethical mantle as creators, to take care of all our progeny, even of the artificial variety. The love she appears to propose is not of the romantic, infatuated ilk—classic technophilia—but of the familial variety, with its overtones of responsibility, ethical guidance and life-long commitment."[20]

Jane Alexander—Humananimals/Animalshuman

Jane Alexander's work[21] is more explicitly political than Patricia Piccinini's. Or, to be more precise, the reference to politics in Alexander's work is more directly addressed to the effects of human interaction on other humans. In contrast to Piccinini's creatures, who have been engineered to serve humans, while retaining as much of their animality, in Alexander's hybrid creatures we encounter humananimals, creatures who are part human and part animal. There is a preponderance of birds. The repertoire of hybrid humananimals is not extensive, which makes her creations the more strikingly compelling and provocative One of the central themes of Alexander's work is "vulnerability" and "violence." In fact, most of her installations evoke scenes where violence has taken or is about to take place. Both themes, however, are intricately entwined, and they get to be embodied in her wounded, precarious, ailing creatures. Even those humananimals that would inspire terror in us are themselves past victims of torture and violence. Note, for instance, the image titled *Butcher Boys 1985–86*. This installation shows three humanoids with heads of what appear to be a cross between cattle and some sort of feline. They have horns, but some are broken. They do not appear to have mouths. Their eyes are dark and terrifying. Deep creases, traces of long endured frowns of worry and intense concentration, mark their brows. Their body postures of expectant ease contrasts with their otherwise mangled faces. But what is almost missed is that the three *Butcher Boys* bear scars across the center of their chests, as if they had undergone open-heart surgery. Their genitals are concealed/sealed behind what seemed to be metal or hard plastic cups that appear surgically affixed to their groins. Their muscularity stands in contrast with the absence of genitalia. These are emasculated hypermasculine killing machines, who are themselves victims of violence.

Butcher Boys seems to say that we humans become beastly through our

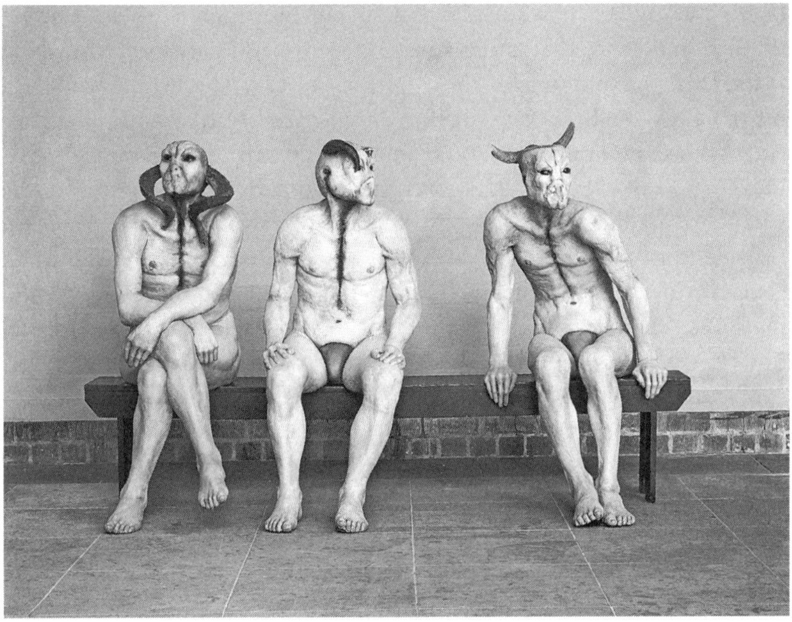

Butcher Boys 1985–86. © Jane Alexander/DALRO.

The sacrifices of God are a troubled spirit. 2004. © Jane Alexander/DALRO.

own violence. We are the animals that can turn themselves into beasts through a violence that is excessive and exorbitant. This logic of making vulnerable, or exploiting human vulnerability, to render beastly works the other way as well. In her installation *The sacrifices of God are a trouble spirit. 2004* we see a humanoid with an animal head, which is wearing a mask. This mask, however, does not cover the entire face but is placed over the snout, leaving the eyes visible. On closer inspection, we see that the mask is made from a skull fragment of a Gemsbok with deformed horns, when generally they are straight. The humanoid is walking with the aid of walking sticks over a floor covered entirely with glossy red gloves. The installation is surreal and haunting. There is a postapocalyptic character to it. The figures to the right of the altar standing with their arms spread out as if scarecrows have humanoid faces.

Another installation is titled *Bom Boys. 1998;* "Bom" is the Afrikaans word for "bomb," and the title of the installation is taken from local graffiti. *Bom Boys* shows a series of boys wearing masks. Some of the masks are of animals, but some are simply covering hoods. One of the children, which someone with a US racial imaginary could take for a black child, is wearing

Bom Boys. 1998. © Jane Alexander/DALRO.

a blindfold. Some of the kids are naked; others are wearing rags. They are all in some state of disarray and disregard. *Bom Boys* is an installation that arranges chaos; it represents disconnection while, at the same time, degrees of violence. The fact that the masks span the range of hoods, blindfolds to animal masks, suggests that we put animal masks on humans to do violence to them: we animalize to render vulnerable, to render injurable.

Like Piccinini, Alexander indexes her work with respect to apartheid and postapartheid South Africa. Her work references not simply racial violence, but the violence of a society marked by the devices of racial separation. Thus, for Alexander becoming animal is being racialized. Racialization is a technology of bestialization, which in turn is a way to make injurable. Crossing the animal/human divide means also crossing racial boundaries in postcolonial contexts. The subtext, then, is that the history of becoming animal, of being turned into a beast, is part and parcel of the history of racial/colonial violence. We cannot innocently become transhuman or posthuman. At the same time, Alexander urges us to recognize that this racial/colonial violence has had a profound impact on the animals of the frontier land, the land that had to be eviscerated and evacuated of other humans and autochthonous animal species so that it could be colonized.

Guillermo Gómez-Peña: Ironic Naftazteca Borgs of the Haunted Border

Guillermo Gómez-Peña is without question the best-known Latino, Chicano, Mexican performance artist in the United States, Latin America, and the art world of the North Atlantic corridor.[22] His performances are legendary, unforgettable, and visually assaulting. His dioramas project a distinct and unmistakable aesthetic that is part street pastiche, part postcolonial neobaroque, and part Silicon Valley cybernetic funk. Entering one of his many installations and performance settings, it is hard to be sure whether you are walking through a curiosity shop in an intergalactic outpost or through a hilarious ethnographic museum reconstructed from the sarcastic notes of a colonial informer. From afar, the commentary on the performances that booms from speakers in electronically remastered voices sounds like Derridian, or Spivakian—accented by the idioms of Octavio Paz and Subcomandante Marcos—as it might be rapped by the Mission Street homeboys. Indeed, Gómez-Peña's performances are theatricalizations of poststructuralist and postcolonial theorizations. His performances are also aurally eerie, as echoes of languages no longer spoken or yet to be spoken, or in the process of dying or of being born, reverberate in the background. Gómez-Peña is a virtuoso of the visual and the uttered. He does not speak his performances. He languages them, in Martin Heidegger's sense of language being more than a mere speaking. Language is a way of opening up a world. It is world disclosing. First there is the speaking, the performance of the word, laying before us a possible world, and then there is language, a way to communicate information through words.

Among the many themes that run through his books, performances, and communiqués there are two that I want to foreground here vis-à-vis the work of Piccinini and Alexander. The first has to do with Gómez-Peña's uses of technology, which I would like to call ironic—in the sense meant by Don Ihde's ironic technics.[23] His characters and personas are built from the recycled and reappropriated technologies left over as debris in the post-imperial and postcolonial junkyards of postindustrial cities. The "Cyber-Aztec," one of Gómez-Peña's ironic creatures, for instance, is pastiche made up of pre-Colombian lore and postindustrial junk. He is what I call an ironic cyborg in the sense that technology is now used for aims forgotten or no longer relevant. At the same time, this very technology is possessed. It has had and will have unintended consequences. *Cyber-Aztec* is an allegorization of the ways in which we have become victims of our own inventions and technological accoutrements. We aim to shield ourselves with our technologies, but through them we make ourselves vulnerable in new and unforeseen ways. Through our technologies we aim to make our worlds fully available to us, operable, manipulable, calculable, to use Nietzsche's terminology, but in the process of making our world operable, we turn into operable man ourselves, as Peter Sloterdijk notes.[24] Homo naturans became Homo laborans, and Homo laborans became Homo technologicus. As we became the creature user of technology, however, we became a cog in our own machines—we became the Homo operabilis: operable man. Operating others requires that we operate us, as we operate other of our technologies. This task always has unintended consequences; it is as if our technologies have spirits of their own. Technology thus appears as if possessed by evil spirits that must be exorcised. The "Ethnograpic loco," another of Gómez-Peña's ironic personas, performs this haunting of and by technology. The "Border Brujo," thus, is the other Janus face of the Cyber-Aztec. In Gómez-Peña's work, then, technologies are deployed to draw boundaries: boundaries between the animal and the human, but also between humans, between cultures, and between races. The border is a technology, and a complex one with its own ghosts and incubuses. But in drawing these borders, we unleash unintended consequences.

"The Border Brujo." Permission by Guillermo Gómez-Peña.

"El Ethnographic Loco." Permission by Guillermo Gómez-Peña.

The second theme that I want to foreground from Gómez-Peña's work is his relentless commenting, analyzing, demythologizing, and desacralizing of the border. Borders are human inventions, but in the United States in particular, the border has become something like an 'a priori,' a given, an uncontested force of nature. Since his earliest work, already more than two decades old, Gómez-Peña has been performing, writing, and commenting on the US-Mexico border. One of his earliest performances had to do with "exorcising" the border, and thus was born the "Border Brujo." The border was also the scene of many a "crucifixions"—in particular of the Tex-Mariachi and the warriors for Gringostroika. The border is not a line, nor is it simply a wall. It is a technological dispositif that extends in both directions, forming less an impermeable membrane and more a passageway. The border is a haunted technological dwelling. It is the scene of much human suffering, indeed; the place where both nations meet and bleed, like Gloria Anzaldúa puts it, but it is also the place where two cultures meet and give birth to new culture. Naftazteca, child of this encounter, shows us not simply how a new type of culture is born from the tatters of an old one, but how all culture is always hybrid, part pastiche, part failed attempt at

"Cyber-Aztec." Permission by Guillermo Gómez-Peña.

authenticity. All culture is cultivated hybridity, which is why we are always haunted by the memory of suffering, of conquest, but also resistance.

Conclusion

The background assumption of this chapter is that we need to develop a post-Kantian, and even post-Levinasian, version of cosmopolitanism that overcomes both notions that the human is a rational animal that occupies a place on the "chain of being" above all other animals, and that this human animal's worldhood is to be thought merely in terms of historicity, à la Heidegger.[25] The aim, thus, is to develop an "interspecies cosmopolitanism," or "cosmopolitanism of animal companions" to use Haraway's felicitous expression. This chapter focuses on three contemporary artists who have been exploring the region that marks the divide among humans and animals from the standpoint of biotechnology and biopolitics. Whereas Piccinini and Gómez-Peña engage the uncanny dimensions of a posthuman, postanimal future through ironic and sardonic iterations of technologies utopias turned demonic chimeras, Alexander engages the somatic or corporeal dimension of embodied animal existence. If Gómez-Peña engages ambiguous technology as a postcolonial and postimperial 'American' subject, Alexander challenges the racist dimension of how technology is used for social manipulations in South Africa. Thus, the chapter focuses on two axes of analysis when engaging these unsettling artists: the techno-scientific that points in the direction of what Sloterdijk has called the "operable man," and the corporeal or embodied dimension of animal existence. I argue that a form of "interspecies cosmopolitanism" demands attention to the phylogeny of the *scala naturae*, which also demands attention to how we became human/animals on this planet.

I began this chapter, however, by invoking a certain reading of Nietzsche that places him within what Norris calls a "biophilic tradition" and what the Acamporas call becoming feral, or wild. Nietzsche's injunction that we should aim to transcend the human, in these readings, does not mean that we should cease to be animal. On the contrary, to become Übermensch means also to become an Übertier, that is, an animal that has refused the technologies of docility and domestication that had turned us into slaveanimals (*Sklaventiere*). We are the slaveanimals of our biopolitical technologies of subjection and normalization. Along with Foucault, we could say that the grotesquely calculative height of these biopolitical technologies was the National Socialist project to eradicate the Jewish people, which allegedly sought to save us from a vermin, a plague.[26] For Foucault, however, the concept of biopolitics was a placeholder, or more precisely the name for a distinct regime of governmentality. I suggest, in fact, that we ought to understand this new regime as the opening up of a new horizon of the political in which possibility and necessity are determined according to a certain imaginary. In Foucault's work we cannot talk about biopower, biohistory, and biopolitics[27] without also considering that for him the "power" that circulates because of biopolitics is both generated and contested by the interaction among: forms, or games, of veridiction, procedures of governmentality, and pragmatics or technologies of the self.[28] In other words, biopolitics is a particular configuration of knowledge/power/subjection in which we allow ourselves to be defined as members of a species, whose biological existence is to be monitored, analyzed, and determined by the life sciences in such a way that our truths are disclosed either by the color of our skins or the deep grammar of DNA strings. Yet, as pastoral power was contested by the counterconducts of heretics, biopolitical power is contested by those who refuse to submit to the imperatives of a governmentality that aims to maximize their blood, flesh, and semen, for the alleged benefit of the nation, the "superior" race, the human "race."

My three artists, dwellers of border cultures, performers of counterconducts, skeptics of the regimens of governmentality that forces us to continuously to mark, build, and police boundaries among ourselves, with our technologies, and against our "companion species," incite us with their powerful aesthetic creations to invent, fashion, live out our own counterconducts against our modern biopolitical form of governmentality. Rather than talk of a "negative" biopolitics or a contestational biopolitics, I think we should think of counterconducts. This is what Nietzsche means when he beseeches us to leap over the abyss that yawns under the human. This is what Norris and the Acamporas mean when they invite us to dwell in the worlding of "feral ontologies."[29]

Notes

The author would like to thank William Glenn, Kristen Nyitray, and F. Jason Torre, librarians at Stony Brook University, for help in locating some of the images here reproduced. The author also expresses his deep gratitude to the artists who allowed their works to be reproduced.

1. Friedrich Nietzsche, *The Gay Science*, trans. Josefine Nauckhoff (Cambridge: Cambridge University Press, 2001), 114. This comes from bk. 3, sec. 115.
2. Friedrich Nietzsche, *Thus Spoke Zarathustra. A Book for All and None*, trans. Adrian del Caro (Cambridge: Cambridge University Press, 2006), 5–7. This citation comes from the first part, "Zarathustra's Prologue," secs. 3 and 4.
3. Friedrich Nietzsche, *Beyond Good and Evil. Prelude to a Philosophy of the Future*, trans. Walter Kaufmann (New York: Vintage, 1966), 154, sec. 225 (italics in original).
4. See the outstanding work by Vanessa Lemm, *Nietzsche's Animal Philosophy: Culture, Politics, and the Animality of the Human Being* (New York: Fordham University Press, 2009).
5. For a discussion of the bestiary, see my essay "Political Bestiary: On the Uses of Violence," *Insights*. 3, no. 5, http://www.dur.ac.uk/ias/insights/volume3/article5/.
6. The classic study on this tradition is Arthur O. Loveyjoy, *The Great Chain of Being: A Study of the History of an Idea* (Cambridge, MA: Harvard University Press, 1964).
7. Margot Norris, *Beasts of the Modern Imagination: Darwin, Nietzsche, Kafka, Ernst, & Lawrence* (Baltimore: Johns Hopkins University Press, 1985).
8. Ibid., 22.

9. Christa Davis Acampora and Ralph R. Acampora, eds. *A Nietzsche Bestiary: Becoming Animal Beyond Docile and Brutal* (Lanham, MD: Rowman & Littlefield, 2004), 2.
10. On the Übertier, see Friedrich Nietzsche, *The Will to Power*, trans. Walter Kaufmann and R. J. Hollingdale (New York: Random House, 1967), 531, sec. 1027 (from the Spring-Fall 1887).
11. Acampora and Acampora, *Nietzsche Bestiary*, 7.
12. For a presentation of "interspecies cosmopolitanism," see my essays: "Interspecies Cosmopolitanism: Towards a Discourse Ethics Grounding of Animal Rights," in "Recenterings of Continental Philosophy," ed. Cynthia Willett and Leonard Lawlor, special issue, vol. 35, SPEP Supplement 2010 of *Philosophy Today* 54, 208–16; and "Interspecies Cosmopolitanism," in *Handbook of Cosmopolitan Studies*, ed. Gerard Delanty (New York: Routledge, 2012), 276–87.
13. Michel Foucault, *The History of Sexuality*, vol. 1, *An Introduction*, trans. Robert Hurley (New York: Pantheon, 1978), 143.
14. I recommend the beautifully illustrated "Animals," special issue, *Lapham's Quarterly* 6, no. 2 (2013). One could take *Lapham's* special issue as a twenty-first-century biopolitical bestiary.
15. Donna Haraway, *When Species Meet* (Minneapolis: University of Minnesota Press, 2007).
16. Many of her works and essays are found at http://www.patriciapiccinini.net/.
17. See Frances Bartkowski, "All That Is Plastic . . . Patricia Piccinini's Kinship Networks," *Insights* 4, no. 7 (2011), http://www.dur.ac.uk/ias/insights/volume4/.
18. See Peter Hennesay, "Patricia Piccinini. Plastic Realist" in *Photo Files: An Australian Photography Reader*, ed. Blair French (Sydney: Power, 1999), 247–54.
19. See Donna Haraway, "Speculative Fabulations for Technoculture's Generations. (Tender) Creature Exhibition Catalogue. Artium." (2007), http://www.patriciapiccinini.net/writing/30/242/68.
20. Jacqueline Millner, "Patricia Piccinini: Ethical Aesthetics," *Artlink* (2001), http://www.patriciapiccinini.net/printessay.php?id=4.
21. I have relied on Jane Alexander, *On Being Human*, ed. Pep Subirós (Durham: Institute of Advanced Study, Durham University, 2009). There is now a comprehensive catalog of her work: *Jane Alexander: Surveys (From the Cape of Good Hope)*, ed. Pep Subirós (New York: Museum for African Art and ACTAR, 2011). There is also a very useful overview of a 2013 exhibition of her work at the Cathedral of St. John the Divine in Manhattan, at http://africasacountry.com/strange-cargo-jane-alexander-at-the-cathedral-of-st-john-the-divine/. Some images and additional information appear on Alexander's university home page at http://www.michaelis.uct.ac.za/staff/alexander/.
22. Some works and texts of Guillermo Gómez-Peña are found at http://www.pochanostra.com/antes/jazz_pocha2/index.htm.
23. See Don Ihde, *Ironic Technics* (Copenhagen: Automatic Press, 2008).
24. Peter Sloterdijk, *Das Menschentreibhaus. Stichworte zur historischen und prophetischen Anthropologie* (Weimar: Verlag und Datenbank für Geisteswissenschaften, 2001), sec. 4 in particular. See also his *You Must Change Your Life*, trans. Wieland Hoban (Cambridge: Polity, 2013).
25. I have analyzed Heidegger's use of animals in terms of his own political-ontological bestiary; see "El Bestiario de Heidegger: El Animal sin Lenguaje ni Historia," *Revista Filosofía UIS* (Colombia) 11, no. 1 (2012): 17–43.

26. See Jeffrey Herf, *The Jewish Enemy: Nazi Propaganda During World War II and the Holocaust* (Cambridge, MA: Harvard University Press, 2006).
27. See my entries on "biopolitics," "biopower," and "biohistory" in *The Cambridge Foucault Lexicon*, ed. Leonard Lawlor and John Nale (Cambridge: Cambridge University Press, 2014).
28. For a discussion of these themes and the way in which Foucault thought about biopolitics, see Dianna Taylor, ed., *Michel Foucault: Key Concepts* (Durham, NC: Acumen, 2011).
29. Norris, *Beasts of the Modern Imagination*, 101.

PART THREE

Medicine and Sexuality: The Question of the Body

NINE

Patient Activism and Biopolitics: Thinking through Rare Diseases and Orphan Drugs

CARLOS NOVAS

This chapter examines the forms of political activism that developed around orphan drugs from the late 1970s to the present. The chapter begins by discussing how the problem of developing drugs aimed at treating rare diseases was made knowable to political authorities in the United States through a combination of patient group activism, congressional hearings, surveys, academic conferences, and media reports. Through the formation of an effective coalition, patients' associations were able to aid the passage of legislation specifically concerned with altering the economic and regulatory circumstances that had previously prevented pharmaceutical companies from developing treatments aimed at rare diseases. In the second section of the chapter, I discuss how the US Orphan Drug Act (1983) has been adopted as a policy model in a number of countries such as Singapore, Japan, Australia, and, most recently, the European Union. The chapter examines the impact that this legislation has had on the biotechnology industry, which has benefited considerably from the provisions contained within orphan drug legislation. The chapter concludes by examining some contemporary biopolitical problems that orphan drugs pose as health-care authorities attempt to reconcile providing expensive treatments for a small number of patients while still trying to finance the health-care needs of the general population. This chapter aims to examine how patient activism has shifted from shaping the economic and regulatory circumstances that prevented orphan drugs from being marketed to more recent efforts that focus on challenging the calculative practices that restrict patients from having access to orphan drugs.

Orphan Drugs, Patient Activism, and the Pharmaceutical Industry

Orphan drugs provide an interesting case study to explore the normative character of contemporary biopolitics. Historically, the qualifier "orphan drug" referred to medically significant chemical substances or therapeutic categories that by virtue of their clinical, legal, proprietary, or market characteristics were not invested in by governmental, pharmaceutical, or scientific groups. The term now refers to treatments intended for rare diseases or small patient populations. Current debates around orphan drugs tend to focus on their cost, as they are generally more expensive than treatments aimed at larger population groups. The clinical and economic features of orphan drugs spur debate about how well existing cost-effectiveness methodologies assess these drugs' funding by national healthcare systems. Orphan drugs provide a historical and contemporary example through which to explore the biopolitical dimensions of the drug development process and the calculative practices used to assess the cost-effectiveness of medicines.

The highly regulated trajectory that a chemical or biological substance must follow to become an authorized medication requires that it pass through relays of power concerned with the administration of life (Lakoff 2008). In this chapter, I draw on Michel Foucault's concept of biopolitics (Foucault 1978, 1994, 2001a, 2001b) to analyze the forms of power relations that shape the development of pharmaceuticals and that govern access to medicines. What Foucault claimed to be distinctive about this form of power is its concern with addressing governmental problems in relation to phenomena characteristic of human populations (Dean 1999; Foucault 1991). Biopolitics is concerned with phenomena that affect the health and vitality of a population such as birth and death rates, levels of morbidity, fecundity, sanitary practices, hygiene, and the safety of consumer goods. As a governmental practice, biopolitics is concerned with the range of social, cultural, economic, environmental, and geographic conditions that limit or foster the health of the human species (Rose 1999). As Foucault (1978) explains in the *History of Sexuality*, the development of this novel form of power gave rise to forces that resisted it. He claims that the "forces that resisted relied for support on the very thing it invested, that is, on life and man as a living being" (Foucault 1978, 144). On the one hand, we can understand the regulation of pharmaceuticals as being a key site of biopolitical intervention as it relates to the governance of human health. On the other hand, the regulation of pharmaceuticals (i.e., the drug development process) and their use within health-care systems have given rise to numerous resistance movements. The kinds of struggles that emerge around the circulation and regulation of pharmaceuticals, as Andrew Lakoff (2008) usefully characterizes them, relate to struggles over the supply of pharmaceuticals (the drug development process; which illnesses are targeted as being of market interest) and the demand for pharmaceuticals (pricing, access to medicines, who pays for medicines). Orphan drugs provide a unique vantage point from which to consider these types of biopolitical struggles as in the past they raised questions about the supply of drugs for particular patient populations, while in the present they pose questions about access to medicines. I propose that the types of struggles around orphan drugs are biopolitical in orientation as they are concerned with including the therapeutic needs of particular population groups into the investment decisions of pharmaceutical and biotechnology companies, as well as statist calculative practices relating to access to medical treatments. As I attempt to show, these struggles are normative in orientation as they concern the administration of the lives of persons with rare diseases.

Throughout the twentieth century, states have actively developed regulations to protect national populations against life's contingencies and health risks (Timmermans and Berg 2003). Regulations are particularly prominent in the field of consumption, most notably relating to food and drugs (O'Malley 2004). In the case of pharmaceutical products, regulatory institutions such as the US Food and Drug Administration (FDA) were established in the early part of the twentieth century to minimize threats posed to the public's health by unsafe or ineffective medicinal goods (Marks 1997). The institutional forms and practices associated with pharmaceutical regulation have been shaped by a number of disasters (Daemmrich and Krücken 2000). The greatest one, which has shaped drug regulation from the 1960s onward, relates to the drug thalidomide. Thalidomide, a sedative, was administered to women to treat pregnancy-related nausea. One of the tragic consequences associated with thalidomide was that it led to the birth of a great number of deformed babies. In the United States, the thalidomide disaster led to the passage of the Kefauver-Harris Amendments to the US Food and Drug Act (1962). These amendments required pharmaceutical companies to provide greater safety and efficacy data based on adequately conducted clinical trials prior to the approval of a drug for sale (Abraham 1995). What the sweeping reforms to pharmaceutical regulation in the United States and in other jurisdictions accomplished was a new series of standards for approving chemical or biological substances as authorized medications.

While the implementation of new regulatory standards safeguarded the public's health, it also served to alter the techno-economic network of drug development. In the 1970s, the pharmaceutical industry claimed that the Kefauver-Harris Amendments had increased drug development costs, extended the time required for a drug to gain regulatory approval, and eroded the duration of patent terms (Mitchell and Link 1976). Representatives from the pharmaceutical industry asserted that the focus on drug safety had shifted their investment decisions toward drugs intended for short-term use and for large patient populations. This view was echoed in a number of conferences and publications in the early 1980s that focused on the orphan drug problem. These studies showed that the rising costs of drug development, combined with the prospect of limited economic returns associated with drugs targeted at small patient populations and the greater difficulties associated with conducting clinical trials for rare diseases, served to marginalize this therapeutic category from drug development (Brewer 1983; Karch 1982; Scheinberg and Walshe 1986; Van Woert and Chung 1985). Over the course of the 1970s, the orphan drug problem began to be understood as one shaped by the unwillingness of pharmaceutical companies to conduct the requisite clinical trials needed to sponsor medically promising drugs through the FDA approval process (Asbury 2007).

One side effect of the new drug safety standards was a new model of drug development that served to exclude particular sets of patient populations. As Carolyn Asbury has carefully documented, this novel drug development model excluded these drug and therapeutic categories: chronic diseases, vaccines, women of childbearing age, children and the elderly, rare diseases, unpatentable substances, substance abuse treatments, and drugs for developing countries (Asbury 1985, 2007). In the early 1970s, growing recognition of what was then known as "drugs of limited commercial value" began to problematize the government of drug regulation and the investment decisions of the pharmaceutical industry. This problem was biopolitical in orientation given that the stakes involved concerned the inclusion of the medical needs of a previously neglected population into the institutional and economic relays of drug development.

In the United States, two interdepartmental task forces were convened in 1974 and 1978 to examine the problem of drugs of limited commercial value (Asbury 1985; US Department of Health, Education, and Welfare 1979). Both task forces recognized that the government had an obligation to provide medically important drugs to populations that needed them (Asbury 1985). Drugs of limited commercial value problematized existing regulatory policies and industry drug development practices by documenting the range of medically significant chemical substances that pharmaceutical companies were refusing to fund. Although a number of recommendations identified potential financial, legal, and regulatory measures that might remedy the problem, they did not garner the support of the pharmaceutical industry (Asbury 1985). It was not until the late 1970s and early 1980s, through the advocacy efforts of patients' organizations and academic scientists, that the orphan drug problem became known beyond a few specialists and identified exclusively with the plight of persons affected by rare diseases.

In the late 1970s and early 1980s, the orphan drug problem emerged as a significant biopolitical issue that concerned the *inclusion* of the therapeutic needs of rare disease patients into the biopolitical paradigm of drug development. Steven Epstein's (2007, 18) work is useful here as it draws attention to how socially marginalized subpopulations come to be known and acted on within the institutional and economic circuits of drug development. The needs of rare disease patients began to be articulated during this period by Sharon Dobkin, the National Myoclonus Foundation, and Dr. Melvin Van Woert (Meyers 1988). For many years, they had tried—unsuccessfully—to find a commercial sponsor for L-5HP, a then-promising therapy for myoclonus. This failure convinced them that legislative action was needed to change regulations concerning drug development and the investment decisions of the pharmaceutical industry. They contacted Representative Elizabeth Holtzman, who after hearing of their experiences introduced a bill in Congress in June 1980 (Asbury 1985). This led to hearings on this proposed bill conducted by Representative Henry Waxman. These hearings were historically significant in that they created a forum through which the orphan drug problem came to be known and acted on. These struggles to legislate orphan drugs can be understood, following Foucault (1978, 145), as being concerned with "life more than the law" in that they aimed to reshape the biopolitical paradigm of drug development.

Congressional hearings (US Congress 1981) on orphan drugs narrowed the scope of the problem from a broad range of therapeutic categories to the treatment of rare diseases (Asbury 1985). Pushed by the advocacy of patients' groups for myoclonus, Tourette syndrome, Huntington's disease, Wilson's disease, and other rare diseases, the hearings documented how the medical profession, the FDA, and the pharmaceutical industry were not adequately meeting the therapeutic needs of the rare disease population. Press coverage of the hearings led the popular television series *Quincy M.D* to air an episode portraying the unique problems faced by patients with rare diseases.[1] The subsequent media attention that followed this episode helped to transform the orphan drug problem into a national issue (Asbury 1985; Henkel 1999). Of further significance is how the hearing brought into being a novel patient population. While patients affected by rare diseases were previously isolated by the very rarity of their illnesses, the forms of activism that they engaged in over the course of the late 1970s and early 1980s helped to bring the needs of this particular patient population to the attention of the pharmaceutical industry, as well as to medical and governmental authorities.

Because the hearings included testimony from all the major players—the pharmaceutical industry, the FDA, academic science, and patients' groups—they created an important forum through which the newly constituted patient population could begin to get its needs met. Nonetheless, defining the appropriate scope of governmental intervention created intense political struggle. This intensity had to do with the stakes at hand: the regulatory framework and the economic premises underpinning the drug approval process in the United States. At the hearings, the Pharmaceutical Manufacturers Association (PMA) argued that the FDA standards relating to drug safety and efficacy significantly increased the time and expense of developing new medicines, which created disincentives for industry to invest in drugs targeted toward small patient populations. The PMA proposed the development of a two-tiered drug approval system: one system that employed the existing standards for drugs intended for large segments of the population and a second, fast-track system for orphan drugs. By way of contrast, the FDA argued that orphan drugs should follow the same regulatory provisions as drugs aimed at common conditions. The FDA suggested that to do otherwise would unfairly subject the rare disease population to a different standard. Patients' testimony fit into the dominant "drug lag" discourse present in the United States from the early 1970s to the mid-1980s (Daemmrich and Krücken 2000). This discourse suggested that stringent FDA standards led to lengthy and expensive delays in marketing new drugs in comparison to other countries, causing unwarranted suffering and loss of life to US patients and further discouraging pharmaceutical companies from investing in the development or marketing of drugs intended for rare diseases (US Congress 1981: statements by Abbey Meyers, Jack Klugman, and MS Society). Over the course of the hearings, the highly regulated practices of drug development were problematized in terms of their failure to meet the therapeutic needs of patients with rare diseases. The question then became one of how the existing standards could be adapted promote the development of treatments for rare diseases.

The hearings made apparent that the existing system of drug regulation created disincentives for firms to invest in disease treatments thought to be unprofitable. Given the FDA's concerns that existing drug safety standards be maintained, the problem was posed in terms of the regulatory and legal incentives that would encourage pharmaceutical companies to sponsor orphan drugs through the FDA approval process. The Orphan Drug Act (ODA), in its final formulation, included a range of such legal and economic incentives, and helped to create a novel trajectory of drug development through its various provision (e.g., seven years of marketing exclusivity for orphan drugs, tax credits, drug application assistance, and a grant program).[2] The series of struggles that took place around orphan drugs contributed to the reconfiguration of the biopolitical paradigm of drug development through the inclusion of a previously unrecognized patient population. It further altered this biopolitical paradigm in an economic sense by modifying some of the risks and uncertainties associated with the development of drugs intended for rare disease patient populations. As the following section describes, the ODA has successfully increased the number of therapies available to this population.

Markets, Firms, and the Orphan Drug Act

Foucault's multiple investigations into biopolitics often refere to the salience of economic discourses in shaping the various technologies of power concerned with the administration of life. He draws attention to the work of scholars such as François Quesnay, Jean-Baptiste Moheau, Johann Peter Sussmilch, and William Petty, who had first begun to reflect on the relationship between economic resources and populations (Foucault 1978, 140). Foucault further claims that biopower is indispensable to the development of capitalism: first, in terms of disciplining bodies to fit into the machinery of capitalist production, and second, on a more general scale, facilitating the regulation of various population trends to economic processes (Foucault 1978, 140–41). As I have attempted to show in the previous section, the passage of the ODA concerned the health and well-being not only of the rare disease patient population but also of the capitalist biopolitical paradigm of drug development. In this section, I highlight how the emergence of a novel population (i.e., rare disease patients) has contributed to the development of a niche global economic sector focused on meeting specific medical needs (i.e., those of the rare disease patient population).

The advocacy of patients' groups during the fight for orphan drug legislation illustrates that the struggle concerned not only patients' health but also the drug development process. And by many accounts, the ODA is considered a success (Haffner 2006; Pulsinelli 1999; Tambuyzer 2000). First, this legislation has shaped the conduct of firms as can be judged by the dramatic increase in the number of therapies marketed to treat rare conditions. Second, as a governmental technology for encouraging the development of rare disease treatments, the ODA has been adopted as a policy model by other countries and regions. Last, the impact of the ODA in shaping drug development markets can be evaluated by the growing numbers of firms that specialize in the development of treatments for rare diseases. What is significant about the provisions contained within the ODA is that they have effected a revaluation of rare diseases: previously these illnesses were considered to have limited commercial value; now they are considered a potentially profitable market sector.

The ODA has influenced the research and investment decisions of the drug development industry. Through removing some of the risks and uncertainties previously associated with the development of drugs targeted at small patient populations, the ODA has positively shaped the kinds of illnesses that the pharmaceutical industry decides to treat. One measure used to evaluate the success of this legislation is the number of orphan drugs marketed since 1982. In contrast to the 35 orphan drugs on the market in the 1970s, since the passage of the ODA, 285 of these drugs are now marketed (Haffner, Whitley, and Moses 2002). The significant growth in the number of orphan drugs marketed can be attributed to the growth of the biotechnology industry from the 1980s onward. As many rare diseases are genetic in origin, the legal and economic incentives contained within the act provide a stimulus to an emergent biotechnology industry (Haffner, Whitley, and Moses 2002; Meyers 1992; Tambuyzer 2000). The ODA not only has shaped the conduct of markets and firms, but has contributed significantly to the health of populations affected by rare diseases.

By acting on the economic circumstances surrounding the development of drugs for small patient populations, the ODA has augmented the health of the estimated 20 to 25 million Americans affected by rare diseases. It is suggested that 14 million Americans have benefited from new treatments for their rare illnesses (Haffner, Whitley, and Moses 2002). What I believe to be of further significance is how the act has extended the clinical reach of medicine to conditions where no viable treatment option previously existed. Patients' groups, industry, and the FDA argue that the success of the ODA in promoting the development of new treatments has led to the adoption of this legislation in locales outside the United States.

The ODA has further served as a model for the development of similar legislation in Singapore, Japan, Australia, and most recently, the European Union. As a biopolitical paradigm of drug development, the ODA is valued in jurisdictions outside of the United States as it combines a program for the achievement of public health objectives in addition to promoting innovation in the biotechnology sector. The adoption of this legislation in other countries involves recognition that the market fails to develop treatments for rare diseases in the absence of economic incentives. Furthermore, it indicates how the health needs of rare disease patient populations have been included into the biopolitical paradigms of drug development in jurisdictions outside of the United States.

Although the ODA was designed to provide a range of incentives to the pharmaceutical industry, its greatest impact has been on the biotechnology industry. The ODA is often credited with fomenting the growth of this industrial sector. For many start-up biotechnology companies, orphan product designation is instrumental to securing venture capital investment, which is assured by the prospect of seven years of market exclusivity and the tax credits associated with clinical trial–related expenses. The significance of the ODA to this nascent industry is best captured in an article in *Nature Biotechnology* that reported that "56% of biotechnology products launched between 1982 and 2000 were first approved for orphan drug indications" (Ashton 2001; see also Haffner, Whitley, and Moses 2002). The ODA has effectively shaped the conduct of particular companies that have found a niche in satisfying the medical needs of rare disease patients.

Despite the rather rosy picture I have painted, the creation of market conditions for rare diseases is not without its problems. The ODA was initially developed to encourage firms to develop therapies that had very limited prospects for profits. However, some orphan drugs are among the most expensive in the world. Take the example of a drug developed by Genzyme known as Cerezyme. The average annual cost of treating a patient with Cerezyme varies from $70,000 to $300,000, depending on dosage and body weight. Considering that a treatment such as Cerezyme is required for the remainder of a patient's life, it generates substantial revenues. Genzyme is not alone in producing profitable orphan drugs: Genentech's Propotin, Amgen's Epogen, and Novartis's Gleevec have all managed to make it onto the list of the top 100 selling drugs. It is important to emphasize that the vast majority of orphan drugs have sales in the region of $15 to $20 million and are not profitable.

The topic I discuss next is how the high cost of a small number of orphan drugs raises the more serious problem of how national health-care systems are able to pay for these treatments.

Pharmaco-economics, Orphan Drugs, and Contemporary Patient Activism

In his lecture "The Crisis of Medicine or the Crisis of Anti-Medicine," Foucault (2004) argued that states took on the responsibility of providing healthy lives to their citizens in the aftermath of World War II. From this moment onward, he suggested, health became a significant item of state expenditure and site of political struggle. Although the circumstances have changed, these two axes of Foucault's analysis are relevant to considering how states over the course of the past three decades have become concerned with reducing health-care expenditures. The growth of health-care budgets is due in part to advances in the biomedical sciences that have expanded the field of health-care intervention in terms of scope and cost. Orphan drugs are representative of this trend since they extend the domain of clinical intervention to illnesses that were previously untreatable. Furthermore, orphan drugs, in comparison to treatments intended for large population groups, are generally more expensive. These characteristics of orphan drugs have led states to question whether they should fund these types of therapies. High-cost orphan drugs pose a financial problem in the sense that the provision of these treatments to a small number of patients may lead to budgetary constraints in meeting the health-care needs of the general population. In the remainder of this section, I concentrate on the forms of struggle around access to orphan drugs. As I show, the high cost of some orphan drugs renders them problematic to agencies concerned with evaluating the cost-effectiveness of pharmaceuticals. I suggest that these forms of cost-effectiveness analysis work to exclude many high-cost orphan drugs. The types of patient group struggles that are taking place around cost-effectiveness analysis are concerned with altering this form of analysis to accommodate the particularities of orphan drugs. These struggles are biopolitical in orientation as they are concerned with reshaping the forms of economic calculation that shapes access to medicines.

As states became concerned with health-care spending from the 1970s onward, the field of cost-effectiveness analysis gained prominence in shaping health-care decisions. In the late 1980s and early 1990s, cost-effectiveness evaluations gained an institutional foothold through the creation of bodies such as the National Institute for Health and Clinical Excellence (NICE, UK), Australian Health Technology Assessment Committee,

Canadian Coordinating Office for Health Technology Assessment, and the Swedish Council on Technology Assessment in Health Care. This field of knowledge and practice is concerned with developing a range of principles to evaluate health practices so that the most efficient use is made of limited health resources. In the course of making these types of decisions, health economics has developed standardized methodologies for evaluating the clinical- and cost-effectiveness of drugs. On one register, cost-effectiveness evaluations entail an analysis of the clinical effectiveness of a therapy in actual practice to evaluate whether it produces significant health benefits. On a second register, the field of health economics has invented a range of standardized metrics by which to assess the cost-effectiveness of therapies. The most known and widely used metric is Quality Adjusted Life Years (QALYs). This standardized metric enables comparison of widely differing health interventions on clinical and monetary bases. QALYs make it possible to compare the cost-effectiveness of different therapies that compete for the same budget. Although QALYs do not constitute the sole metric used to evaluate drugs, they figure prominently in establishing health-care funding guidelines. For example, in the United Kingdom, drugs that exceed a threshold of £20,000 to £30,000 per QALY are generally not reimbursed (National Institute for Health and Clinical Excellence 2006). In the case of orphan drugs, the unique clinical and economic features of these therapies are often difficult to accommodate within the methodologies employed in most forms of cost-effectiveness analysis. As I attempt to show, the atypical qualities of orphan drugs problematizes existing methodologies used to evaluate the funding of medicines within European health-care systems. The forms of struggle that take place around these standards are politically significant since they are concerned with the inclusion of the therapeutic needs of rare disease patients into the regulatory regimes that shape access to medicines.

As health-care systems throughout Europe struggle to contain health-care expenditures, cost-effectiveness evaluation has become another regulatory hurdle that a drug must pass to enter human bodies (Simon 2006; Simon and Kotler 2003). Through assessing drugs on the bases of clinical- and cost-effectiveness, the methodologies that health economists employ seek to depoliticize the process through which decisions are made over which treatments are funded by national health-care systems. The impulse behind cost-effectiveness analyses is to ensure that therapies funded by health-care systems are based on scientific evidence and that the most efficient use is made of limited health budgets.

To illustrate how high-cost orphan drugs do not readily fit into the methodologies used to evaluate the clinical- and cost-effectiveness of pharmaceuticals, I draw on the experience of the West Midlands Specialised Commissioning Unit (WMSCU). The forms of cost-effectiveness evaluation that this group employs is analogous to that which takes places across most European countries (my case study builds on the work of Moreira 2007; Mulkay, Ashmore, and Pinch 1987; Sjögren and Helgesson 2007; these researchers have examined cost-effectiveness analysis in the United Kingdom and Sweden). The WMSCU is a group composed of more than thirty primary care trusts in the West Midlands region of England that provide specialized health services to a population of 5 million people. One of the decisions that the WMSCU had to make was whether to fund a series of high-cost orphan drugs used for the treatment of lysosomal storage disorders known as Gaucher's disease, Fabry's disease, and mucopolysaccharidoses type 1. To establish a consistent and principled basis for making these types of decisions, the WMSCU commissioned a series of studies that examined the economic, social, and legal aspects of funding these types of therapies. Toward this end, the group commissioned cost-effectiveness studies of Imiglucerase, Fabryzyme, and Migulsat (Aggressive Research Intelligence Facility 2004a, 2004b) and conducted focus group studies to ascertain public opinion on this topic (McIver 2004). Strategically, the WMSCU was attempting to move away from a system of health-care rationing driven haphazardly by political expediency, patient group influence, and industry lobbying. Alternatively, the WMSCU sought to develop a rational, transparent, and fair means of determining health-care priorities and the allocation of limited health budgets.

Similar to the experience of other countries, one of the problems that orphan drugs present to the methodologies used to assess the cost-effectiveness of a medicine is limited clinical trial data. As many orphan drug trials involve small patient populations and have not been in use for lengthy periods, there is often insufficient data to evaluate their clinical and cost effectiveness. Second, the high cost of many orphan drugs means that their cost per QALY considerably exceeds established guidelines. In the case of the WMSCU, the cost per QALY for Imiglucerase was £400,000, £252,000 for Agalsidase beta, and £450,000 for Laronidase (Burls, Austin, and Moore 2005). Based on their evaluation of the clinical evidence and the cost-effectiveness of these therapies, the WMSCU decided not to fund treatments for lysosomal storage disorders. In making this decision, the WMSCU highlighted that based on existing methodologies, there are no adequate cost-effectiveness measures to account for the unique features of orphan drugs. Second, the WMSCU justified their decision using the economic concept of opportunity costs to address the normative question of the social distribution of health-care resources. In an article describing the experience of the WMSCU, Amanda Burls and colleagues (2005) explain that when resources are devoted to high-cost treatments that affect a small percentage of the population, the opportunity costs involved mean that fewer resources are available to fund health interventions that offer better value for money and have a greater impact on the health of the population. Shortly after this decision, the UK Department of Health decided to fund treatments for lysosomal storage disorders and asked NICE to review existing methodologies for evaluating high cost orphan drugs (Department of Health 2004). After reviewing existing methodologies, NICE concluded that in the case of ultra-orphan drugs (to treat conditions that affect fewer than 1,000 people in the United Kingdom), further decision rules would need to be developed. NICE stated that it would undertake further work on appraising ultra-orphan drugs and recommended the establishment of an Ultra-Orphan Drug Evaluation Committee (National Institute for Health and Clinical Excellence 2006). The UK Department of Health has not yet established this committee. At present, the cost-effectiveness assessment of orphan drugs remains uncertain in the United Kingdom and throughout Europe.

The success of orphan drug legislation in promoting the development of a range of treatments for rare disorders has generated a novel interrelated problem: how can national health-care systems afford to pay for the increasing number of high-cost orphan drugs? On an individual basis, expensive orphan drugs do not affect national health-care budgets given that they treat a small number of patients; collectively, however, they can have a significant impact on these budgets (Hollis 2005). At its core, this problem involves a conflict between attempting to increase the capacity to act on human health while attempting to contain increasing health-care costs.

Conclusion

The orphan drug problem has changed the scale and organizational form of rare disease patients' groups. The process of attempting to enact legislation relating to the orphan drug problem led to the formation of a coalition of patients' groups known as the National Organization for Rare Disorders (NORD). Through organizing themselves as a population of rare disease patients, this coalition was able to overcome the rarity of their conditions and effectively struggle to have their therapeutic needs included in the institutional and economic circuits of drug development. In the 1980s, a different coalition of patient activists, those organized around AIDS, were able to further effect changes to the regulation of pharmaceuticals and the conduct of clinical trials (Epstein 1996). This model of activism has grown in scope, as witnessed by the increasing involvement of patients' organizations in biomedicine. In thinking about this in relation to contemporary biopolitics, it could be said that by changing the scale of their organizational efforts, patients' organizations have managed to integrate themselves into the relays of power through which matters of health are thought about and acted on. Through their formation into coalitions, patients' organizations have been able to assume of number of important functions in relation to the government of health.

Notes

1. After reading the *Los Angeles Times* report of testimony on this bill, Jack Klugman and his brother Maurice decided to do an episode of *Quincy* about persons affected by rare diseases (Henkel 1999). This episode was incredibly popular and was followed by testimony from Jack Klugman and Adam Seligman in Congress in relation to a bill that was reintroduced by Representative Henry Waxman.
2. The Orphan Drug Act (ODA) provides a number of incentives to pharmaceutical and biotechnology companies to develop rare disease treatments. The greatest incentive consists of seven years of market exclusivity. Under the terms of this provision, the FDA cannot license any other drug or product for the same indication, with the exception of products that demonstrate a greater safety profile or clinical effectiveness. The ODA provides tax credits for up to 50 percent of clinical trial–related expenses. Another incentive consists of the waiver of application fees relating to new drug applications (application fees are in the region of $500,000). To encourage the development of treatments for rare diseases, sponsors of a drug can request written protocol assistance from the FDA relating to the design of clinical trials that are most likely to satisfy the drug approval process. This provision was put in place since the conduct of clinical trials on small patient populations is often difficult and it is not always possible to conduct randomized control trials. In the United States, the ODA provides a number of research grants to encourage the development of orphan drugs or medical devices.

References

Abraham, John. 1995. *Science, Politics, and the Pharmaceutical Industry: Controversy and Bias in Drug Regulation.* New York: St. Martin's Press.

Aggressive Research Intelligence Facility. 2004a. "Recombinant Human Alphagalactosidase for the Treatment of Patients with Fabry's Disease." Report commissioned by the West Midlands Specialised Services Agency, Birmingham.

———. 2004b. "Recombinant Human Alpha-Liduronidase for the Treatment of Patients with Mucopolysaccharidosis Type 1." Report commissioned by the West Midlands Specialised Services Agency, Birmingham.

Asbury, Carolyn H. 1985. *Orphan Drugs: Medical versus Market Value.* Lexington, MA: Lexington.

———. 2007. "Orphan Drugs." In *Enclyclopedia of Pharmaceutical Technology*, edited by James Swarbrick, 2468–74. New York: Informa Healthcare.

Ashton, Gabby. 2001. "Growing Pains for Biopharmaceuticals." *Nature Biotechnology* 19 (4): 307–11.

Brewer, George J., ed. 1983. *Orphan Drugs and Orphan Diseases: Clinical Realities and Public Policy.* New York: Alan R. Liss.

Burls, Amanda, Daphne Austin, and David Moore. 2005. "Commissioning for Rare Diseases: View from the Frontline." *British Medical Journal* 331 (7523) (2005): 1019–21.

Daemmrich, Arthure, and Georg Krücken. 2000. "Risk versus Risk: Decision-Making Dilemmas of Drug Regulation in the United States and Germany." *Science as Culture* 9 (4): 505–34.

Dean, Mitchell. 1999. *Governmentality: Power and Rule in Modern Society.* London: Sage.

Department of Health. 2004. "National Designation and Funding of Treatment for Patients with Lysosomal Storage Disorders." On file with author.

Epstein, Steven. 1996. *Impure Science: AIDS, Activism, and the Politics of Knowledge.* Berkeley: University of California Press.

———. 2007. *Inclusion: The Politics of Difference in Medical Research.* Chicago: University of Chicago Press.

Foucault, Michel. 1978. *The History of Sexuality.* Vol. 1, *The Will to Knowledge.* London: Penguin.

———. 1991. "Governmentality." In *The Foucault Effect: Studies in Governmentality*, edited by Graham Burchell, Colin Gordon, and Peter Miller, 87–104. London: Harvester Wheatsheaf.

———. 1994. "On the Government of the Living." In *Ethics: Subjectivity and Truth*, edited by Paul Rabinow, 81–85. New York: New Press.

———. 2001a. "*'Omnes et singulatim'*: Toward a Critique of Political Reason." In *Power: Essential Works of Foucault, 1954–1984*, edited by James D. Faubion, 298–325. London: Allen Lane.

———. 2001b. "The Politics of Health in the Eighteenth Century." In *Power: Essential Works of Foucault, 1954–1984*, edited by James D. Faubion, 90–110. London: Allen Lane.

———. 2004. "The Crisis of Medicine or the Crisis of Antimedicine?" *Foucault Studies* 1 (1): 5–19.

Haffner, Marlene E. 2006. "Adopting Orphan Drugs: Two Dozen Years of Treating Rare Diseases." *New England Journal of Medicine* 354 (5): 445–47.

Haffner, Marlene E., Janet Whitley, and Marie Moses. 2002. "Two Decades of Orphan Drug Development." *Nature Reviews Drug Discovery* 1 (10): 821–25.

Henkel, John. 1999. "How TV Launched the Orphan Drug Law." *FDA Consumer Magazine*, May–June.

Hollis, Aiden. 2005. "Drugs for Rare Diseases: Paying for Innovation." Paper presented at "Health Services Restructuring: New Evidence and New Directions," Queen's University, Canada, November 17–18.

Karch, Fred. 1982. *Orphan Drugs.* New York: M. Dekker.

Lakoff, Andrew. 2008. "The Right Patients for the Drug: Pharmaceutical Circuits and Codification of Illness." In *The Handbook of Science and Technology Studies*, edited by Edward J. Hackett, Olga Amsterdamska, Michael Lynch. and Judy Wajcman, 741–59. Cambridge, MA: MIT Press.

Marks, Harry M. 1997. *The Progress of Experiment: Science and Therapeutic Reform in the United States, 1900–1990*. Cambridge: Cambridge University Press.

McIver, Shirley. 2004. "The Views of the Public about Funding Enzyme Replacement Therapy." Report commissioned by the West Midlands Specialised Services Agency, Birmingham.

Meyers, Abbey. 1988. "Working toward Passage of Orphan Drug Act: An Example of Determination." *American Medical Writers Association Journal* 3 (1): 3–7.

———. 1992. "The Impact of Orphan Drug Regulation on Patients and Availability." *Food and Drug Law Journal* 47:9–14.

Mitchell, Samuel A., and Emery Link, eds. 1976. *Impact of Public Policy on Drug Innovation and Pricing: Proceedings of the Third Seminar on Pharmaceutical Public Policy Issues.* Washington, DC: American University.

Moreira, Tiago. 2007. "Entangled Evidence: Knowledge Making in Systematic Reviews in Healthcare." *Sociology of Health & Illness* 29 (2): 180–97.

Mulkay, Michael, Malcolm Ashmore, and Trevor Pinch. 1987. "Measuring the Quality of Life: A Sociological Intervention Concerning the Application of Economics to Health Care." *Sociology* 21 (4): 541–64.

National Institute for Health and Clinical Excellence. 2006. "Appraising Orphan Drugs." On file with author.

O'Malley, Pat. 2004. *Risk, Uncertainty and Government.* London: Glass House.

Pulsinelli, Gary A. 1999. "The Orphan Drug Act: What's Right with It." *Santa Clara Computer and High Tech Law Journal* 15:299–345.

Rose, Nikolas. 1999. *Powers of Freedom: Reframing Political Thought.* Cambridge: Cambridge University Press.

Scheinberg, I. Herbert, and J. M. Walshe, eds. 1986. *Orphan Diseases and Orphan Drugs.* Manchester: Manchester University Press.

Simon, Françoise. 2006. "Market Access for Biopharmaceuticals: New Challenges." *Health Affairs* 25 (5): 1363–70.

Simon, Françoise, and Philip Kotler. 2003. *Building Global Biobrands: Taking Biotechnology to Market.* New York: Free Press.

Sjögren, Ebba, and Claes-Fredrik Helgesson. 2007. "The Q(u)ALYfying Hand: Health Economics and Medicine in the Shaping of Swedish Markets for Subsidized Pharmaceuticals." *Sociological Review* 55 (s2): 215–40.

Tambuyzer, Erik T. 2000. "The European Orphan Medicinal Products Regulations and the Biotechnology-Based Industry in Europe." *Journal of Commercial Biotechnology* 6 (4): 340–44.

Timmermans, Stefan, and Marc Berg. 2003. *The Gold Standard: The Challenge of Evidence-Based Medicine and Standardization in Health Care.* Philadelphia: Temple University Press.

United States. Department of Health, Education, and Welfare. 1979. *Significant Drugs of Limited Commercial Value.* Washington, DC: US Department of Health, Education, and Welfare. http://rarediseases.info.nih.gov/files/1979_interagency_task_force_report_on_significant_drugs_of_limited_clinical_value.pdf.

United States. House of Representatives. 1981. *Hearings before the Subcommittee on Health and the Environment of the Committee on Energy and Commerce, House of Representatives, Ninety-Seventh Congress, Orphan Drugs, HR 1663, March 9, 1981*. Washington, DC: US Government Printing Office.

Van Woert, Melvin H., and Eunyong Chung, eds. 1985. *Cooperative Approaches to Research and Development of Orphan Drugs.* New York: Alan R. Liss.

TEN

The Biopolitics of HIV Prevention Discourse

DAVID M. HALPERIN

Contemporary practitioners of queer studies sometimes speak of gay male "subjectivity," but we avoid delving into the subject in any kind of depth or detail. There are many reasons for this avoidance: a wish to escape the pitfalls of essentialism, a psychoanalytically informed reluctance to believe that subjectivity itself is differentiated along the lines of social identity, a new triumphalism about the progress of gay rights and the imminent integration and assimilation of gay people into mainstream society—a prospect threatened by the notion that gay men might have a distinctive subjectivity of their own. But there is another reason that the question of gay male subjectivity remains a minefield for queer theory. Despite much conceptual work, *subjectivity* is still hard to disentangle from *psychology*, and psychology has long represented a tainted category for lesbians, gay men, and other sexual dissidents.

For a couple of centuries, any deviation from very strict standards of normative gender presentation and heterosexual behavior was considered, and treated, as the sign of a psychological illness—as a symptom of a diseased state, variously described as "moral insanity," "sexual perversion," "personality disorder," "mental illness," or "maladjustment," but characterized in any case as a kind of abnormal psychology. The rise of the various psy-disciplines and the sciences of man that accompanied the growth of what Michel Foucault called "biopower," the state administration of human life at both the level of the population and the individual body, produced new political hierarchies and forms of social stratification grounded in a system of statistical and descriptive norms, such as "health" and, ultimately, "sexuality," that worked against the life chances of queer people. Even today, some reputable psychologists, psychiatrists, and psychoanalysts maintain in all seriousness that homosexuals are sick. In the wake of this continual medical and forensic treatment of homosexuality as a psychological pathology or aberration, lesbians and gay men have directed much political effort to undoing the presumption that there is something fundamentally wrong with us. In this context, it has seemed necessary to close off the entire topic of gay subjectivity to respectable inquiry, so as to prevent gayness from ever again being understood as a sickness.

Foreclosing the question of gay subjectivity, in short, has been a small price to pay for scuttling a psychological model of homosexual difference premised on sexual abnormality and replacing it with a political program grounded in a nonpsychological notion of gayness as a quasi-ethnic social identity.

A small price—but a necessary one? Foucault himself famously quipped (in a 1981 interview with the German filmmaker Werner Schroeter) that the whole art of life consists in killing off psychology ("L'art de vivre, c'est de tuer la psychologie"). But does that spell the death of subjectivity as well? What if psychology were not the only means of access to the supposed truth of subjectivity? What if the psychic, so central to the current deployments of biopower in our society—to the army of therapists, doctors, experts, public health authorities, forensic pathologists, developmental psychologists, and educational theorists who regulate our lives—were not in fact the royal road to knowledge about the internal workings of the gay subject?

Even during the period of their ascendancy, psychology and psychoanalysis were shadowed by an alternate tradition, perhaps not a consistent tradition so much as a persistent impulse, visible conspicuously (though not exclusively) in the work of a number of gay male writers. In their different ways, Walter Pater, Oscar Wilde, André Gide, Marcel Proust, Jean Genet, and Roland Barthes all attempted to imagine and to represent human subjectivity without recourse to psychology.

If we are to keep that queer project alive, we will have to forge alternatives to the modern, scientific culture of the self and its psychological hermeneutics of the subject. Queer culture has shown itself in fact to be prolific in inventing concrete escape routes from self-analysis, ranging from gay churches to gay bathhouses, just as a number of gay male writers have elaborated various antipsychological approaches to representing the queer subject. So queer people have a multitude of both social and intellectual resources to draw on when it comes to figuring out how to have a subjectivity, and a sexual subjectivity in particular, without conceptualizing or experiencing it in terms of either psychology or psychoanalysis. We should use them. After all, our lives depend on it.

If there was ever any doubt about the urgent need to find ways of representing gay male subjectivity without necessary or automatic recourse to psychology and psychoanalysis, the merest glance at contemporary discourses about why some gay men have risky (or "unsafe") sex would dispel it. The topic of gay men's sexual risk taking has opened new perspectives onto gay male subjectivity and occasioned a multitude of inquiries—by scientists, journalists, community leaders, and activists—into what gay men want. Nearly all of those inquiries have taken the form of psychological speculation about gay men's motives for engaging in risky sex. That speculation has led in turn to a revival of medical thinking about homosexuality: a style of reasoning that distinguishes "healthy" from "unhealthy" behavior and thereby tends to smuggle into an ostensibly scientific analysis many stealth assumptions about good and bad sex, functional and dysfunctional subjectivity, proper and improper human subjects. Starting from the premise that no sane person would ever put his life at risk to obtain sexual pleasure—a dubious premise at best, which acquires a specious plausibility by being grounded in unexamined normative notions about psychological health—most efforts to understand gay men's sexual risk taking complicate their task by setting themselves the impossible goal of explaining behavior that has already been defined as deeply irrational or incomprehensible.

Next, the causes of such behavior tend to be sought in various psychological "deficits" that impair gay men's mental health and interfere with normal functioning. Some of this speculation aims to be gay-friendly: it portrays gay men as victims of hostile social forces, diagnoses the various ills from which we suffer as a consequence, and offers to help us into recovery by providing therapeutic remedies for the damage we have sustained. Other accounts are more straightforwardly homophobic. But whatever the intention, the result is to portray gay men as beset by a number of serious psychological conditions, ranging (on the "victim" end of the scale) from internalized homophobia, survivor guilt, and posttraumatic stress disorder to (on the pathological end) low self-esteem, addictive personality syndrome, sexual compulsiveness, and lack of self-control.

It is not clear exactly how many gay men engage in sex that carries a real risk of transmitting HIV, how many of us take what kinds of risks, to what extent we succeed in minimizing those risks, or what the demographic distribution of gay risk takers is, either in the United States or around the world. Formal monitoring of the HIV/AIDS epidemic has often been spotty or inconsistent, especially in the United States, where the Centers for Disease Control and Prevention (CDC) did not receive confidential name-based reporting of cases of HIV infection from all fifty states and six dependent areas until 2008. Even the most sophisticated statistical analyses of gay men's risk taking tend to be based on partial, incomplete, unrepresentative, or flawed information.

Nonetheless, certain consistent patterns have emerged. (Note that all the data that follows was accurate, to the best of my knowledge, in 2008, when this study was last updated; I have not been able to take account of more recent epidemiological findings or to alter my conclusions in the light of them, except in my concluding remarks, below, which date to 2013.) David Nimmons, who surveyed more than sixty behavioral studies published during the 1990s, reports that "the proportion of [gay] men primarily behaving safely [which means different things in different studies, but tends to equate somewhat misleadingly with using condoms] commonly hovers between 60 and 70 percent"; by contrast the percentage of heterosexual women and men practicing safe sex rarely attains a third of the sample.[1] Those percentages are not in themselves reliable indicators of actual risk, however, because not all *unprotected* (i.e., condomless) sex is necessarily *unsafe* sex. Unprotected sex cannot enable the transmission of HIV unless it takes place between an infected and an uninfected partner, and even then the exact degree of risk involved depends on the specific sexual acts performed and a complex array of secondary factors. In any case, what does seem to be clear is that only a minority of self-identified gay men in the industrialized world currently put themselves at significant risk of contracting HIV/AIDS in their sexual practices.

To be sure, the last two decades *have* witnessed a growing alarm in both popular and scientific circles about increased sexual risk taking by gay men. The production of panic scenarios about gay men's sexual behavior in the context of the HIV/AIDS epidemic is, of course, nothing new, but a number of recent developments have provided a fresh motive for particular concern. Chief among them is the emergence of "barebacking," the deliberate, premeditated practice of unprotected anal sex with casual or anonymous partners, which has garnered considerable attention from journalists and epidemiologists alike. Still, there is reason to believe that barebacking, however worrisome, does not represent the terrifying menace to public health that some have taken it to be. By saying this I do not mean to minimize the gravity of the situation. Concerns about a rise in risky sex among gay men are hardly groundless: reported HIV/AIDS diagnoses in men who have sex with men, which had declined annually on average from 2001 to 2003 in a majority of the states in the US, rose more than 3 percent in 2004, followed by a 6 percent jump in 2005, and another increase of nearly 4 percent in 2006—all in all, a sharp rise of 13.3 percent over the course of three years from 2003 to 2006 (though it is not clear how much of that rise should be put down to increased HIV testing or the duplication of diagnoses; moreover, the estimated number of newly diagnosed cases of AIDS among gay and bisexual men in the entire United States did not significantly increase from 2002 to 2006). In any case, the alarm over barebacking is certainly motivated by some quite real and ominous developments, but it may turn out to have been overblown—or, even, in some instances, to be mistaken.

For one thing, initial reports of barebacking often referred to condomless sex among men who had *already* been infected with HIV. All subsequent research has consistently shown that barebacking is practiced more frequently by men who are HIV-positive than by men who are HIV-negative. Furthermore, those HIV-positive men who bareback do it most often with men who either are or are assumed to be HIV-positive, or who are assumed to know that their partner is or is likely to be HIV-positive (very few HIV-positive men are willing to bareback with men they know to be HIV-negative).[2] Medical authorities have cautioned against unprotected anal sex between HIV-positive partners, fearing the possibility of "superinfection," the reinfection of an already infected person with a different and possibly more virulent or drug-resistant strain of HIV; accordingly, safe-sex campaigns in the United States and Europe have often urged HIV-positive men to use condoms in anal sex with one another. But no hard evidence has so far materialized to justify that conservative prevention message: the authors of a recent literature review conclude that "the true rates and consequences of HIV superinfection have yet to be well delineated," and they point out that superinfection, though it can occur and has been documented, especially among newly infected individuals (at annual incidence rates as high as 5 percent) and among individuals who have interrupted treatment, is not well attested in the scientific literature (only sixteen cases reported worldwide between 2002 and 2005). In fact, exposure to different strains of HIV-1 regularly fails to produce expected occurrences of superinfection.[3] To the extent that it takes place among HIV-positive partners—or among HIV-negative partners, for that matter—barebacking carries no established risk of transmitting HIV and so provides no immediate cause for alarm on that score.

Indeed, the conscious and deliberate practice of confining unprotected sex to relations between seroconcordant gay men (gay men who have the same HIV-serostatus), a practice that is coming to be favored by gay men in cities where the numbers of HIV-positive individuals allow for sufficient latitude and ease in the selection of sexual partners, is now being hailed as a new "safe-sex" technique called "serosorting." For example, serosorting has been credited with sharply reducing the rate of HIV transmission among gay men in San Francisco since 2001 (though the claimed decline in the transmission rate may misrepresent the available data, and the ascription of that supposed decline to serosorting is a matter of interpretation). And on November 6, 2006, the Department of Public Health in San Francisco unveiled a new social marketing campaign to promote serosorting as a means of HIV/AIDS prevention. Called "Disclosure" and produced by Better World Advertising, the campaign consists of bus shelter and billboard posters that feature the message, "disclosure is HIV prevention," explaining that "Status sorting is a prevention strategy" and urging gay men to "test & tell."

As Kane Race has pointed out, much of what was vilified as "barebacking" in the mid-1990s, and is often denounced today as evidence of gay men's pathological self-destructiveness, is in fact precisely what is now being celebrated under the revamped designation of "serosorting." The different valences attached to the two terms do not necessarily result from any difference in the nature of the sexual behaviors to which they refer. After all, as Race observes, in many cases "barebacking" and "serosorting" are simply alternate names for the same practice. The two words differ less in what they describe than in their respective interpretations of what condomless sex is all about. If "barebacking" alarmingly presents unprotected sex as a sign of irresponsible, hedonistic, reckless abandon on the part of gay men, Race argues, that is because it ignores the sexual protocols being worked out within HIV-positive communities and treats the rejection of condoms as an expression of unconstrained, libertine individualism. "Serosorting," by contrast, reassuringly implies a set of communal, hygienic arrangements for containing the virus within stable relationships, and it generates an image that foregrounds not the pleasure-seeking individual but the prudent, conjugal couple.[4]

Serosorting can reduce the risk of HIV transmission for individuals, but it certainly does not eliminate it, and it may even introduce new risks. For example, many young men in the United States who have been infected with HIV wrongly believe themselves to be HIV-negative. When the practice of serosorting leads men who actually *are* HIV-negative to have condomless sex with men who sincerely but mistakenly *believe* they are HIV-negative, such HIV-negative men put themselves at a very significant risk of infection with HIV—in which case serosorting turns out to be a snare and a delusion. Also, HIV-negative men may not always have the insider knowledge of HIV-positive gay men's sexual cultures that they require to pick up the tacit, subtle hints that HIV-positive men may think they are clearly dropping about their HIV-positive status. And both HIV-negative and HIV-positive men may overestimate the degree to which their partners can correctly infer their true serostatus from the kinds of sexual behaviors they engage in or the sorts of risks they are willing to take. The new emphasis on disclosure as a prevention strategy, as Race points out, may even backfire, insofar as it could give HIV-negative men the false impression that HIV-positive men routinely announce their serostatus and that those who do not identify themselves unambiguously as HIV-positive are not infected. Much serosorting, especially among young HIV-negative men, has therefore been redescribed as "seroguessing." So the efficacy of serosorting as a safe-sex technique is not assured: it varies in accordance with frequency of HIV testing, accuracy of knowledge of one's serostatus, explicitness of disclosure among sexual partners, and the sheer amount of HIV present within a particular network or population of individuals who have sex with one another. And the only kind of serosorting that is sure to be safe as a prevention technique is the serosorting practiced among self-identified HIV-positive men, since they are the ones who are most likely to know what their true serostatus is (and someone who tells you he is positive is probably not lying to you).

Nonetheless, to the extent that "barebacking" may turn out to coincide, or at least to overlap, with "serosorting," it may not represent quite such an alarming development, especially in sexual cultures in which gay men are well informed about their true serostatus and are willing (or empowered) to disclose it. By contrast, barebacking that involves *deliberate, intentional unprotected sex between HIV-positive and HIV-negative men,* though it may be fantasized about or even celebrated among a few fringe groups,[5] seems to be extremely rare. Race notes that "the only quantitative study to strictly define and describe barebacking so far found that 229 (91%) of the 252 HIV-negative and untested men surveyed who had heard of the term *barebacking* had *not* 'intentionally set out to have unprotected anal sex with someone other than a primary partner' in the previous 2 years."[6] And the broadest statistical and demographic measures confirm that nearly 90 percent of all HIV-negative men who have sex with men in the United States continue to avoid unprotected anal sex with potentially HIV-positive men.[7] This generalization holds even for men advertising for condomless sex on special Internet sites catering to barebackers: contrary to what the mass media would often lead us to believe, the vast majority of men who use such websites to meet sexual partners for unprotected anal intercourse do *not* set out to be infected with HIV or to infect others, but rather use serosorting or other techniques to minimize the risk of HIV transmission.[8] There is ample reason, in short, for reining in some of the recent panic about the supposed outbreak of unrestrained, irrational, and self-destructive risk taking on the part of gay men.

In this context, it is important to recall that "safe sex" (a set of practical guidelines for sexual behavior designed to reduce or eliminate the transmission of HIV), which remains the best hope for stopping the epidemic, was originally a gay, grassroots invention. It was created by gay communities in North America and propagated throughout them even before an infectious agent associated with AIDS was discovered; it has continued to undergo periodic renovation ever since, as gay men informally but collectively revise their procedures for protecting themselves and their partners from infection in the light of changing medical technologies, increased experience of them, and more sophisticated mutual understanding. Safe sex has proven to be overwhelmingly effective in preventing HIV transmission and reducing the prevalence of HIV infection in many parts of the world. Already by the end of the 1980s epidemiologists noted that documented changes in gay men's sexual behavior constituted "the most profound modifications of personal health-related behaviors ever recorded."[9] Recently, the authors of an authoritative epidemiological report on sexual practices in Australia observed, "In general, the majority of homosexually active men have sustained a 'safe sex' culture [since the advent of the HIV/AIDS epidemic] even though sustaining safe sex over such a long period is difficult."[10] That poker-faced final clause is a masterpiece of understatement.

In the United States alone, white gay men—once the main vector of HIV transmission—now account for less than a quarter of new HIV infections.[11] In fact, barely more than a third of the *entire* new AIDS caseload reported in 2006 (the latest year for which statistics were available in the United States in 2008) can be traced specifically to gay sex.[12] Summing up the data for Europe, Canada, Australia, and the United States, one researcher has recently noted that, despite the evidence of an increase in sexual risk taking, "the majority of gay men continued to report safer sex practice" in the period since the introduction of the new generation of antiretroviral therapies in 1996. And a number of recent studies indicate that sexual risk taking by gay men may now be leveling off or even declining.[13]

Serosorting represents merely the latest chapter in this ongoing saga. Indeed, as Kane Race has lately reminded us, the evolution of safe-sex practices indicates that "risk" and "safety" are not opposites or alternatives. Much of what now qualifies as "safe sex" is the result of gay men's spontaneous improvisations, their calculated experimentation with varying degrees of risk. After all, the medical establishment in the United States never told us it was safe to have sex, let alone that sex was good for us. They still have not given us a green light to practice unprotected oral sex (even without ejaculation), though in 2014 the CDC finally admitted that the per-act probability of acquiring HIV from an infected partner through oral sex is so small as to be unquantifiable.[14] Nor did they say we could have as many sexual partners as we wanted so long as what we did with them was safe. They have never endorsed condomless sex among HIV-positive men or agreements among HIV-negative partners in a stable relationship to have unprotected sex with each other so long as they protect themselves from infection in their contacts with casual sexual partners: this technique, a precursor of serosorting, was identified in 1991–92 by Susan Kippax and her team of Australian social researchers into gay men's sexual practices and included by them in their new category of "negotiated safety," but that practice has not been consistently promoted as a prevention technique by public health authorities in the United States.[15] The US medical establishment never even told us it was safe to kiss.

It is gay men themselves who have continued to define, and to redefine, the limits of safety through an ongoing history of sexual experimentation and mutual consultation, and who have thereby produced, over time, workable compromises and pragmatic solutions that balance safety and risk in proportions that have turned out to be both acceptable to a majority of gay men and successful in limiting the transmission of HIV.[16] The upturn in condomless sex that we have been witnessing recently does not signal the end of safe sex, the failure of HIV prevention, or a new indifference on the part of gay men to the risks of HIV infection (after all, as Michael Shernoff points out, there have always been some gay men who, for whatever reason, have preferred to take significant risks in their sexual practices).[17] The current loosening of the condom code may simply be the latest stage in the ongoing evolution of harm reduction techniques among men who have sex with men, the most recent adjustment or refinement in the complex protocols of safe sex.

Some of the probabilistic calculations behind this loosening of safe-sex practices may be merely wishful or careless, even misguided and dangerous, so there is good reason to be concerned about the possibility that the growing experimentation with condomless sex may lead to increased transmission of HIV, especially among the young, the poor, the badly served and badly informed, or those who do not consider themselves gay. We should certainly remain alert to the way that harm reduction techniques, even among relatively canny and self-aware gay men, can sometimes lead to new possibilities for harm *increase*.[18] But there do not seem to be grounds for believing that gay men as a whole have suddenly abandoned the cause of HIV prevention or given up on either the idea or the practice of safe sex. The evidence hardly supports the generalization, popularized by psychoanalytic theorist Tim Dean, that "erotic risk among gay men has become organized and deliberate, not just accidental."[19]

On the contrary. Although there does appear to have been an increase in condomless sex among men who have sex with men, it does not follow that there has been an increase in *deliberate* risk taking. Some of the new harm reduction techniques that have been substituted for the use of condoms may not be as effective as they are intended to be, but the data do not require us to conclude that gay men have massively renounced safe sex as a goal in favor of what Dean calls "purposeful HIV-transmission."[20] Rather, what we may be witnessing is a change in the definition and concrete manifestations of safe sex, which have moved HIV prevention practices beyond the antiquated, unworkable rule of "use a condom every time." While there are new risks associated with this trend—risks that are hardly trivial, as the recent rise in reported HIV/AIDS diagnoses in American men who have sex with men suggests, and that therefore demand to be addressed—there is little reason to presume that gay men no longer consider HIV/AIDS prevention an urgent matter, that they have accepted new levels of risk as a matter of course, or that their behavior reveals the symptoms of a collective psychological affliction.

"How do we measure the success or failure of HIV prevention?" Eric Rofes asked in 1998.

> If we use what appears to be the common standard driving most HIV-education programs, we view each new infection of a gay man as evidence of failure and proclaim there is an expansion of the epidemic. While literally true in aggregate numbers of infected gay men, this kind of evaluation is inappropriate for use with communicable diseases that require long-term, sustained approaches and achievable goals. To mistake the utopian rhetoric that we could end AIDS tomorrow if everyone practiced safe sex 100 percent of the time for appropriate public health strategy is terribly misguided. Instead, I believe the way to assess our prevention efforts is by examining the level of seroprevalence within successive generations of gay men. A realistic aim might be that each successive cohort of gay men in a particular location show a specific decline in level of infection. . . . Rather than a crisis-driven, drama-queen exaggeration of continuing HIV infection among gay men,

> coming to terms with the AIDS epidemic means confronting the authentic reality we face.[21]

The spread of epidemic disease is notoriously difficult to stop by means of behavioral interventions alone. And yet nearly two-thirds of gay men in the United States contrive to remain uninfected over the long term.[22]

Gay men have rarely gotten credit for those lifesaving breakthroughs, however, either in scientific or in popular writing about the epidemic. Nimmons discovered that "a literature search on 'gay men, HIV, and risk factors' finds some 588 studies, yet the same search on 'gay men, HIV and safety' yields only 9. Of those, only 8 papers explicitly set out to chronicle the native strategies gay men have developed, the affirmative and creative ways ordinary people have invented to protect each other."[23] Gay men are more likely to see the continued existence of male-to-male HIV transmission (at annual rates ranging from a stubborn and unacceptably high 1 percent to a truly terrifying 4 percent or more among some younger men and some African American men)[24] used against them, while their scattered departures from perfect adherence to safe-sex protocols typically occasion global judgments about their impaired or abnormal psychology. Only 61 percent of Americans in 1996 consistently reported using seatbelts, according to the National Highway Traffic Safety Administration, as Nimmons pointed out, but that statistic—which, thankfully, has since improved—did not raise comparably grave concerns about the mental health of the heterosexual majority. Similarly, the most recent literature review analyzing the causes of risky sexual behavior among young heterosexual women and men looked exclusively to social factors for explanation, never to psychological ones.[25]

The alleged failure of safe-sex campaigns and the defective psychology of gay men that has been posited to explain it are quickly laid at the door of all gay men without qualification. Particular blame has lately been attached to those non-white gay and bisexual men in the United States among whom HIV/AIDS diagnoses have recently been increasing at the sharpest rates in the nation—even though a careful study of a multiethnic population of gay men in San Francisco determined that "the prevalence of barebacking did not differ by race/ethnicity."[26] In particular, the higher rate of infection among black men who have sex with men does not correlate with higher levels of documented risk taking. As one researcher sums up the current findings, "Black men who have sex with men (BMSM) are disproportionately affected by HIV/AIDS in the United States. . . . Yet the disparity is not explained by higher rates of unprotected anal and oral sex."[27] This crux in the available data makes it particularly hazardous to generalize about the relation between race and risk in black communities, to say nothing of assigning blame. Nonetheless, epidemiological evidence provides a convenient vehicle for ratifying moral and psychological judgments against those who are already devalued on the grounds of sexuality or race (or both); members of socially privileged groups, by contrast, suffer only individual, not collective, discreditation. HIV/AIDS prevention allows such a normalizing strategy to proceed under the protective, politically virtuous cover of enlightened concern for the downtrodden, and stigmatized groups are pathologized on the pretext of victim advocacy.

To explain why some proportion of gay men continue to take certain risks in their sexual practices, even in the third decade of the HIV/AIDS epidemic, both scientists and journalists have to reckon with gay men's motivations for risk taking, which means in turn that they have to address the topic of gay sexual subjectivity—what gay men want. HIV/AIDS prevention has now come to be the one genre of public discourse in which gay male subjectivity, far from being bracketed or sidelined, is a continual subject of discussion. The focus on gay subjectivity is sharpest in the case of white, socially privileged gay men, whose agency and autonomy are not likely to have been compromised by political oppression or external constraint and whose behavior therefore cannot be explained by social factors: that is why much of what I have to say here will refer to them, though my conclusions will have a wider application.[28]

Most often the efforts to account for gay men's risk taking, as Gregory Tomso notes, revolve obsessively around a single, rather unhelpful question—"What makes them do it?"—which then leads directly to an entire series of unwarranted and even crude (if sometimes well-meaning) psychological speculations about the nature of gay male subjectivity itself.[29] What is it that goes on in the minds and hearts and psyches of gay men? What makes them tick? What makes them so different from normal people? Why do they behave so badly, so irrationally, so self-destructively? What is wrong with them? What determines the affective structure of their feelings and impulses? Why do they seem to be so impervious to HIV/AIDS prevention efforts? And what picture can we draw, on that basis, of the peculiar and characteristic features of gay male subjectivity itself?

It is probably pointless to try and explain why anyone has unprotected sex. There are obvious contributing factors to the practice of risk, which it is always useful to analyze,[30] but plumbing the depths of human motivation in search of ultimate causes is a bad idea because it is likely to produce specious answers. The notion that people, once they truly understand what is in their own best interests, always act rationally to maximize them—on the basis of an accurate, long-term calculation of the likely costs and benefits of their behavior—has taken a persistent battering ever since Socrates first proposed it. Rational calculation does not always explain even the sort of human behavior most amenable to cost-benefit analysis, namely economic behavior. (Just think of global warming.) Human beings tend to have a hard time giving up something they badly want, when it is readily available and within easy grasp, for the sake of a less immediate and more long-term benefit. Why should we suppose that sex, of all things, would be the sphere in which people could be depended on to act in a *more* rational or calculating fashion than they typically do?

Especially since sex without condoms is fun.

As Stephen Lyng and his colleagues have shown, modern Westerners have a very complicated relation to risk that can, however, be summed up pretty quickly. Risk is thrilling.[31] It hardly seems surprising that sexual behavior, whose thrills are routinely intensified by various sorts of risk (pregnancy, infidelity, naughtiness, pain, shame, disgust, disease, love), does not readily lend itself to consistent and careful calculation or control.

In fact, what seems truly remarkable is not that some people continue to take some risks in their sexual practices, but that such a large proportion of gay men should have succeeded for so long in adhering so prudently to so many constraining and annoying limitations on their sexual behavior. To speculate about the peculiar subjectivity of gay men on the basis of their occasional sexual risk taking is a misguided project. There is, in all likelihood, nothing in particular to explain beyond the obvious: sex is easier and more pleasurable without condoms, and people have a hard time consistently and categorically forgoing a pleasure they deeply cherish when they have to do it over a very long period of time. It was never realistic to expect that HIV/AIDS alone would "provide the requisite motivation for sweeping behavior changes" or that it would lead people to jettison en masse and without a backward glance the sexual activities central to their senses of themselves.[32]

The narrator of an unpublished novel/memoir by Kirk Read puts it very well:

> I grew up post-AIDS, where I wasn't privy to some collective generational memory of what it was like before the epidemic. I knew it felt better. I mean, that's the dirty little secret of bareback sex, the thing nobody ever says out loud. It feels better. You feel more connected to the person you're with, the

> friction is smoother, there's a sort of abandon that's intoxicating. The Centers for Disease Control never includes that basic, obvious truth in press releases: It feels better.
>
> You know what I'm talking about, where they take a street survey that isn't even remotely scientific—like 250 gay men in one neighborhood of one city randomly selected by young non-profit wage slaves with a clipboard questionnaire. Then they turn it into a huge pronouncement that unprotected anal sex is on the rise by 48% in urban gay men and it's all because gay men have low self-esteem. And all these medical "experts" line up to shake their fingers, especially at the young ones who should know better, those of us who didn't lose all of our friends in the 80s. They say we're suicidal. They never just say *We're really sorry. Condoms suck. We know it feels better. We just have to do this for a while.* Why can't the government say that? Even a lot of HIV positive guys are self-righteous, saying they have no sympathy for anyone who gets infected in this day and age, knowing what we know. Like they had no idea it could happen to them. Like they were all infected in 1978, back before we all got kicked out of Eden. Honey, no one's innocent and everyone's innocent. It's complicated. It's context.[33]

Although not all research into gay men's sexual practices is as inept as the kind of behavioral survey that Read's narrator lampoons, he is right to protest about the way alarmist statistics are used to generate punitive judgments about the sanity of gay men. He is also right to believe that the question of why some gay men have risky sex is itself badly formulated and prejudicial.

I find it unfortunate to have to enter a discussion of gay male subjectivity that has already been framed by the question of why some gay men have risky sex, a discussion that cannot fail to lend the question itself a substance and a dignity it has no right to claim. One of the purposes of my little book, *What Do Gay Men Want?*, was to argue that the very question "What makes them do it?" produces, inevitably and necessarily, bad and irrelevant answers. Far from opening up new perspectives on sex and risk, it simply creates the unfortunate presumption that there is something wrong with particular groups and individuals. In order to confront sexual risk taking both practically and imaginatively, we need to take it out of the realm of morality and normalizing judgments (whatever our personal feelings about it may be). And instead of representing sexual risk in the era of AIDS as a vertiginous high-wire act, a drama of virtue forever teetering on the brink of the abyss, we might begin by trying to de-dramatize the practice of risk taking altogether.[34]

That is precisely what Lauren Berlant proposes to do in a highly original essay on obesity, now included in her book, *Cruel Optimism*. Her analysis of everyday practices and habits "focuses on what's vague and gestural about the subject and episodic about the event. It presumes nothing about the meaning of decision or the impact of an act." Berlant argues that unless we attend "to the varieties of constraint and unconsciousness that condition ordinary activity," we are likely to ascribe an excess of meaning to particular acts, to endow modern subjects with an untrammeled sovereignty, an inflated capacity of cognition, and a powerful, effective, consequential agency that they do not in fact possess, and to ascribe to them on that erroneous basis a heightened responsibility such that all their actions take on the melodramatic intensity of a life-or-death decision. She prefers, instead,

> to rethink some taxonomies of causality, subjectivity, and life making embedded in normative notions of agency. More particularly, I want to suggest that to continue to counter the moral science of biopolitics, which links the political administration of life to a melodrama of the care of the monadic self, we need to think about agency and personhood not only in normative terms but also as activity exercised within spaces of ordinariness that does not always or even usually follow the literalizing logic of visible effectuality, bourgeois dramatics, and lifelong accumulation or fashioning. . . . I recast these within a zone of temporality we can gesture toward as that of ongoingness, getting by, and living on, where the structural inequalities are dispersed, the pacing of their experience intermittent, often in phenomena not prone to capture by a consciousness organized by archives of memorable impact. I want to prompt a thought about a kind of interruptive agency that aspires to detach from a condition or to diminish being meaningful. Crisis management produces dramas that obscure the motives and temporalities of these aspects of living.[35]

Despite the somewhat enigmatic language in which it is couched, Berlant's project is an admirable one. Its importance lies in its attempt to move us beyond the familiar binary model that positions the hypercognitive, well-disciplined, rational, and calculating neoliberal subject over against his shadowy opposite, the pathological, defective, victimized, reason-impaired subject. That is precisely the set of false alternatives that we need to get past if we are to rethink sex and risk productively in the context of HIV/AIDS prevention.

The only problem with Berlant's approach for the purpose of dedramatizing the practice of sexual risk is that her model of "ongoingness" does not lend itself to the phenomenology of HIV infection. There is nothing "episodic" about the event of HIV transmission: it happens once, and it may even happen suddenly, in a transient instant. HIV infection does not occur according to the slow temporalities of obesity; on the contrary, nothing could be more calculated to restore to "the event" its singularity and narrative prestige. Sexual risk and HIV transmission would seem to have a kind of melodrama built into them. Although the degree of risk involved in any one act of unprotected sex between an infected and an uninfected partner is very slight, a single, momentary contact *can* have life-changing consequences. The stark realities of sexual risk may tend, if anything, to intensify the melodramatics of responsibility, intentionality, agency, and moral seriousness that already attach to the scene of gay male sex in the context of the HIV/AIDS epidemic and that complicate the work of prevention.

Nonetheless, if there is nothing episodic about HIV transmission, there may well be much that is episodic about condomless sex. So that is where the usefulness of Berlant's approach may lie. In actual practice, sexual risk may partake of precisely the sort of intermittence, ongoingness, and ordinariness that Berlant evokes. That is true especially to the extent that condomless sex can become, within the context of an individual life, a cumulative habit without a specific accompanying consciousness, an unreflective tendency, a gradual or occasional letting go of meaning, agency, will, or cognition.

In any case, it will be important to find some way to evade the choice between the rational subject and the pathological subject, as well as some way to take the drama out of the practice of risk. To do that, we may need to move HIV/AIDS away from the center of all thinking about gay men's sexual health. As the late Eric Rofes argued eloquently for more than a decade, the health of gay men ought to be conceptualized in an affirmative, holistic, politically imaginative fashion. It should not be narrowly defined by purely medical thinking about HIV/AIDS, reduced to the treatment of pathologies, or overwhelmed by panic scenarios about the epidemic.[36] For all of these reasons, it is crucial to detach our models of gay male subjectivity from discourses of mental health, the high moral drama of the individual sexual act, the dichotomous opposition between rational agency and pathology, and the epidemiology of risk.

Unfortunately, the perennial need for effective HIV/AIDS prevention strategies has once again made it possible, as well as politically palatable, to ask in all seriousness a battery of psychological and psychoanalytic questions about the nature of gay male subjectivity that had long been considered quaint, pointless, or prejudicial, and that had in any case been discredited by their implication in the protracted and shameful history of pseudoscientific homophobia. It is as if the very indecorousness of the topic of unprotected anal intercourse, coupled with the undoubted epidemiological imperative to address its occurrence among gay men, had emboldened researchers to break the various taboos that, in progressive circles at least, had previously prohibited any substantive discussion of the subjective life of male homosexuality.

It is, of course, legitimate to guard against increases in sexual risk taking by gay men. To the extent that such increases are accurately documented, they should arouse real concern and evoke a vigorous and sustained response. The minority of gay men who put themselves at significant risk of HIV infection deserve a disproportionate amount of attention from epidemiologists and prevention activists since those gay men are the ones whose behavior presents a challenge to public health and threatens to enlarge the scope of the epidemic both within gay communities and beyond them. It is therefore understandable that so much interest should be devoted to the percentage of gay men who fail to protect themselves adequately from the danger of HIV infection, nor is there anything particularly sinister about the fact of such interest in itself. The current panic over the alleged failure of safe-sex education, however, goes beyond the scope of the actual problem and produces a number of unfortunate results.

First, it exaggerates the extent of the danger (which is real but not desperate). Next, it intensifies the obsession with risky sex, making it seem central to the definition of gay male identity in the present—and therefore harder to resist. The sensationalistic treatment of gay male risk takers portrays gay men themselves as the sole source of the problem, as if they were to blame for the absence in the United States of any public support for a vibrant, sophisticated, and safe gay sexual culture as well as for the lack of publicly available, explicit, precise, reliable, appropriate, and practical information about how to prevent HIV infection. It substitutes for these social and political factors a multitude of moral and psychological ones, and it treats as inexplicable lapses gay men's principled refusals to adhere unswervingly to outdated, unrealistic, or excessively stringent safe-sex protocols. The moral and psychological defects from which gay male risk takers supposedly suffer are then generalized, categorically extended to lesbians, bisexuals, and gay men as a group. In this way, public discourses about "the return of unsafe sex" have contributed to the repathologizing of homosexuality. And so they affect queer people as a whole.

Today the gay male subject of unsafe sex, as Barry Adam maintains, has come to represent the cultural antithesis of the "calculating, rational, self-interested subject" who constitutes the presumptive subject of free-market neoliberalism and, thus, the new model of the "autonomous, self-regulated individual" in our disciplinary societies—the notional norm of a responsible, self-governing Everyman.[37] It is at least partly because the gay male subject of unsafe sex represents such a scandalous rebuke to neoliberal models of individual rationality that, in the last fifteen years or so, a vast scientific and popular literature has come into being that focuses with a kind of horrified fascination on the minority of gay men who have risky sex.

One effect of that literature has been to produce a new and exotic bestiary of barebackers, bug chasers, gift givers, and other adepts of extreme sensation, whose behavior is explained by reference to their intellectual or emotional "deficits." Such deficits include the "preset psychological variables" of "low self-esteem, sexual identity problems, or general sexual impulsivity," which are then compounded by "a host of irrational intervening factors: complacent AIDS optimists, reason-impaired drug users, personality-defective sensation seekers, and so on," culminating in "the panic icons of the popular imagination: demon infectors . . . , monster AIDS transmitters" as well as "barebackers and their mirror image, bug chasers, nearly always reported third hand or as seen on the internet."[38] (It is the researchers and writers who pursue these creatures, more than gay men themselves, who deserve to be labeled "bug chasers," as Michele Morales has pointed out.)[39]

Gregory Tomso, who has also studied the various discourses surrounding this topic, observes that "the list of reasons" given in the recent literature for gay men's engaging in risky sexual behavior is "expansive: low self-esteem; the physical pleasures of condomless sex; a 'culture of disease' created by glossy HIV-medication ads that equate infection with 'popularity and acceptance'; childhood sexual abuse; drug use; rebellion against authority; 'sexual self-control deficits'; and the eroticization of risk itself, to name just a few."[40]

In short, just when sexual difference was finally getting distinguished from pathology, just when homosexuality was ceasing to be considered a sickness, just when HIV/AIDS was at last starting to be recognized as a global pandemic, a public health emergency, and a terrible historical accident instead of being portrayed as a gothic tale of sexual crime and punishment—just at that moment a new syndrome, that of the psychologically afflicted gay individual who knowingly seeks to be infected or to infect others with HIV, has restored to homosexuality its venerable identity as a disease.[41]

Out of the vast and repellent literature on the supposed irrationality and self-destructiveness of gay men who take risks in their sexual practices, I will present just one trivial if typical example. I have chosen it not for its outlandish rhetoric or sexual demonology—the prize for that could probably go to the notorious 2003 *Rolling Stone* feature, "In Search of Death"[42]—but, on the contrary, for its banality and relative inoffensiveness (though I would hardly want to minimize its power to offend). On September 24, 2002, the *New York Times* published an article by Dr. Richard A. Friedman entitled, "A Clue to Why Gays Play Russian Roulette with H.I.V." The occasion for that typically sensationalistic representation of gay men's relation to risk turns out to have been a study presented at the 14th International Conference on HIV/AIDS in Barcelona, which found that 77 percent of a group of fifteen- to twenty-nine-year-old HIV-positive "gay and bisexual" men were unaware that they had been infected with HIV, despite supposedly having engaged in sexual behavior that might have exposed them to the virus. The authors of the study freely admitted that their figure of 77 percent was "upwardly biased to some unknown extent," but you would never know that from reading the article in the *New York Times*, which presented that excessively high figure at face value without further qualification—and as if the prevalence of unrecognized HIV infection did not vary wildly from one country to another, clearly correlating with differences in local social and political conditions (which should warn us not to seek its causes in individual gay men's psychology).[43] Undeterred by such considerations, Dr. Friedman went on to ask, "How can one explain such potentially fatal self-destructive behavior? . . . Why would anyone knowingly expose himself to a potentially lethal infection?"

In response to his own questions, with their gratuitous ascriptions to queer men of extreme self-destructiveness and conscious death-seeking, Dr. Friedman appealed to the work of Columbia psychiatrist Richard C. Friedman (no relation). The latter, identified as the coauthor of a promising new book entitled *Sexual Orientation and Psychoanalysis*,[44] located one cause of "this dangerous behavior" in "a phenomenon called internalized homophobia." That phenomenon is characterized by both Drs. Friedman as "a common and often serious psychological problem in gay men and women that lies at the root of many self-destructive behaviors, including risky sex." Note that the explanation for the irrational behavior of the risk-taking subgroup of gay men is generalized here and immediately made to apply to "gay men *and women*" without qualification, who are then described as being given to "*many* (other, unspecified) self-destructive behaviors, including"—but by no means limited to—"risky sex."

What is enlightening about this article is obviously not the insight it purports to offer into what it called "an aspect" of gay male sexual life "that has eluded explanation" until now. After all, "internalized homophobia" has not been news for at least forty years, and the invocation of that notion here merely provides a politically palatable cover for the continued, insistent association of queer people with psychopathology. (Would the *New York Times* ascribe "serious psychological problems" to heterosexual women as a class and diagnose them all as suffering from "internalized misogyny" because some of them sometimes "play Russian roulette" with pregnancy? It is an intriguing idea, in its way, even if it is misguided for the reasons I have already mentioned, but in any case we should not expect to see it expressed very soon in the *New York Times*.) As lesbian psychologist Celia Kitzinger argued thirty years ago, "internalized homophobia" is the gay-affirmative version of homosexual pathology: it is not their homosexuality but their homophobia that now makes queer people sick.[45]

More recently, Peter Hegarty has established that nearly one-quarter of current research into homophobia by professional psychologists focuses not on straight people's aversion to queers but on queers' own supposed aversion to themselves. "In the last five years examined," he wrote in 2006, "24 per cent of the articles that mentioned homophobia concerned the ways that gay men, lesbians and bisexuals (in that order) internalised prejudice rather than the ways that heterosexuals endorsed or enacted it."[46] As Hegarty noted, summarizing a number of recent studies, internalized homophobia has become a favorite explanation among psychologists for why gay men have condomless anal sex, even though at least one literature review concluded that "the evidence for a link between unsafe sex among bisexual and gay men and internalised homophobia was weak at best."[47]

What the *New York Times* article does show is that the topic of "unsafe sex" opens up an inviting public space for respectable if misguided speculation about what gay men want, about the subjective life of male homosexuality. The article also dramatizes the urgent need for a counterdiscourse of gay male subjectivity: not a wholesale avoidance of the subject, and not a more gay-friendly brand of psychology attentive to the disabling effects of homophobia—that, as the article itself illustrates, is already part of the problem—but a discourse free from psychology itself and from psychology's tainted opposition between the normal and the pathological.

That is what I have tried to offer, tentatively and experimentally, in my recent work—chiefly in the body of my 2009 book, *What Do Gay Men Want?* (from whose introduction these remarks have been extracted), and more recently, as well as at greater length, in my 2012 book, *How to Be Gay*. There I focus on gay male cultural practices, and I look to them as sources for the study of gay male subjectivity, considered as a collective, transpersonal, social phenomenon, rather than as an individual, individualizing, and therefore potentially pathologizing deviation. My aim has been to frustrate some of the deployments of contemporary forms of biopower, which seek to extend the processes of social control into ever more intimate spheres of personal life, by investing our relations to our own bodies and our practices of pleasure with new normalizing forms of self-surveillance disguised as HIV/AIDS prevention and public health.

Meanwhile, the epidemiological situation remains dire, and the news seems to be getting worse. The latest *HIV Surveillance Report* (vol. 23) from the CDC, published in February 2013 but reporting "Diagnoses of HIV Infection in the United States and Dependent Areas" through June 2012, indicates that nearly two-thirds of new HIV infections (65 percent) in 2011 were attributable to male-to-male sexual contact. Men who have sex with men now represent one of the few transmission categories in which the annual number of diagnosed HIV infections actually increased from 2008 to 2011.

In this context, the last thing I want to do is to minimize the gravity of the situation or to imply that gay men are doing just fine on their own, as if we only need the public health establishment to stop harassing us in order to defend our ways of life and to be healthy. At the same time, if public health interventions—grounded in such biopolitical procedures as surveillance, screening, testing, contact tracing, "high-impact prevention," responsibilization, risk elimination pedagogy, the condom code, the pathologization of risky sex, and the criminalization of HIV through laws requiring the disclosure of HIV infection before sexual contact—were actually forms of care instead of forms of regulation, if they really qualified as good medical science, good psychology, or good social policy, we would expect them to be considerably more effective at preventing new HIV infections among queer adolescents and adults than they have proved to be.

We would do well to recall in this context that safe sex is a gay, grassroots invention. Every single behavioral technique that has proven to be effective in stopping or slowing the spread of HIV was spontaneously discovered and disseminated by gay male communities themselves—with the exception of the "zero grazing" campaign in Uganda. Other than that, not one means of preventing the transmission of HIV that actually works was invented by the authorities or by experts in public health and then prescribed to gay men. All the safe-sex techniques that are currently approved were invented by gay men, acting collectively. They were actually considered risky by the public health establishment at the time they were devised—and that includes the use of condoms for anal sex. Accordingly, grassroots prevention guidelines, back in 1983, warned that anal sex with condoms was "possibly unsafe" (scientists did not complete the study showing that latex condoms could prevent the spread of HIV until 1985). It was only later, as the techniques that gay communities had invented and promulgated to preserve our sexual lifestyles while reducing the risk of HIV transmission gradually revealed their efficacy, that public health authorities jumped on them and transformed them into public health policy, turning around and imposing them on gay men (and others) as obligatory behaviors.

This creates a particular problem for gay men's health. When gay men's indigenous safe-sex inventions are turned around by public health authorities and then imposed on gay men as obligations, as behavioral norms, as duties, or as emblems of good hygiene, moral virtue, and patriotic citizenship, they cease to be weapons that we use to defend ourselves and our pleasures from a homophobic society, on the one hand, and from an intolerant public health establishment and other biopolitical institutions, on the other, and become instead a new kind of discipline to which we must submit, if we want to show our straight friends (and some gay ones) that we are *good* gay men, not those evil, irresponsible, sex-obsessed sluts that everyone thinks we are—not unjustifiably.

The dangers of turning risk reduction into a compulsory form of hygiene are vividly illustrated by the fate of safe sex itself, which has been alienated from gay men by the public health establishment (including some gay prevention activists) and has been deployed against us: our failure to adhere completely to the very practices of risk reduction that we invented, and that we promoted in the teeth of authoritative public health exhortations not to practice them, now serves to measure the extent of our deviance, of our pathology, of our sickness. And so those safe-sex practices have come to produce strenuous resistance from some gay men, precisely because they seem to be imposed on us from without. They no longer represent to us the strategies we have adopted ourselves, out of a determination to fight back, to survive, and to thrive. That is at least one reason why they are being rejected.

The latest statistics from the CDC, in short, show that the current prevention strategies are not working with gay and bisexual men. Or rather, they work very well when it comes to strengthening the institutions and technologies governing the administration of life at the level of the population and the individual, but they do not work well at stopping the spread of HIV among queer adolescents and adults. What that population needs is a different, nondisciplinary approach to HIV prevention, not a new deployment of biopower, but a prevention strategy as effective as safe sex used to be during the first fifteen years of the epidemic in the United States.

Notes

Excerpts from What Do Gay Men Want? An Essay on Sex, Risk, and Subjectivity, *by David M. Halperin (Ann Arbor: University of Michigan Press, 2007; rev. ed. 2009), 11–37.*

1. David Nimmons, *The Soul Beneath the Skin: The Unseen Hearts and Habits of Gay Men* (New York: St. Martin's, 2002), 59–61 (quotation on 59). Later studies have echoed those statistics.
2. See Jonathan Elford, "Changing Patterns of Sexual Behaviour in the Era of Highly Active Antiretroviral Therapy," *Current Opinion in Infectious Diseases* 19 (2006): 26–32, esp. 28–29. On the last point mentioned, see, e.g., P. Keogh, S. Beardsell, and Sigma Research, "Sexual Negotiation Strategies of HIV-Positive Gay Men: A Qualitative Approach," in *AIDS: Activism and Alliances*, ed. Peter Aggleton, Graham Hart, and Peter Davies (London: Taylor & Francis, 1997), 226–37, cited, discussed, and corroborated by Ian Hodges and Eamonn Rodohan, "Living with Homophobia: Exploring Accounts of Communication and Disclosure from London Gay Men Diagnosed with HIV," *Lesbian and Gay Psychology Review* 5, no. 3 (2004): 109–17; similarly, "High-Risk Sexual Behavior by HIV-Positive Men Who Have Sex with Men—16 Sites, United States, 2000–2002," *Morbidity and Mortality Weekly Report* (Centers for Disease Control and Prevention) 53, no. 38 (2004): 891–94.
3. Davey M. Smith, Douglas D. Richman, and Susan J. Little, "HIV Superinfection," *Journal of Infectious Diseases* 192, no. 3 (2005): 438–44 (quotation on 442). See, further, Matthew J. Gonzales, Eric Delwart, Soo-Yon Rhee, Rose Tsui, Andrew R. Zolopa, Jonathan Taylor, and Robert W. Shafer, "Lack of Detectable Human Immunodeficiency Virus Type 1 Superinfection during 1072 Person-Years of Observation," *Journal of Infectious Diseases* 188, no. 3 (2003): 397–405; Rose Tsui, Belinda L. Herring, Jason D. Barbour, Robert M. Grant, Peter Bacchetti, Alex Kral, Brian R. Edlin, and Eric L. Delwart, "Human Immunodeficiency Virus Type 1 Superinfection Was Not Detected Following 215 Years of Injection Drug User Exposure," *Journal of Virology* 78, no. 1 (2004): 94–103; Michael Shernoff, *Without Condoms: Unprotected Sex, Gay Men & Barebacking* (New York: Routledge, 2006), 222–23.
4. Kane Race, "Engaging in a Culture of Barebacking: Gay Men and the Risk of HIV Prevention," in *Gendered Risks*, ed. Kelly Hannah-Moffat and Pat O'Malley (London: Routledge-Cavendish, 2007), 99–126.
5. See, e.g., Christian Grov, "'Make Me Your Death Slave': Men Who Have Sex with Men and Use the Internet to Intentionally Spread HIV," *Deviant Behavior* 25, no. 4 (2004): 329–49. Grov does not consider the possibility, later documented by Shernoff, *Without Condoms*, 162–63, that some men who advertise as HIV-negative bug chasers are actually HIV-positive men who seek to heighten or intensify the thrills of sex with other HIV-positive men by pretending to be HIV-negative.

6. Kane D. Race, "Revaluation of Risk among Gay Men," *AIDS Education and Prevention* 15, no. 4 (2003): 369–81 (quotation on 374), citing Gordon Mansergh et al., "'Barebacking' in a Diverse Sample of Men Who Have Sex with Men," *AIDS* 16, no. 4 (2002): 653–59.
7. Only 12.1 percent of HIV-negative men in a San Francisco study conducted between 1999 and 2001 reported practicing unprotected anal intercourse (UAI) with "potentially serodiscordant" partners, according to Sanny Y. Chen, Steven Gibson, Darlene Weide, and Willi McFarland, "Unprotected Anal Intercourse between Potentially HIV-Serodiscordant Men Who Have Sex With Men, San Francisco," *Journal of Acquired Immune Deficiency Syndromes* 33, no. 2 (2003): 166–70, even though other reports indicate that UAI is coming to be a community norm among some gay men in San Francisco. More recent studies have put the percentage of HIV-negative gay men who have UAI with potentially HIV-positive men at around 11 percent.
8. See Alvin G. Dawson Jr., Michael W. Ross, Doug Henry, and Anne Freeman, "Evidence of HIV Transmission Risk in Barebacking Men-Who-Have-Sex-With-Men: Cases from the Internet," *Journal of Gay & Lesbian Psychotherapy* 9, nos. 3–4 (2005): 73–83, esp. 81, whose conclusion, about the lack of deliberate intent to infect or be infected with HIV but the high rate of indifference to the possibility of HIV infection on the part of bareback website users, has been massively confirmed by Shernoff, *Without Condoms*, 157–64 and 318–19, and by Richard Tewksbury, "Bareback Sex and the Quest for HIV: Assessing the Relationship in Internet Personal Advertisements of Men Who Have Sex with Men," *Deviant Behavior* 24, no. 5 (2003): 467–82. Similar statistics have been reported for gay and bisexual men both on and off the Internet. "In an online survey of more than 2500 MSM, only 1.9% reported that they had ever had anal sex without a condom because they wanted to get HIV," according to Patrick S. Sullivan and Richard J. Wolitski, "HIV Infection among Gay and Bisexual Men," in *Unequal Opportunity: Health Disparities Affecting Gay and Bisexual Men in the United States*, ed. Richard J. Wolitski, Ron Stall, and Ronald O. Valdiserri (Oxford: Oxford University Press, 2008), 220–47 (quotation on 228), citing an unpublished paper by Wolitski and his collaborators.
9. Ron D. Stall, Thomas J. Coates, and Colleen Hoff, "Behavioral Risk Reduction for HIV Infection among Gay and Bisexual Men: A Review of Results from the United States," *American Psychologist* 43, no. 11 (1988): 878–85, cited and quoted by Nimmons, *Soul Beneath the Skin*, 61. Eric Rofes, however, who also quotes this passage, has argued with some plausibility that the change in gay men's sexual behavior during the 1980s may have been exaggerated: see Eric Rofes, *Reviving the Tribe: Regenerating Gay Men's Sexuality and Culture in the Ongoing Epidemic* (New York: Harrington Park Press, 1996), 145–51, 173–79, and *Dry Bones Breathe: Gay Men Creating Post-AIDS Identities and Cultures* (New York: Harrington Park Press, 1998), 199–200.
10. Patrick Rawstorne, Carla Treloar, and Juliet Richters, eds., *Annual Report of Behaviour 2005: HIV/AIDS, Hepatitis and Sexually Transmissible Infections in Australia*, National Centre in HIV Social Research, Monograph 3/2005 (Sydney: National Centre in HIV Social Research, 2005), 4.
11. Of the estimated 54,230 new HIV infections among whites, blacks, and Hispanics in 2006, 13,230 or just 24 percent were among white gay and bisexual men who were not also at risk of infection from injecting drugs, according to the CDC's calculation: see J. Prejean, R. Song, Q. An, and H. I. Hall, "Subpopulation Estimates from the HIV Incidence Surveillance System—United States, 2006," *Morbidity and Mortality Weekly Report* 57, no. 36 (2008): 985–89, esp. 987, table. This implies that white

gay men represented an even smaller proportion of the overall number of new HIV infections in the United States as a whole.

12. The Centers for Disease Control and Prevention (CDC) report that men who have sex with men accounted for 49 percent of HIV/AIDS cases diagnosed in the United States in 2006: see Centers for Disease Control and Prevention, *HIV/AIDS Surveillance Report*, vol. 18, *Cases of HIV Infection and AIDS in the United States and Dependent Areas, 2006* (Atlanta: US Department of Health and Human Services, 2008), 6 (referring to 11, table 1, which seems, however, to yield a figure of 47.5 percent, whereas A. Mitsch et al., "Trends in HIV/AIDS Diagnoses Among Men Who Have Sex with Men—33 States, 2001–2006," *Morbidity and Mortality Weekly Report* 57, no. 25 [2008]: 681–86, put the figure at 46 percent [681]). This proportion has been increasing annually since 2001, when it was 40.5 percent, according to the CDC's annual surveillance report for 2005. Even so, those figures are based on the thirty-three states and five territories that have provided the CDC with confidential, name-based reporting of HIV infection since at least 2003; the figures represent, by the CDC's own estimate, 63 percent, or less than two-thirds, of the total epidemic in the United States. When the CDC estimates the annual number of cases of AIDS in the entire United States and "dependent areas," not just in states and territories that furnish it with name-based reporting (51), it finds that *male-to-male sexual contact accounted for only 35 percent of total AIDS cases diagnosed in 2006* and only 44 percent of all AIDS cases diagnosed in the United States from the start of the epidemic through 2006: see 37, table 17. Similarly, the CDC attributed more than 43 percent of new HIV infections in forty-five states and five territories of the United States in 2006 to male-to-male sexual contact (36 percent of the total from the start of the epidemic to 2006): see 38, table 18. Nonetheless, the CDC *estimates* that male-to-male sexual contact accounted for 42 percent of AIDS cases diagnosed in the entire United States and its territories in 2006 (a proportion that has steadily increased from just under 39 percent in 2001), and 46 percent of the cumulative total of such cases from the beginning of the epidemic through 2006: see 12, table 3. A recent report from the CDC estimated that from 2001 to 2004 "44% of new HIV infections [in the United States] were in MSM": see "The Global HIV/AIDS Pandemic, 2006," *Morbidity and Mortality Weekly Report* 55, no. 31 (2006): 841–44 (quotation on 842). But the CDC's later estimate of HIV incidence in the United States hypothesizes that men who have sex with men accounted for a massive 53 percent of all new HIV infections in 2006: see H. Irene Hall, Ruiguang Song, Philip Rhodes et al., "Estimation of HIV Incidence in the United States," *JAMA* 300, no. 5 (2008): 520–29, esp. 524. That figure, though a statistical projection, has proved to be prophetic. Furthermore, the CDC now conjectures that "HIV incidence has been increasing steadily among gay and bisexual men since the early 1990s" (*Estimates of New HIV Infections in the United States*, CDC HIV/AIDS Facts, August 2008, 3), by which it seems to mean total numbers of new infections each year. An independent literature review finds that true HIV incidence (i.e., rates of HIV transmission) among MSM did not either rise or decline between 1995 and 2005 (Ronald Stall, "What's Driving the US Epidemic in Men Who Have Sex with Men?" [paper delivered at the 15th Conference on Retroviruses and Opportunistic Infections, Boston, February 3–6, 2008, podcast accessed on August 26, 2008]).

13. Elford, "Changing Patterns," 27; see also Jonathan Elford, Graham Bolding, Lorraine Sherr, and Graham Hart, "High-Risk Sexual Behaviour among London Gay Men: No Longer Increasing," *AIDS* 19, no. 18 (2005): 2171–74; Juliet Richters, ed.,

HIV/AIDS, Hepatitis and Sexually Transmissible Infections in Australia: Annual Report of Trends in Behaviour 2006 (Sydney: National Centre in HIV Social Research, 2006), 2, 9–16.

14. On this much-debated topic, see J. Campo, M. A. Perea, J. del Romero, J. Cano, V. Hernando, and A. Bascones, "Oral Transmission of HIV, Reality of Fiction? An Update," *Oral Diseases* 12 (2006): 219–28, who emphasize the relatively low risk of HIV transmission in oral-genital contact and describe the factors that elevate or diminish the risk. For the CDC's current estimates, updated last on July 1, 2014, see "HIV Transmission Risk" (accessed September 24, 2014), http://www.cdc.gov/hiv/policies/law/risk.html.
15. Susan Kippax, June Crawford, Mark Davis, Pam Rodden, and Gary Dowsett, "Sustaining Safe Sex: A Longitudinal Study of a Sample of Homosexual Men," *AIDS* 7, no. 2 (1993): 257–63.
16. Race, "Engaging in a Culture of Barebacking."
17. Shernoff, *Without Condoms*, 12.
18. I owe this point to Kane Race, specifically to his workshop at the 2006 University of Michigan conference "Against Health." See also Barry D. Adam, Winston Husbands, James Murray, and John Maxwell, *Renewing HIV Prevention for Gay and Bisexual Men: A Research Report on Safer Sex Practices Among High-Risk Men and Men in Couples in Toronto*, A Report to Health Canada (2003), 32, who warn that "'risk reduction' messages" may "ultimately affirm the 'boundary pushing' practices of some men, and indeed may help consolidate such practices as viable forms of risk reduction." In this way, they observe, some harm reduction strategies may "simply function[] as a cover for harm increase." For a detailed and subtle analysis that supports this point, with specific reference to the Australian model, see Juliet Richters, Olympia Hendry, and Susan Kippax, "When Safe Sex Isn't Safe," *Culture, Health & Sexuality* 5, no. 1 (2003): 37–52. For a careful and nuanced assessment of risk reduction messages and their effects, see Sean Slavin, Juliet Richters, and Susan Kippax, "Understandings of Risk among HIV Seroconverters in Sydney," *Health, Risk & Society* 6, no. 1 (2004): 39–52.
19. Tim Dean, Robert Caserio, Heather Love, and Jean-Michel Rabaté, "On Bareback Subcultures and the Pornography of Risk," *Slought Foundation Online Content* (October 6, 2006; accessed November 30, 2006), http://slought.org/content/11332/.
20. Ibid.
21. Rofes, *Dry Bones Breathe*, 129–30, summarizing a position he had staked out at greater length and in greater detail in *Reviving the Tribe*, 203–24. It may be worth observing in this connection that even the CDC at one point set a goal that stopped considerably short of ending the epidemic: its HIV Prevention Strategic Plan aimed to halve the number of yearly HIV infections in the United States from forty thousand to twenty thousand. See the CDC's *HIV/AIDS Surveillance Report*, vol. 16, *Cases of HIV Infection and AIDS in the United States, 2004* (Atlanta: US Department of Health and Human Services, 2005), 5.
22. Sullivan and Wolitski, "HIV Infection," 221–22, citing the CDC's National HIV Behavioral Surveillance survey for 2004–5, report a prevalence rate for HIV infection among men who have sex with men, based on a study of 1767 MSM in five large US cities, of 25 percent overall, rising to 37 percent in the forty- to forty-nine-year-old age group.
23. Nimmons, *Soul Beneath the Skin*, 62.
24. See Sullivan and Wolitski, "HIV Infection," 222, 224.

25. See Nimmons, *Soul Beneath the Skin*, 62, for the quip about seatbelts and heterosexual mental health. For sexual risk taking among heterosexual youth, see Cicely Marston and Eleanor King, "Factors that Shape Young People's Sexual Behaviour: A Systematic Review," *Lancet* 368, no. 9547 (2006): 1581–86.
26. Mansergh et al., "Barebacking," 653; also 655, 657. Mitsch et al., "Trends in HIV/AIDS Diagnoses," 683, observe that whereas the increase between 2001 and 2006 in the number of HIV/AIDS diagnoses among all black men who have sex with men was 12.4 percent in the thirty-three states with established confidential, name-based reporting (compared to 8.6 percent among MSM overall), diagnoses among black MSM aged thirteen to twenty-four years increased during the same period by 93.1 percent, twice as fast as among white MSM of the same age group (and, among Asian/Pacific Islander MSM aged thirteen to twenty-four years, the increase in HIV/AIDS diagnoses was 255.6 percent—though the database for this demographic group is too small to generate entirely reliable figures). The CDC estimates that in 2006 "the HIV incidence rate [in the United States as a whole] was 7 times as high among blacks (83.7; 95% CI, 70.9–96.5) as among whites (11.5; 95% CI, 9.6–13.4)": see Hall, Song, Rhodes et al., "Estimation of HIV Incidence," 524.
27. David J. Malebranche, "Black Men Who Have Sex with Men and the HIV Epidemic: Next Steps for Public Health," *American Journal of Public Health* 93, no. 6 (2003): 862–65 (quotation on 862). Malebranche offers a number of possible, though speculative, solutions to this puzzle, none of which can be confirmed until more research is completed. See, further, Gregorio A. Millett, John L. Peterson, Richard J. Wolitski, and Ron Stall, "Greater Risk for HIV Infection of Black Men Who Have Sex with Men: A Critical Literature Review," *American Journal of Public Health* 96, no. 6 (2006): 1007–19, who also conclude that "high rates of infection for Black MSM . . . were not attributable to a higher frequency of risky sexual behavior" (1007). See, further, Sullivan and Wolitski, "HIV Infection," 228–29, and Robert E. Fullilove, *African Americans, Health Disparities and HIV/AIDS: Recommendations for Confronting the Epidemic in Black America* (Washington, DC: National Minority AIDS Council, 2006).
28. Cf. Tim Dean, *Beyond Sexuality* (Chicago: University of Chicago Press, 2000), 134n: "I focus primarily on gay men's responses to safe-sex education, not because other demographic groups remain unaffected by the issues I discuss (far from it), but because the problem is most acute for gay men and most demonstrably involves complex psychical factors." For a classic study of sexual risk taking by gay men who belong to an ethnic minority, a study that invokes a complex set of cultural and political constraints on the agency and autonomy of these men to explain why they sometimes fail to protect themselves from infection despite their desire to do so and their clear understanding of the risks, see Rafael M. Díaz, *Latino Gay Men and HIV: Culture, Sexuality, and Risk Behavior* (New York: Routledge, 1998).
29. Gregory Tomso, "Bug Chasing, Barebacking, and the Risks of Care," *Literature and Medicine* 23, no. 1 (2004): 88–111: "What makes them do it? It seems impossible today to have a conversation about bug chasing or barebacking without hearing someone raise the question, 'What makes them do it?'" (90). The most ambitious, comprehensive, and exhaustive effort to answer that question has been provided by Shernoff, *Without Condoms*, esp. 65–100, who approaches the problem from a psychological angle but who at least tries to offer a nonpathologizing solution to it.
30. For a comprehensive overview, see Ron Stall, Mark Friedman, and Joseph A. Catania, "Interacting Epidemics and Gay Men's Health: A Theory of Syndemic Production among Urban Gay Men," in *Unequal Opportunity: Health Disparities Affecting*

Gay and Bisexual Men in the United States, ed. Richard J. Wolitski, Ron Stall, and Ronald O. Valdiserri (Oxford: Oxford University Press, 2008), 251–74.

31. Stephen Lyng, ed., *Edgework: The Sociology of Risk-Taking* (New York: Routledge, 2005).
32. Rofes, *Reviving the Tribe*, 162.
33. I quote, with Kirk Read's kind permission, from his work in progress called *This Is the Thing*. One of the few researchers to have anticipated Read's call and to be willing "to validate the negative feelings" about condoms that most gay men have is Díaz, *Latino Gay Men and HIV*, 15.
34. See Rofes, *Dry Bones Breathe*, esp. 198–205, for a good example of an effort to dedramatize sexual risk taking. Shernoff, *Without Condoms*, 267, proposes a similar strategy.
35. Lauren Berlant, "Slow Death (Sovereignty, Obesity, Lateral Agency)," *Critical Inquiry* 33, no. 4 (2007): 754–80 (quotations on 757–59).
36. See Rofes, *Reviving the Tribe*, esp. 227–82; *Dry Bones Breathe*, esp. 241–92. Cf. Susan Kippax, "A Public Health Dilemma: A Testing Question," *AIDS Care* 18 no. 3 (2006): 230–35, who makes a similar point about the current promotion of HIV testing for the purposes of prevention. And compare Sullivan and Wolitski, "HIV Infection," 235: "Achieving sustainable long-term gains in HIV prevention efforts for MSM may require a more holistic approach to the health and well-being of MSM."
37. Barry D. Adam, "Infectious Behaviour: Imputing Subjectivity to HIV Transmission," *Social Theory and Health* 4 (2006): 168–79 (quoting Barry Smart's *Economy, Culture and Society* and Deborah Lupton's *Risk* on 169–70). For a similar account of how "modern systems of risk administration" create a context in which "risk rationalities" give rise to "risk identities," see Mark Davis, "HIV Prevention Rationalities and Serostatus in the Risk Narratives of Gay Men," *Sexualities* 5, no. 3 (2002): 281–99 (also cited by Adam), esp. 285: "it flows from this system of identity production that to act in an irrational manner is deviant and, therefore, such conduct becomes a sign of defectiveness."
38. Adam, "Infectious Behaviour," 170, who provides plentiful references to this literature (some of the quoted phrases I have embedded in my text are from Adam himself and some are from the literature he cites, though I admit that the latter are hard to tell apart from Adam's withering rephrasings of them). The most egregious example of third-hand, anecdotal, undocumented discussions of barebackers and bug chasers can be found in DeAnn K. Gauthier and Craig J. Forsyth, "Bareback Sex, Bug Chasers, and the Gift of Death," *Deviant Behavior* 20, no. 1 (1999): 85–100, an article whose complete lack of solid evidence for its sensationalistic claims has not prevented it from becoming a staple of the scholarly literature on gay men and risky sex.
39. Michele Elaine Morales, "Persistent Pathologies: The Odd Coupling of Alcoholism and Homosexuality in the Discourses of Twentieth Century Science" (PhD diss., University of Michigan, 2006), 179. For a painstakingly detailed review and refutation of the scapegoating of some gay men by other gay men for allegedly lapsing from perfect compliance with safe-sex protocols, see Rofes, *Dry Bones Breathe*, 123–97.
40. Tomso, "Bug Chasing," 90. Similarly, Morales, "Persistent Pathologies," 179, who has surveyed the epidemiological literature about homosexuality and alcoholism, comments that "the end result [is] that LGBT people can never 'get well.'"
41. See the brilliant, eloquent, incisive, and moving essay by Tony Valenzuela, "Killer

Gay Sex!" *POZ Magazine*, May 7, 2008, http://www.poz.com/articles/killer_gay_sex_hiv_401_14539.shtml.

42. Gregory A. Freeman, "In Search of Death," *Rolling Stone*, February 6, 2003, 44–48. For a detailed critique of this article and of the claims advanced in it, see Shernoff, *Without Condoms*, 172.
43. The study in question is by Duncan A. MacKellar, Linda A. Valleroy, Gina M. Secura, Stephanie Behel, Trista Bingham, David D. Celentano et al., "Unrecognized HIV Infection, Risk Behaviors, and Perceptions of Risk Among Young Men Who Have Sex with Men: Opportunities for Advancing HIV Prevention in the Third Decade of HIV/AIDS," *Journal of Acquired Immune Deficiency Syndromes* 38, no. 5 (2005): 603–14, whose sample was drawn from six US cities; for the acknowledgment about upward bias, see 612. A 1996 investigation noted, "Whereas 90% of Australian gay men know their antibody status, US estimates place the percentage at 65% for gay men": see Thomas J. Coates et al., "HIV Prevention in Developed Countries," *Lancet* 348, no. 9035 (1996): 1143–48, at 1146, citing D. C. Berrios, N. Hearst, T. J. Coates et al., "HIV Antibody Testing among Those at Risk for Infection: The National AIDS Behavioral Surveys," *JAMA* 270, no. 13 (1993): 1576–80.
44. Richard C. Friedman and Jennifer I. Downey, *Sexual Orientation and Psychoanalysis: Sexual Science and Clinical Practice* (New York: Columbia University Press, 2002).
45. Celia Kitzinger, *The Social Construction of Lesbianism* (London: Sage, 1987), 52–57.
46. Peter Hegarty, "Where's the Sex in Sexual Prejudice?," *Lesbian & Gay Psychology Review* 7, no. 3 (2006): 264–75 (quotation on 266).
47. Ibid., citing Theo G. M. Sandfort, "HIV/AIDS Prevention and the Impact of Attitudes toward Homosexuality and Bisexuality," in *AIDS, Identity, and Community: The HIV Epidemic and Lesbians and Gay Men*, ed. Gregory M. Herek and Beverly Greene (Thousand Oaks, CA: Sage, 1995), 32–54.

ELEVEN

Precarious Life: Butler and Foucault on Biopolitics

JANA SAWICKI

> It seems to me that contemporary political thought allows very little room for the question of the ethical subject.[1]
>
> —Michel Foucault

> I am one of those people . . . trying to figure out how to live.[2]
>
> —Judith Butler

What role does Michel Foucault's analysis of biopolitics play in Judith Butler's work? What are her most salient and important creative deployments of Foucault? Where does she bend his work to a different purpose? Extend his project? Supplement it?

Whereas in her earlier books, most notably *Gender Trouble, The Psychic Life of Power,* and *Undoing Gender,* Butler consistently appealed to Foucault's concepts of power, normalization, and subjection to address sexual and gendered life, in her more recent post–9/11 work she takes up broader questions concerning the politics and ethics of human life itself. What makes a life livable? Which lives count as human? How do norms of the human make some lives possible and render others culturally unintelligible, and hence dispensable? And how might we resist this normative violence?

Indeed, Butler's effort to develop an ethical response to contemporary forms of normative violence in global politics dovetail remarkably with Foucault's shift from a focus on the regime of sexuality in his earlier biopolitical writings, to his later preoccupations with broader questions concerning neoliberal governmentality, the government of individualization and possible ethical responses to its more pernicious effects. They also appear to deviate from Foucauldian orthodoxy, especially since they also rely on a revised psychoanalytic framework.

In what follows, I focus principally on Butler's more recent book, *Frames of War*, and her essay "What Is Critique?" to explore her relationship to Foucault's reflections on the emergence of biopolitics. Yet I also briefly address her account of subjection in *The Psychic Life of Power* since there we see the unique way in which she uses psychoanalytic theory, placing herself between Foucault and Sigmund Freud, to give an account of the subject as it is produced within relations of power. Rather than dialectically oppose them to one another to reconcile their differences, I want to position Butler *beside* Foucault—an exercise in nonbinary thinking introduced by queer theorist Eve Kosofsky Sedgwick. "*Beside* comprises a wide range of desiring, identifying, representing, repelling, paralleling, differentiating, rivaling, leaning, twisting, mimicking, withdrawing, attracting, aggressing, warping, and other relations" (Sedgwick 2003, 8).

As I will show, Butler's project leans on, imitates, differentiates itself, and parallels Foucault's. One also discovers echoes of Foucault's idea of the "tactical polyvalence of discourse" in Butler's notion of resignification. Her use of Foucault could be considered a form of citation that repeats Foucault with a difference. The effect of his work in her thought becomes a new event in thinking—one that arises within a different context, a different, queer space and time.[3] What appears initially as a recourse to the very biopolitical discourses that Foucault rendered suspect—humanism and psychoanalytic thought—represents instead a creative and tactical redeployment of them to address global problematics that are remarkably Foucauldian in spirit.

The Biopolitical in Foucault

It will be helpful at the outset to summarize Foucault's understanding of biopower to understand how Butler creatively adapts the "biopolitical" within her own work. The key ideas associated with Foucault's analytics of biopower are as follows:

1. Biopower targets and attends to life and displaces the centrality of juridical sovereign power—a form of power that operates through prohibition, repression, and taking life away.
2. Sexuality is the target and anchor of both of its poles—the disciplinary and regulatory—insofar as it involves the individual body, norms of develop-

ment, sexual behaviors, family and reproductive life, and the life of the population.

3. Whereas juridical power associated with sovereignty "always refers to the sword," and "does not have to draw the line that separates the enemies of the sovereign from his obedient subjects, . . . [a] normalizing society is the historical outcome of a technology of power centered on life." (Foucault 1980, 144)
4. Scientific racism emerges as an alibi for biopower. Foucault observes: "never before did . . . regimes visit such holocausts on their own populations. . . . Entire populations are mobilized for the purpose of wholesale slaughter in the name of life necessity: massacres have become vital" (1980, 136–37). Here the question arises: "How can a power such as this kill, if it is true that its basic function is to improve life, to prolong its duration, to improve its chances, to avoid accidents, and to compensate for failings?" (Foucault 2003, 254). The answer is scientific racism. Scientific racism divides the population into good and inferior, normal and monstrous or incorrigible, healthy and ill. As a normalizing biopolitics of the human species, it isolates internal threats to the vitality and health of the population and thereby sanctions death.
5. A new set of rights emerges as a counterdiscourse once life becomes a political objective—"right to life, to one's body, to health, and to happiness, to the satisfaction of needs, and beyond all oppressions or alienations, the right to rediscover what one is and all that one can be" (Foucault 1980, 145).
6. Ethical techniques of self-transformation and desubjectivation associated with the aesthetics of existence might become a resource for bolstering communities of resistance to biopolitics, opening up spaces for other ways of living where different possible futures might be prefigured.[4]

Butler on Critique, Normalization, and Subjection: The Eclipse of the Biopolitical?

In an essay on Butler's discussion of Foucault's *Herculine Barbin*, Finnish scholar Jemima Repo argues that Butler evades a treatment of biopolitics (Repo 2014). Repo argues that her preoccupations are ontological, not genealogical. Butler wants to explain how one comes to appear as gendered, and how gender varies from culture to culture as a particular manifestation of sex. Whereas a Foucauldian might ask how relations of power/knowledge make possible the emergence of the discourse and practices that produce and normalize gender (even though Foucault did not distinguish sex from gender), Butler wants to understand how the ontology of a binary gender system that produces sex develops within heterosexist and patriarchal cultures. Thus she focuses on a different set of normalizing discourses, namely feminist and nonfeminist theoretical accounts of gender, rather than the psychiatric, biological, criminological, and medical discourses that Foucault addressed in his genealogical writings on biopower.

To account for the production of sexed and gendered subjects, Butler turns to Gayle Rubin's (1984) account of sex/gender systems, a model of gender production within different cultural frameworks that draws upon both psychoanalytic theory and Claude Lévi-Strauss's theory of kinship structures. Both Rubin and Butler explain the emergence of gendered desire as an effect of the prohibition of desire and exogamy within kinship systems. Butler does not engage genealogy of gender to show how the development of theories of gender and gender norms arose within biopolitical strategies aimed at governing individual bodies and regulating the population. Indeed, as Repo implies, perhaps unlike Butler, Foucault would have been more likely to target the psychoanalytic and structuralist anthropological theories to which Butler turns as forms of power/knowledge than use them to explain variations in sex/gender systems.

So, at this earlier stage of her work, Butler is more interested in developing a theory of gender identification, of normative femininity and masculinity, than she is in a genealogical study of the emergence of the regime of sexuality that forms the conditions of possibility of particular experiences of sex and gender within already operating hierarchical relations of power between men and women, parents and children, teachers and students, and so on. This may account in part for why the biopolitical has received less attention within queer theory over the past two decades.

Butler elaborates more fully on how subjects are produced and subjected through normalization in *The Psychic Life of Power*. Here she provides an account of the subject's attachment to its subordinating identity. She asks: What does it mean for a norm to be internalized? Her answer is that the process of internalization produces a psychic interior—a psychic interior already marked, indeed constituted, by power but not determined by it. The psyche bears the marks of the social, but it exceeds the conditions of its own possibility. She asks: What about the individual makes it susceptible to subjection? To the desire for subjection? And what makes it possible for the individual to rework and dismantle the conditions of its subjection?

She answers these questions in *The Psychic Life of Power*. Here Butler positions herself between Foucault and psychoanalytic thought to bring into relief the dynamics of affect and desire at work in our attachments to particular understandings of ourselves, most notably, our subjection within the normalizing regime of sexuality. This attention to affective life might be understood as an important supplement to Foucault's approach to critique, even though, as I have already indicated, it raises questions about the desirability of recourse to what Foucault tended to regard as normalizing psychological sciences.[5] To be sure some of the most problematic features of psychoanalytic theory are substantially revised or jettisoned in Butler's use of it. She rejects anthropological universals—the Oedipus complex, the incest taboo—that inform Jacques Lacan's psychoanalytic theory. Nonetheless, her recourse to psychoanalytic discourse raises important questions about whether and how to appeal to theories that Foucault linked with biopolitical normalization in his genealogies of biopower. Yet surely we must not all be genealogists. And surely it is possible that concepts available in these theories can be redeployed in strategic ways to resist normalization. Butler's appeals to Freudian and Lacanian concepts might be understood as indicative of the ongoing eventalization of discourses, as events created by using materials available in the culture to open up other ways of thinking and doing. Still, insofar as she is interested in explaining the psychodynamic process of subjection, as we will see below, she appears to be departing from Foucault considerably.

Adopting Foucault's usage, Butler describes "subjection" as both a process of subordination and a process of becoming a subject. She wants to explain the psychodynamics of gender identification. Borrowing from Freud and Friedrich Nietzsche, she describes the reflexive subject as formed by a desire that, in the face of another, turns back on itself. This turning back represents the process through which a self-conscious subject emerges—a subject that comes to desire the reflexive circuit itself. In effect, our sense of self is premised on a desire to be—a desire that depends on the recognition of others for its continued existence. Because the terms of that recognition are not of our own choosing, we can become attached to our own subordination. We become passionately attached to those who care for us even under condition of exploitation or abuse. Where the terms of our existence involve internalizing norms or social categories that are subordinating, we prefer them to no social existence whatsoever. Our very survival depends on it. As a result, our primary dependency coupled with a desire to survive renders us vulnerable to subordinating forms of subjection.

Psychic subjection is a particular form of subjection. It represents a potentially more insidious power dynamic than explicit coercion or force—

namely an exercise of power and aggression over oneself. An effect of reflexivity formed through prohibition of desire, conscience emerges in Butler's account as a vehicle through which the subject becomes an object for itself. In response to prohibitions, one makes oneself an object of reflection, measures oneself against external ideals, and berates oneself for failing to measure up. Berating and censoring the self become outlets for the very aggression and libido that they thwart.

At the same time, desire can be regulated in ways that escape self-reflection. Butler adopts Freud's distinction between two types of prohibition, namely, repression and foreclosure. Unlike repressed desires, desires that may once have existed apart from their prohibition, foreclosed desire is barred from consciousness altogether, thereby "constituting the subject through a certain kind of preemptive loss" (1997, 23). What is foreclosed represents the "constitutive outside of the subject"—that which cannot become conscious without threatening dissolution of the subject.

In addition, Butler contends, foreclosure represents a constitutive melancholia at the heart of subject formation. Our identities are established on the ground of loss—an "incomplete and irresolvable grief" (1997, 23). Melancholia is an expression of that grief that also signals the limits of coherent subject formation. In an intriguing move, she connects this concept of foreclosure to Foucault's idea of a normalizing power that establishes the boundaries between possible and impossible forms of love—"a foreclosure of love that becomes the condition of possibility for social existence" (1997, 24). The sociality achieved by such a mechanism becomes, in her account, a melancholic sociality insofar as it produces one set of possibilities for intelligible social existence and bars others—the lost objects. One transgresses the limits of intelligible ways of being at risk of social death.[6]

Insofar as we are reflexive subjects with the capacity to relate to ourselves, we can also turn against the conditions that enable our existence. We can risk "social death" to "expose and open to transformation the hold of social power on the conditions of life's persistence" (1997, 29). Furthermore, for a given set of relations of power to be reproduced, subjects must repeat them through time. This temporal gap between repetitions, or reiterations, of power secures the possibility of their alteration because we can fail to repeat them "in the right way." Such failure reveals the contingency of established conditions of intelligible being and implies there might be other possibilities for meaningful existence. In this way, we surpass the power by which we are enabled, even if we do not entirely escape it.

In this account of subjection, one can hear echoes of several Foucauldian motifs. Although Butler's appeal to time and repetition relies more on speech act theory and the Derridean concepts of iteration and citation, one can also detect traces of Foucault's Nietzschean idea of the eventalization of concepts through the use of genealogy as well as an appeal to desubjectivation (dissolution and desubjugation of the subject?) as counters to normalizing power. Moreover, Foucault once described himself as "living in society's shadows"—an example of a kind of social death that resonates clearly with Butler's account of resistance in *The Psychic Life of Power*.[7] Ultimately, at this stage of her work, Butler envisions a form of political agency premised on the possibility that social subjects can fail to reinstate norms, to "repeat" them in ways that risk social death and challenge the dominant organization of life, the boundaries of intelligibility.

Foucault and Butler are both interested in how power/knowledge regimes constitute the conditions of possibility for thinking and being. Both also appeal to reversal as a viable tactic within biopolitical struggles. Additionally, insofar as these regimes do not totalize the social field, it would seem that the only "outside" available must be found within a historical field. How else can one make sense of Foucault's idea of heterotopic spaces? Or his idea that we use materials available within the culture to invent new ways of thinking and being—in effect, relying on possibilities for resignification (tactical polyvalence) and cultural innovation within the historical conditions of possibility that are available? Furthermore, Foucault's antihumanism targets fixed understandings of human nature and transcendental arguments that place the constitutive subject outside history. Butler eschews both. Her philosophy of the subject strikes me as thoroughly historical and, hence, to some extent compatible with Foucault's approach.

Finally, has the biopolitical really dropped out of Butler's analysis here, as Repo claims it does? Although Repo may be right that a more Foucauldian approach would involve doing a genealogy of gender rather than offering a revised psychoanalytic account of subjection that appeals to a historical unconscious, in a broader sense Butler's analysis of subjection and normalization represents a counterdiscourse to the politics of gender and sexual life found within anthropological and psychoanalytic discourses (Lacan, Lévi-Strauss) that might be understood as further developments within the regime of sexuality.[8] And to the extent that Foucault's analysis of biopower concerned quasi-scientific discursive apparatuses that perpetuate normalization and impose unnecessary constraints on ways of being and living, it would seem that at this stage Butler was remarkably attuned to further developments within the biopolitical regime of sexuality.

Butler and Biopolitics Post–9/11: Resignifying the Biopolitical

Feminist political theorist Moya Lloyd notes that although preoccupation with normative violence (normalization) and embodiment are persistent themes throughout Butler's writings, within the past decade she has moved on to talk not only about normalization associated with sexed and gendered corporeality, but also the occlusions and exclusions associated with ableism, racism, ethnocentrism, and, I would add, nationalism (2007, 138). Lloyd also suggests that Butler's shift toward a contingent and contested if quasi-universal humanism could be understood as her way to avoid exclusion in her own critical theory. In any case, Butler's post–9/11 writings do involve a significant shift of focus. Concerns about sexual minorities persist, but in this later work she focuses less on normalizing subjection in the regime of sexuality and more on the ways in which feminist and queer projects have either been enlisted in the service of homonationalism or been used to legitimate war and the abjection of Islamic cultures as intolerant and uncivilized. Thus, for example, she calls attention to the George W. Bush administration's strategic use of feminist concerns about the plight of the Taliban's treatment of women to justify its incursions into Afghanistan, and Dutch immigration policies that make intolerance of homosexuality a litmus test for Muslim citizenship.

In *Frames of War* Butler returns to a project that she began in her earlier post–9/11 book *Precarious Life,* namely, "reimaging the possibility of community on the basis of vulnerability and loss" starting with the "human" (2004, 20). In this later work certain themes from her earlier work (grief, loss, normalization, attachment, subjection, foreclosure, and recognition, to name the most pertinent) reappear in the context of her critique of the war on terror. She asks: "what constitutes the being of a life, and "what are the mechanisms of power through which life is produced?" (2009, 6). What makes some lives more visible and recognizable, more or less precarious, than others despite the fact that the condition of precariousness is shared by all of us? And how do existing norms of the human "allocate recognition differentially" (2009, 6)? Again, whereas Foucault's approach to the biopolitical involves descriptive genealogical work on the emergence of life sciences that normalize, regulate, and order it, Butler moves within a different dimension of Foucault's analysis insofar as she considers the relationship between the biopolitical construction of life and waging war. Like Foucault she accords racism a central role in making war possible. In this case, "racism" refers not to the biologically based species racism found in Foucault's account, but the more familiar racism based on ethnic and racial differences. In what she calls the "twisted logic" of this contemporary version of racism, "the loss of [certain] populations is deemed necessary to protect the lives of 'the living'" (2009, 31). The necessity here is not the necessity associated with the logic of scientific racism described by Foucault and exemplified in the Nazi regime. It would seem that in some respects Butler has reverted to an earlier logic alluded to by Foucault—one that he calls "the relationship of war" that does not require scientific biopolitical racism to function. Even if within this earlier rationality killing the other might protect me, it does not make life in general "healthier and purer" (2003, 255).

On the other hand, Foucault does suggest that if the "power of sovereignty . . . that has the right of life and death wishes to work with the instruments, mechanisms, and technology of normalization, it too must become racist" (2003, 255). Accordingly, Butler claims that the war on terror involves a resurgence of sovereignty operating within a racist technology of normalization—in this case, the normalization of "humanity." This new technology of normalization makes it easier to wage war insofar as it exploits current forms of racism and ethnocentrism to exclude certain populations from its normalizing frame of the human. For example, media and governmental "frames of war" exclude thousands of Iraqi civilians killed in the second Iraq war. Some lives are rendered invisible while other lives become normatively human. Thus connections between human beings are severed, occluding the fact of our dependency and reliance on one another, perpetuating inequality in the conditions of existence within which different lives are lived.

In Butler's reading of Foucault, governmentality displaces sovereign power. Whereas Foucault describes state power as a dispersed array of state and nonstate institutions that target populations, conduct conduct, and use both political economy and apparatuses of security to manage and administer populations, Butler emphasizes the way in which the Bush administration's suspension of law in the case of indefinite detention of alleged terrorists in Guantanamo produced an illegitimate and lawless sovereignty—its "resurgence within the field of governmentality" (2004a, 53).

Furthermore, Butler uses the term "sovereignty" to denote not only a certain understanding of state power, but also "mastery" in a more specific sense—the mastery of the meanings of one's words, one's intentions, and the effects of one's deeds and utterances. And given her emphasis on our dependency on others not only for survival but also for carrying on the meanings of our utterances and carrying forth our projects, exerting some form of control on the future, one can see why, given her sensitivity to affective dynamics, she imagines that the fear of loss of sovereignty might invite compensatory mechanisms. Accordingly, she reads the post–9/11 demand for retaliation in a preemptive war with Iraq as a form of indignant and retaliatory rage stemming from grief, and a desire to both escape a sense of vulnerability and to regain a sense of mastery.[9]

Butler's Ethical Turn

Foucault once remarked that liberation movements in his time "need an ethics," but not one "founded on so-called scientific knowledge of what the self is, what desire is, what the unconscious is, and so on" (1994a, 255–56). Butler appears to be responding to Foucault's call for a new ethic—one that operates outside the confines of a scientific, hence biopolitical, understanding of the self, desire, and the unconscious. While she certainly departs from Foucault insofar as she draws on psychodynamic theories, like Foucault she is sensitive to the dangers of normalization and unnecessary constraints on ways of living within biopolitical relations of power/knowledge. She clearly wants to avoid these dangers. Her attention to the dynamics of affective life represents a potentially important supplement to Foucault's analytics of power, for she speculates that we are conditioned and regulated not only in our conduct and self-understandings, but also at the level of our "affective and moral responsiveness." Even though we may take our enlarged capacities for sympathy and other regarding behavior as marks of our humanity, we often fail to recognize that "the humanity in question is, in fact, implicitly divided between those for whom we feel urgent and unreasoned concern and those whose lives and deaths simply do not concern us" (2009, 50).

Butler notes that her task is not to provide "a genealogy of the concepts of life or death," but to think about precariousness "as something both presupposed and managed by . . . [biopolitical] discourse, while never being fully resolved" (2009, 18). She draws our attention to the injustice in current biopolitical modes of administering life. And though she proposes grounding an ethics on the shared precariousness of our lives, her starting point is quite abstract and, hence, not based on an account of human nature. Her resources are existential-phenomenological, not scientific. Her ethics builds on our lived experience of embodiment. In her quasi-Hegelian understanding of the self, we are ek-static subjects who depend on others (past and present) for our very sense of who we are. We also depend on normalizing cultural terms of recognition for survival. What's more, our bodies expose us to the touch and gaze of the other. We attach ourselves to the available cultural possibilities for self-understanding to survive. Consequently, there are limits to our ability to account for ourselves.[10] At the same time, she asserts, our capacity to be responsive to (if not entirely responsible for) others is of paramount importance. Just as we are not masters of our own thoughts, deeds, or affective responses, so too do others face the same limitations. The impossibility of complete self-transparency and control, hence any strong form of self-determination or autonomy, counsels both humility and a responsibility not to rush to judgment and condemnation.[11] Butler urges us to critically reflect on the obstacles to recognizing inequalities in the distribution of precariousness, to engage in cultural translation as well as inquiry into the forces that operate to justify or dismiss inequality within a global political frame.

As we saw above, it is not inevitable that grief, loss, and the fact of dependency will lead to an awareness of shared vulnerability. It can lead instead to righteous indignation and retaliatory violence to reassert the illusion of mastery. Butler observes:

> We are bound to others not only through networks of libidinal connection but also through modes of unwilled dependency and proximity that may well entail ambivalent psychic consequences, including binds of aggression and desire . . . the shared condition of precariousness leads not to reciprocal recognition, but to a specific exploitation of targeted populations . . . cast as threats to human life *as we know it*. (2009, 30–31; emphasis mine)

To avert the potential for "cycles of violence" in this circuit of affect, Butler turns to ethics:

> This fundamental dependency on anonymous others is not a condition that I can will away. No security measures will foreclose this dependency; no violent act of sovereignty will rid the world of this fact. . . . To be injured means that one has the chance to reflect upon injury, to find out the mechanisms of its distribution, to find out who else suffers from permeable borders, unexpected violence, dispossession, and fear, and in what ways. (2004a, xii)

Butler believes that recent challenges to US national sovereignty present us with an opportunity to imagine a different world, one in which "interdependency becomes acknowledged as the basis for global political community," and we become responsive as global citizens to inequality in distributions of the conditions necessary to sustain life (2009, xii–xiii). Here we find an implicit appeal to a certain version of the new right to life that Foucault invokes as a counterdiscourse within the apparatus of biopower. In the end she responds to the war on terror not with condemnation or justifications of violence (though she does not rule out the possibility that violence may be necessary), but rather public critical questioning and an ethical appeal to others to reflect critically on the global inequalities operating within the politics of life that sever our connections rather than resort to responses to perceived threats to our lives that might blunt or erase "our capacity to feel and apprehend" (2009, xxi). Thus she draws on Foucault's understanding of critique as virtue.

Conclusion: Critique as Virtue, Butler beside Foucault

On the occasion of the creation in Geneva of the International Committee against Piracy, Foucault made a public statement in which he asserted:

> There exists an international citizenship that has it rights and its duties, and that obliges one to speak out against every abuse of power, whoever its author, whoever its victims. After all, we are all members of the community of the governed, and thereby obliged to show mutual solidarity. . . . The suffering of men must never be a silent residue of policy. It grounds an absolute right to stand up and speak to those who hold power. (2000, 64)

Here Foucault invokes a "community of the governed" outside national boundaries, a right not to be governed in a particular way. Hence he reveals the anarchistic strand in his ethico-political standpoint. Foucault advocated for popular sovereignty. In an interview about her own relationship to anarchism, Butler notes that she too wants to preserve the idea of popular sovereignty. Radical political change cannot rely solely on state and other institutional practices since "power is not always exercised through law" (Heckert 2010). Like Foucault, Butler has aligned herself with the "community of the governed" in her political activism and public statements (challenging human rights violations and Israeli state policies toward Palestinians, advocating for undocumented workers in the United States, squatters rights, etc.). Like Foucault, she does not regard anarchism as a way of life but as a strategy for contesting established legal policies, whether by state or nonstate actors.

In her essay "What Is Critique?" Butler asks: What grounds the legitimacy of anarchist revolt? What establishes the legitimacy of this sort of critique? For it is not one based on an argument about the illegitimacy of a given policy with respect to a specific rationality within the present. Instead what it does, according to Butler, is mark "the limit of . . . validity"—in other words, mark the limits of the governmental rationality within which particular policies are justified. She elaborates: "The critical attitude is not moral according to the rules whose limits that very critical relations seeks to interrogate" (2004c, 313). In suspending the ontological ground of a particular regime of truth, the critic effectively places herself at the limit of such a regime. And this, she suggests, is what Foucault means in an essay with the same title when he refers to the "desubjugation of the subject in the play of . . . the politics of truth" (2004c, 315). The virtue of this sort of critique involves "acting without guarantees, risking the subject at the limits of its ordering," positing a value without a foundation within the established order of things (2004c, 319). It is a form of critique that refuses to be bound by the established order in the face of intolerable human suffering—one that "breaks the habits of judgment in favor of a riskier practice" of desubjectivation (2004c, 321).

In Foucault's final volumes of the *History of Sexuality*, he draws our attention to ancient Greco-Roman ethical practices of self-formation that did not involve obedience to prohibitive moral laws or codes, but rather an individual choice about how to live—how to style oneself within the framework of established norms. According to Foucault, these subjects engaged in practices of the self using materials available in their culture (e.g., a plurality of ethical schools) to cultivate a particular self-relation, to create a way of living, and thereby make their lives works of art. In her reading of Foucault's ethical writings, Butler emphasizes that subjects are always formed, and form themselves, in relation to available norms. Thus subject formation becomes a form of poeisis. As she puts it, there is no self-forming outside of norms available for subject formation. As I read her, Butler's ethics aims to cultivate a responsiveness to suffering by drawing on our cultivated capacities for recognizing shared suffering and precariousness and adapting psychoanalytic discourses to address the affective dynamics that occlude the structures that establish unequal access to the conditions necessary to secure life in the biopolitical age. Despite this apparent rejection of Foucault's wariness about the biopolitical discourses of psychoanalysis and the humanism associated with them, Butler's creative use of these biopolitical cultural resources are remarkably Foucauldian in spirit insofar as they creatively repeat his moves, redeploy his terms, and extend and adapt his work to respond to the dangers she perceives within the present.

Notes

1. Foucault (1994b, 256).
2. Kirby (2006, 157).
3. Consider Foucault's description of the question he raises in his critique of power and knowledge in "What Is Critique?" "The question [is] . . . how can the indivisibility of knowledge and power in the context of interactions and multiple strategies induce both singularities, fixed according to their conditions of acceptability, and a field of possibles, of openings, indecisions, reversals and possible dislocations which make them fragile, temporary, and which turn these effects into events, nothing more, nothing less, than events?" (1994c, 278).
4. See also Lemke (2011, esp. 35–52) for a reliable summary of Foucault on biopower.
5. See Huffer (2010) for a provocative treatment of the affective dimension (thinking-feeling) of Foucault's archival work, but also for a discussion of Butler's somewhat twisted reading of the body in *Discipline and Punish* as a stand-in for the psyche. Huffer suggests that in this reading Foucault *becomes* Butler—a relation of identification that, as I have noted, Eve Sedgwick associates with the "beside."
6. Consider Foucault's own remarks about racist normalization and political death in what follows: "When I say 'killing,' I obviously do not mean simply murder as such, but also every form of indirect murder: the fact of exposing someone to death, increasing the risk of death for some people, or quite simply, political death, expulsion, rejection and so on" (2003, 256).
7. Quoted in Huffer (2010, 23).
8. See Butler's *Bodies That Matter*, in which she rejects Lacan's idea of an invariant law of the father and clarifies the sense of "outside" that she finds acceptable (1993, 206–7). Unlike Lacan, Butler adopts a notion of foreclosure that does not determine once and for all the boundaries of social intelligibility. Consider the following remarks: "Which signifiers qualify to unravel the subject and to threaten psychosis remains unfixed in this analysis, suggesting that what constitutes the domain of what the subject can never speak or know and still remain a subject remains variable, that is, remains a domain variably structured by contingent relations of power" (1993, 204). Here she appeals to Foucault's idea of the historical a priori to revise Lacan.
9. For an intriguing analysis of the limits of current radical democratic theory, see McAfee (2008). McAfee calls for political rituals that address the political unconscious, the wounded attachments that perpetuate cycles of violence between groups.
10. "After all, if my survivability depends on a relation to others, to a "you" or a set of "yous" without whom I cannot exist, then my existence is not mine alone, but is to be found outside myself, in this set of relations that precede and exceed the boundaries of who I am" (2009, 44).
11. Foucault does in fact value autonomy defined more narrowly as a reflexive capacity that enables critical distance and self-transformation. Of course, any way of being can become dangerously ensnared within dominating power relations, but it might also be done within communities of resistance to dominant ways of living, and thereby enable some to sustain themselves under conditions that offer limited resources for sustenance (Foucault 1994d).

References

Butler, Judith. 1990. *Gender Trouble: Feminism and the Subversion of Identity.* New York: Routledge.

———. 1993. *Bodies That Matter: On the Discursive Limits of Sex.* New York: Routledge.

———. 1997. *The Psychic Life of Power: Theories in Subjection.* Stanford, CA: Stanford University Press.

———. 2004a. *Precarious Life: The Powers of Mourning and Violence.* London: Verso.

———. 2004b. *Undoing Gender.* New York: Routledge.

———. 2004c. "What Is Critique? An Essay on Foucault's Virtue." In *The Judith Butler Reader*, edited by Sara Salih with Judith Butler, 302–22. Malden, MA: Blackwell.

———. 2009. *Frames of War: When Is Life Grievable?* London: Verso.

Foucault, Michel. 1980. *The History of Sexuality.* Vol. 1, *An Introduction*. Translated by Robert Hurley. New York: Vintage Books.

———. 1994a. "Friendship as a Way of Life." In *Essential Works of Foucault, 1954–1984*, vol. 1, edited by Paul Rabinow. New York: New Press.

———. 1994b. "Questions of Method." In *The Essential Foucault: Selections from Essential Works of Foucault, 1954–1984*, edited by Paul Rabinow and Nikolas Rose, 246–58. New York: New Press.

———. 1994c. "What Is Critique?" In *The Essential Foucault: Selections from Essential Works of Foucault, 1954–1984*, edited by Paul Rabinow and Nikolas Rose, 263–78. New York: New Press.

———. 1994d. "What Is Enlightenment?" In *Ethics: Essential Works of Foucault, 1954–1984*, vol. 1, edited by Paul Rabinow, translated by Robert Hurley, 303–19. New York: New Press.

———. 2000. "Confronting Governments: Human Rights." In *Essential Works of Foucault 1954–1984*, vol. 3, edited by James D. Faubion. New York: New Press.

———. 2003. *"Society Must Be Defended": Lectures at the College de France 1975–1976.* Translated by David Macey. New York: Palgrave Macmillan.

———. 2008. *The Birth of Biopolitics: Lectures at the Collège de France 1978–1979*. Translated by Graham Burchell. New York: Palgrave Macmillan.

Heckert, Jamie. 2010. "On Anarchism: An Interview with Judith Butler." http://www.academia.edu/886666/On_Anarchism_An_Interview_with_Judith_Butler.

Huffer, Lynne. 2010. *Mad for Foucault: Rethinking the Foundations of Queer Theory.* New York: Columbia University Press.

Kirby, Vicky. 2006. *Judith Butler: Live Theory.* London: Bloomsbury Academic Press.

Lemke, Thomas. 2011. *Bio-Politics: An Advanced Introduction*. New York: New York University Press.

Lloyd, Moya. 2007. *Judith Butler*. Cambridge: Polity Press.

McAfee, Noelle. *Democracy and the Political Unconscious*. New York: Columbia University Press, 2008.

Repo, Jemima. "*Herculine Barbin* and the Omission of Biopolitics in Judith Butler's Gender Genealogy." *Feminist Theory* 15, no. 1 (2014): 72–87.

Rubin, Gayle. 1984. "Thinking Sex: Notes for a Radical Theory of the Politics of Sexuality." In *Pleasure and Danger*, edited by Carol Vance. New York: Routledge.

Sedgwick, Eve Kosofsky. 2003. *Touching Feeling: Affect, Pedagogy, Performance.* Durham, NC: Duke University Press.

PART FOUR

Neoliberalism and Governmentality: The Question of the Population

TWELVE

Who's Being Disciplined Now? Operations of Power in a Neoliberal World

TODD MAY AND LADELLE McWHORTER

Things seem different. To those who have spent many years studying Foucault, who have immersed ourselves in his treatment of modern forms of power, and who have seen discipline everywhere around us, the world appears to have changed as we have gazed at it through the lenses he offered us. Normalizing discipline is not as clearly evident as it once was. Psychology and its allied sciences, once the centerpiece of modern power/knowledge, have moved to its edges. At times our reliance on *Discipline and Punish* and the first volume of the *History of Sexuality* feel outdated. The world has changed.

We should have expected this—especially as students of Foucault. After all, Foucault shows us the contingencies of what holds us in place. A central lesson of his work is that there is no overarching trajectory or underlying logic to history, that social arrangements are held in place by relations and forces that can, and do, change. The task, then, for those who take his view seriously, is to understand our world as it has arrived at the place it *now* occupies. Not the place it occupied when Foucault's major genealogical works were published, but *now*. Foucault published the first volume of his *History of Sexuality* more than thirty years ago. Things not only *seem* different now. They *are* different. And we must think through this difference from within the confines of our present—we must take up Foucault's task, not to repeat it, but to extend it. Even for this task, Foucault offers us tools.

Is our time shaped by forces entirely different from those that shaped the 1970s? Yes and no. Some thinkers do see a radical break between the contemporary world and the world of thirty-five or forty years ago, Jean Baudrillard being, perhaps, the best known. His thesis, that we live in an age of hyperreality rather than reality, that our lives are more virtual than actual, more mediated than immediate, seemed radical when it appeared.

"Disneyland is presented as imaginary in order to make us believe that the rest is real, when in fact all of Los Angeles and the America surrounding it are no longer real, but of the order of the hyperreal and of simulation" (Baudrillard 1983, 25). Subsequently, the reality of events like the war in Iraq and the Great Recession have consigned his thesis to the dustbin of history, right next to Francis Fukayama's claim that history has reached its end.

More measured is Gilles Deleuze's hypothesis in "Postscript on Control Societies." Reflecting on Foucault's disciplinary society, he writes, "We're in the midst of a general breakdown of all sites of confinement—prisons, hospitals, factories, schools, the family. . . . *Control societies* are taking over from disciplinary societies" (Deleuze [1990] 1995, 178). Control societies are digital rather than analogical, modulating rather than molding, continuous rather than discrete, metaproductive rather than productive. Whereas disciplinary societies move individuals from position to position (student then worker) and from place to place (school then factory), control societies subject people to continuous oversight that seeks not so much to produce particular types of beings as to channel people into generalized market participation.

Details of Deleuze's proposal need not concern us here. What is of moment is his thesis that one society is being replaced by another. We were once disciplined beings; we are now becoming controlled beings. Does Deleuze believe (or did he believe in 1990) that discipline would disappear? His prose suggests some seismic change is in store. Foucault is more cautious. Although at times, particularly in his earlier work, he describes deep changes with relatively clean breaks, his genealogical writings describe multiple fractures and overlaps; the new does not replace but rather develops alongside the old. In *Discipline and Punish* he writes, "The power of the Norm appears throughout the disciplines. Is this the new law of modern society? Let us say rather that, since the eighteenth century, it has joined other powers—the Law, the Word (*Parole*) and the Text, Tradition—imposing new delimitations upon them" (Foucault 1979, 184).

The task, then, is to understand our situation, not divorced from the past, but not entirely continuous with it either. Discipline still operates, but where and how? And what is new? And where and how does what is new operate? We will call what is new, the new type of power that has arisen over the past forty years or so, neoliberal power—which must not be seen as a tidal wave that has swept the old arrangements of power aside but as a new set of practices and forces that have enmeshed themselves in particular ways and in particular places in our world. Understanding the present requires understanding how we are subject/ed to/by various forms of power in their interrelations and tensions.

Sovereign power, the old binary power of rule, defines two spaces of activity: the permitted and the forbidden. Whatever is permitted is left alone. A step into the forbidden, however, brings merciless punishment, reestablishing the power of the sovereign and demonstrating that power's grandeur. Disciplinary power takes a less spectacular but more certain and more pervasive approach to criminality. "In fact, the shift from a criminality of blood to a criminality of fraud forms part of a whole complex of mechanisms, embracing the development of production, the increase of wealth . . . stricter methods of surveillance, a tighter partitioning of the population . . . the shift in illegal practices is correlative with an extension and a refinement of punitive practices" (Foucault 1979, 77).

Disciplinary power is focused not so much on prohibition and the extraction of signs and goods (fealty and taxes, e.g.) as it is on individual bodies. As Foucault tells the story in *Discipline and Punish*, disciplinary power first arises in monastic settings, where it remains for the most part until the eighteenth century, when population increases in Europe and new technologies (such as rifles and factories) demand changes in the organization of secular institutions and practices. With larger armies made up not primarily of the second sons of nobles but of peasant boys not brought up in the use of weaponry, new modes of training recruits became necessary. At the same time, with new and much faster and therefore more dangerous weapons, recruits had to be disciplined to obey their superiors without hesitation—rather than to turn those weapons on them! With more pupils in classrooms, teachers and schoolmasters had to develop new ways of maintaining disciplinary control while instilling literacy and other skills that were becoming increasingly important in an emerging capitalist economy. Each body had to be trained to be both competent and diligent *and*, at the same time, docile within its respective institutional regime, which meant that each body had to be attended in its specificity and monitored constantly in its progress. Knowledge and measurement became increasingly important in these relations of power. The body of the trainee had to be measured as it developed and assessed against a growing body of information about typical developmental rates—statistical norms, in other words. Disciplinarians were not working to prevent or punish infractions; their job was to produce new capacities according to developmental norms.

In volume 1 of *The History of Sexuality*, Foucault traces the rise of sexuality as a deployment within regimes of normalization. Each body is monitored, its appearance and behavior molded as it matures to conform to gender and sexual norms. But this is only one side of sexuality's functioning. As biopower—the power to make live or to cultivate life, as opposed to the sovereign's right to kill—emerges through the nineteenth century, sexuality functions not only as a tool of normalization (a set of techniques for producing sexual subjects) but also as a tool of population management. Population as an object of study and a target of strategic interventions comes into view as a correlate of biopower in the late eighteenth and nineteenth centuries. (Foucault tells this story in the second and third lectures of his 1978 lectures, *Security, Territory, Population*.) The population is not simply a multiplicity of individuals; it is "a political subject," an emerging "collective subject absolutely foreign to the juridical and political thought of earlier centuries" (Foucault 2007, 42). Statistical techniques are crucial in regard to population management, just as they are in normalizing discipline, but the point is not to measure each individual against a norm but to discern the norms of a population's current existence—its birth rates, morbidity and mortality rates, crime rates, rates of consumption, and so on—and to adjust those norms through what Foucault calls "apparatuses of security" (Foucault 2007, 34).

Apparatuses of security preexist biopolitical population management. They arise as ways of managing the risks of circulation of goods and money in growing market towns. Towns' and then larger cities' well-being depended on constant movement of people and things. But this constant movement increased the risk of epidemic, crime, and economic instability. Sovereigns and bureaucrats needed to understand movement and its risks and guard against them without interfering too much with circulation itself. Thus, risks could not be eliminated. Rather, officials had to determine acceptable risk, to calculate trade-offs, and to do this they began to monitor what we would now call population trends. Within this kind of power regime, what mattered was not that each individual body conform to norms or obey rules, but that conformity reach a level of optimality consistent with free circulation within the given population. Biopolitical population management is a refinement of early security apparatuses that through the nineteenth and much of the twentieth centuries articulated with disciplinary normalization of bodies through deployments such as sexuality.

In many respects, normalizing discipline and population management complemented and reinforced one another. But there were tensions between them as well. Foucault describes discipline as "essentially centripetal" (Foucault 2007, 44); it isolates and encloses. "In contrast," he writes, "apparatuses of security . . . have the constant tendency to expand; they are centrifugal" (Foucault 2007, 45). Indeed, they must expand; circulation without expansion is not sustainable. Innovation and integration of new elements into the circulatory system is necessary for the system's continued functioning. But these two types of power, discipline and security, are not only different; they sometimes actually work against each other. Foucault continues: "By definition, discipline regulates everything. Discipline allows nothing to escape. Not only does it not allow things to run their course, its principle is that things, the smallest things, must not be abandoned to themselves." Everything must be regulated and thus predictable. Not so with apparatuses of security. "The apparatus of security, by contrast, . . . 'lets things happen'" (Foucault 2007, 45). Apparatuses of security must make peace with risk in favor of circulation and momentum, which means they must allow a certain amount of uncertainty, error, and disorganization.

Capitalism requires disciplined bodies. Workers must have skills. Efficiency of movement and concentration of energy are essential to capitalist production. Work occurs on a schedule. Machines have needs that must be met. But capitalism also requires circulation of money and goods, and in some respects—especially as production becomes more efficient—complex, wide, and rapid circuits of exchange become more important than disciplinary control. It is here where we begin to see a shift away from normalizing discipline and the normalized subjects it produces toward apparatuses of security and subjects of exchange that they produce. This prefigures the rise of neoliberalism.

Neoliberalism has its conditions of possibility in both the security framework of the eighteenth century and the classical liberal thought of philosophers like Adam Smith. Regarding the former, Foucault contrasts the normalization of discipline (which in the *Security* lectures he calls *normation*) with the concept of the norm as it is used in practices of security (which he then calls *normalization*). Normation sets a norm and then seeks to develop practices that induce conformity to that norm. The practices of visibility, hierarchical observation, and examination discussed in *Discipline and Punish* are classic examples of normation. By contrast, normalization requires one to understand the normal curve of a given phenomenon (behavior, disease, economic activity), and then to arrange things at the level of the population rather than the individual to maximize efficiency regarding the promotion or diminishing of the phenomenon. "In the disciplines one started from the norm, and it was in relation to the training carried out with reference to the norm that the normal could be distinguished from the abnormal. Here, instead, we have a plotting of the normal and the abnormal, different curves of normality, and the operation of normalization consists in establishing an interplay between these different distributions of normality and [in] acting to bring the most unfavorable in line with the most favorable" (Foucault 2007, 63).

In earlier practices of security, it was assumed that arranging intercessions into the population would converge with the natural tendency of people's interaction to conduce to the most efficient distribution of resources. This innovation was introduced by the Physiocrats, in contrast to the more strictly interventionist policies of the Mercantilists. And it appears in the classical liberal doctrine of Adam Smith in the form of the "invisible hand." The idea was that allowing commerce between people, rather than tending toward an outflow of goods and an impoverishment of a nation's population, would allow for imbalances to be corrected through trade and international interaction generally.

Today we tend to assume that this understanding of the independence of markets is carried over from classical liberalism into neoliberalism. This seems to be the idea behind the Reaganesque mantra that the government is the problem rather than the solution. Allow the markets to function without interference, and everything will work out for the best. Given this mantra, one might wonder why government spending increased so radically during the Reagan administration and the second Bush administration. The mistake is in thinking that neoliberalism follows classical liberalism in holdingthat human interaction will naturally tend toward the greatest efficiency. In fact, in the neoliberal view, according to Foucault, there is no such natural tendency. The government must secure the proper operation of markets, which on their own do not tend toward the efficiency assumed by classical liberalism. The question facing neoliberalism, then, is one of how to establish, maintain, and control a government that is at once essential to markets and at the same time menacing to them.

This problem was addressed by the predecessors of neoliberalism, the ordoliberals of post–World War II Germany and Austria. Writing in the wake of Nazism, they saw capitalist markets as a way of counterbalancing the power of the state, which they saw as the prime cause of the rise of Nazism. At the same time, they jettisoned the belief in natural balance characteristic of classical liberalism. As a result, they required the intervention of the state of which they were otherwise wary. The solution to this problem was to have the state ensure the proper functioning of markets by intervening not into the markets themselves, but into the framework within which markets operate. "What will good interventions act on? Well, on the framework. That is to say, first of all, on the population . . . enabling transfers, migration, and so on. We will also have to intervene at the level of techniques, by making implements available. . . . Third we will also modify the legal framework of farms. . . . Fourth, as far as possible we will modify the allocation of the soil and the extent, natural, and exploitation of the soil available. Finally, if necessary, we will have to be able to intervene on the climate" (Foucault 2008, 140). One intervenes on the market by promoting certain kinds of policies concerning agriculture, population growth and distribution, monopoly regulation, taxes, welfare, and trade. These interventions have as their goal to allow for the support of the competitive mechanisms central to capitalism, since capitalism is not likely to develop them on its own.

Neoliberalism takes up the ordoliberal analysis and expands it.[1] Where ordoliberalism is a theory of markets, American neoliberalism is a theory of living. "American neoliberalism is not—as it is in France at present, or as it was in Germany immediately after the war—just an economic and political choice formed and formulated by those who govern and within the governmental milieu. Liberalism in America is a whole way of being and thinking" (Foucault 2008, 218). For American neoliberalism, the entrepreneurship characteristic of those who enjoy market success is an entire way of approaching all of one's projects. One enters markets to satisfy one's preferences—so likewise one enters schools or training programs, social clubs and religious organizations, and personal relationships to satisfy one's preferences. Utility maximization is a comprehensive way of life. This means that government policy should be directed not only toward developing the framework for the competitive character of markets, but should recognize and where necessary support that type of framework for all aspects of life, from schooling to childcare to penal practices. In fact, in the *Biopolitics* lectures, Foucault explicitly contrasts the neoliberal Gary Becker's proposals for penal reform to those of the disciplinary apparatuses he discussed in *Discipline and Punish*.[2] For Becker, the mechanism of penal reform would not be that of disciplining individual bodies, but rather of creating the kinds of incentives at the level of the population that would make the costs of crime seem to outweigh the costs of lawful behavior. Crime would not disappear but rather would diminish to the most efficient rate, given the costs (including opportunity costs) of deterrence.

Becker's analysis of crime, along with similar neoliberal analyses of other sorts of human behavior and motivation, takes utility maximization as a fact about individuals. People seek to maximize their utility, and each individual is an authority on what constitutes his or her own utility. Not every action undertaken actually succeeds in maximizing one's utility (within a given range of options, one sometimes selects a nonoptimal one), but every action aims to do so, given the information available to the actor at the time. People are better or worse at utility maximization depending on how much they know about the available options.

Among those who may be attempting to maximize their utility in a neoliberal world are the members of the enormous workforce already trained to deliver disciplinary normalization, committed to its goals, and steeped in its assumptions, and there remain a number of institutions around the world designed for disciplinary normalization of various sorts. These include the militaries of most nations (where funding has typically been cut far less than in social services under neoliberal governments), many (if perhaps increasingly fragmented) systems of education, and therapeutic institutions. As long as these networks are not perceived as a drag on the circulation of money and goods, neoliberal apparatuses of security will not challenge them, and they will continue to operate. Accordingly, if they are seen as resource sinks (either because they require inputs beyond their outputs or because they maintain a stock of resources that circulates only internally and remains unavailable to markets), neoliberalism likely will challenge them and attempt to absorb and restructure or simply dismantle them. We might see what has been happening to health care, the medical profession, and locally owned hospitals at the hands of both cost containment–minded US legislators and the global insurance industry over the last few decades as a protracted struggle between techniques of disciplinary normalization and techniques of security in standards of treatment and therapy. At any rate, disciplinary normalization will not disappear as long as there are institutions and well-organized groups whose interests require its continuation. For some time to come, neoliberalism itself may foster its own resistance.

Furthermore, some critics contend that neoliberalism's success as a form of governmentality depends on an established pool of "social capital," which it does not itself produce. This "social capital" includes basic values such as honesty sufficient for minimal trust in one's contractual partners, confidence in the system of contract enforcement, and faith in the market itself.[3] Those values must be produced and cultivated somewhere, along with the ability to make utility-maximizing decisions. As any person who has ever raised or worked with children and young adults (or has clear memories of their own youth) knows, utility maximization involves not only knowledge but also skills. Obviously there are basic skills (such as continence and calculation) that virtually all children acquire early. But there are also much more complex skills of temporal projection and organization that many people do not acquire without a great deal of training, effort, and institutional support. This is why universities offer workshops in time management, for example; time management is crucial for both academic and market success, but it involves a set of skills that are not naturally or easily mastered. Time is one of the few "resources" that all human beings have, yet none of us is born or develops naturally knowing how best to "allocate" it to maximize our utility. If we want to use this language of "social capital," then, we might add to those basic values the skills that are necessary for operating as efficient workers (not only the training that markets can provide, but the basic self-management skills that markets do not provide), that are necessary for translating desire into consumer behavior and for planning for a future in market contexts and terms. The pervasive existence of markets and the possibility of personal failure in them provide incentives for developing some of those skills, but even recognizing the need for high-level skills presupposes that one has already developed some basic skills. In other words, neoliberal government requires subjects who are governable in neoliberal terms, and it is possible that normalizing disciplines are currently among the best, if not the only, ways to produce them.

Normalizing disciplines will survive only where they do not impede circulation of money and goods, then, and they will be supported only where they actually contribute to the production of self-entrepreneurs; where they slow down circulation, they will eventually be superseded. For now, however, it appears that distinctions between the normal and abnormal may still have multiple functions in a neoliberal society. Undoubtedly at present they allow for more or less "safe" disposition of individuals who are extremely unsuccessful in market terms—the severely disabled, the mentally ill, the chronically impoverished. They do a great deal of work in truly educational settings (as opposed to discredited school systems that merely keep children confined for several hours a day), where they are necessary to supplement parental discipline and training for the market. They are also extremely useful in many military settings, combined with older nonnormalizing techniques of military discipline, and it is difficult to see how modern militaries could do without them—self-entrepreneurship being antithetical to unit loyalty and cohesion. Nonetheless, we no longer live in Foucault's disciplinary society. The distinction between the normal and the abnormal is virtually meaningless in the market, where difference can be transformed into fashion statements and expressed through individual consumption choice. In fact, markets thrive on difference, especially in terms of newness and frequent shifts in fashions and tastes, and must generate them if they do not spontaneously occur. The bottom line is that deviance sells. And so it becomes, in the market at least, just a name for a coherent set of consumer choices, a so-called niche.

Despite its long history of disciplinary reinforcement and success, the university is increasingly pervaded by neoliberal values, ideas, and practices. The university's entrepreneurial evolution has been extensively documented.[4] The loss of public financial support for universities has required them to seek funding elsewhere as well as to look for places to cut educational services. The latter has resulted in a particularly harsh squeeze on the humanities in particular; language departments at several state universities have been eliminated, and in 2010 the president of the University of Nevada Las Vegas proposed the elimination of the university's philosophy department. To alleviate financial stress, some universities have secured corporate funding, which they tend to put toward applied sciences, a practice that effectively privatizes research results that under previous funding models would have been public. In addition, the language of investment, consumption, and entrepreneurship has increasingly become the vocabulary of university administrators, as well as students themselves.

Increasingly, students see college as an investment rather than a place to take stock of oneself, become a citizen, or reflect on one's values. For instance, in 1998, the *New York Times* reported that "in the [annual UCLA] survey taken at the start of the fall semester, 74.9 percent of freshmen chose being well off as an essential goal and 40.8 percent chose developing a philosophy. In 1968, the numbers were reversed, with 40.8 percent selecting financial security and 82.5 percent citing the importance of developing a philosophy" (Bronner 1998). If the K–12 years remain within the ambit of a disciplinary apparatus necessary for basic disciplinary training, the post–high school years have been all but captured by the neoliberal paradigm.

Disciplinary training in early life is not the only place one can still see discipline at work, however, if one takes the trouble to look at the periphery rather than the center of neoliberal societies. At their centers, neoliberal societies are not only financialized; they are also postindustrial, for financial investment does not require investment in productive activities. Entrepreneurship is a matter of getting return, and the source of that return does not matter. We have seen this particularly in the events leading up to the financial crisis of 2008. Derivatives of all kinds, many with little or no relation to any productive activity, became a prime source of investment and income. But the trend is of longer standing, tracing back at least thirty years, when neoliberalism began its ascent. John Cassidy, writing in *The New Yorker*, delineates the economic changes regarding finance capital: "Since 1980, according to the Bureau of Labor Statistics, the number of people employed in finance, broadly defined, has shot up from roughly five million to more than seven and a half million. During the same period, the profitability of the financial sector has increased greatly relative to other industries . . . during a period in which American companies have created iPhones, Home Depot, and Lipitor, the best place to work has been in an industry that doesn't design, build, or sell a single tangible thing" (Cassidy 2010, 51). But, if production is declining in areas now dedicated to neoliberal policies and finance capital, it must be intensifying somewhere else. And it is wherever that "somewhere else" is that that one would expect to find discipline at work.

Looking at the periphery, one is not disappointed in this expectation. In Third World countries that produce many of the goods consumed in neoliberal society, discipline is alive and well. Many people are familiar with the case of Nike and its disciplining of female workers in Malaysia.[5] As Naomi Klein documented in her classic *No Logo*, factory discipline is widespread, particularly in the large Third World export processing zones utilized by transnational corporations. Imagine an area of outlet stores of the kind we have all over this country (near where one of the authors live is a town called, no kidding, Commerce, Georgia, which is built around an area of outlet stores). Now imagine it as the place where the goods are produced rather than sold. *Viola*, an export processing zone. As Klein describes them, "The vast majority of workers are women, always young, always working for contractors or subcontractors from Korea, Taiwan, or Hong Kong. The contractors are usually filling orders for companies based in the U.S., Britain, Japan, Germany, or Canada. The management is military-style, the supervisors often abusive, the wages below subsistence and the work low-skill and tedious" (Klein 1999, 205).

Discipline is alive and well in these zones, as well as in the leftover factory areas of First World countries, because it is necessary for production, and without production, there would be nothing to consume, and indeed no material base for neoliberal financial practices. However, discipline has only one foot inside neoliberalism. Since one can invest in nonproductive activities, production and its aligned practices of discipline are a necessary but in some sense external part of the neoliberal apparatus.

Not all areas outside the neoliberal apparatus are productive, however. Both inside countries like the United States and Britain, and outside of them in areas like Africa, there are segments of society or whole societies that are neither neoliberal nor productive. These segments and societies have been abandoned by societies functioning under the neoliberal paradigm, but for distinct reasons. At the risk of oversimplification, we might say that since the end of the Cold War many countries no longer strategically located have been abandoned, cut adrift to be riven by civil strife and genocide.

Indeed, even many segments of neoliberal society have been similarly abandoned. Movement from the welfare state to the neoliberal state brought radical constriction of social services—and its concomitant disciplinary regimes. We see much less disciplinary normalization, for instance, in prisons and inner-city schools, which under neoliberal governmental and fiscal policies have been turned into holding tanks where development is given only minimal attention and docility is not so much ingrained as simply physically enforced. The reason: These arenas are neither ripe for investment (the privatization of prisons has, in many areas, not turned out to be a lucrative endeavor) nor sources of production.

Power still operates in these abandoned regions, but they resemble the premodern operations of power described by Foucault. In *Discipline and Punish,* the reader will recall, power was enacted mainly through *supplice*: torture. The body of the law breaker was subjected to pain both as a retribution for its attack on the body of the sovereign and as a reminder to everyone else what happens in response to such attacks. In societies at the margins of the neoliberal world one can see the operation of such power, although it is often in the hands of local authorities rather than centralized in a sovereign. A vivid, if fictional example of this operation is displayed in the television series *The Wire,* in which the Baltimore drug trade, with rival gangs, enforces its power through torture and murder, often using dead bodies to communicate authority. A nonfictional example is found in Ishmael Beah's memoir of his time as a boy soldier during the 1991–2002 civil war in Sierra Leone. As in many marginalized places, conflicting sides in the war used young boys—Beah was thirteen—as soldiers. These boys were often recruited at gunpoint and then trained to kill and torture mercilessly. Entire villages became subject to the terror of armies of children who had become inured to their own violence. Beah's harrowing descriptions of his own behavior and those around him illustrate the brutality that sought neither profit nor discipline, but only to subdue those who were enemies or potential enemies. In one vignette, Beah writes, "'Where did you get all this ammunition from?' the corporal asked one of the prisoners, a man with an almost dreadlocked beard. He spat at the corporal's face, and the corporal immediately shot him in the head at close range. He fell onto the ground and blood slowly leaked out of his head. We cheered in admiration of the corporal's fierceness and saluted him as he walked by" (Beah 2007, 123). This premodern exercise of power, which, as Foucault points out, has not left our world, often holds a prominent place at neoliberalism's margins.

To conclude, we note an important relationship between the operations of different kinds of power networks and the technologies and material resources available where those kinds of power networks are manifest. In a world without coal and cargo ships, disciplinary normalization might never have developed. Factories and international trade created conditions under which monastic techniques could secularize and swarm, to use Foucault's term. Foucault also alerts us to the invention of the rifle as a significant event in the development of military discipline (Foucault 1979, 163) and to innovations in educational architecture in conjunction with the development of surveillance techniques in schools (Foucault 1979, 143). Bureaucratic institutions developed in tandem with disciplinary normalization because of the increasing need for hierarchical surveillance, information gathering, and data analysis and storage. Without the material resources to produce those storage systems—including the buildings, offices, paper, filing systems, and techniques for cross-referencing and mathematizing data—normalization just would not have been possible. Norms are produced and reinforced in material practices that require a certain set of material things.

Likewise, in a world without electricity and oil, neoliberal networks of power would not be sustainable. Financialization of the economy depends on computers and communication. The rapid circulation of money in a neoliberal world is actually the rapid circulation of digitized and encoded information. In addition to the massive amounts of electricity necessary to power those computers and communications systems, these technologies require rare earths—minerals that are, indeed, rare—and synthetic chemicals that simply did not exist before the twentieth century, many of which may be toxic and most of which require petroleum for some aspect of their production. In other words, the physical infrastructure even for an economy that seems to work in the ether of cyberspace is massive and its inputs are largely nonrenewable. We believe it is crucial to keep these facts in mind in any analysis of power/knowledge networks, and especially in any analysis of neoliberal networks of power. For, as these resources dwindle, neoliberalism will go into crisis and fragment, and we may well see a resurgence of some modified forms of sovereign power or, indeed, something altogether new.

We have tried to offer a glimpse here, however brief, of the operations of different types of power—neoliberal, disciplinary, premodern—in a world where neoliberalism currently holds sway. As we have argued, the power arrangements in such a world are complex. We cannot layer each power over the previous one, but must look instead at the specificities of different areas and their roles in the neoliberal order if we are to understand the concrete operations of power in play.

Notes

1. For a detailed account of the ways in which the Chicago School took up and reworked ordoliberalism, see Van Horn (2009).
2. For a discussion of the extent to which Foucault may have seen his own analysis challenged by Becker's, see Dilts (2008).
3. For a version of this criticism and the use of "social capital" language, see Plant (2010, 168–69).
4. For an early, but still relevant example of this literature, see Reading (1996).
5. See, e.g., http://www.oxfam.org.au/explore/workers-rights/nike/forced-labour-by-nike-supplier and http://www.oxfam.org.au/explore/workers-rights/nike/nike-worker-speaks-out.

References

Baudrillard, Jean. 1983. *Simulations*. Translated by Paul Foss, Paul Patton, and Philip Beitchman. New York: Semiotext(e).

Beah, Ishmael. 2007. *A Long Way Home: Memoirs of a Boy Soldier*. New York: Farrar, Straus, and Giroux.

Bronner, Ethan. 1998. "College Students Aiming for High Marks in Income." *New York Times*, January 12. http://query.nytimes.com/gst/fullpage.html?res=9901E3DE1539F931A25752C0A96E958260&scp=2&sq=college%20students%20%22meaningful%20philosophy%20of%20life%22&st=cse.

Cassidy, John. 2010. "What Good Is Wall Street?" *New Yorker*, November 29, 51.

Deleuze, Gilles. (1990) 1995. "Postscript on Control Societies." In *Negotiations 1972–1990*, translated by Martin Joughin, 177–82. New York: Columbia University Press.

Dilts, Andrew. 2008. "Michel Foucault Meets Gary Becker: Criminality Beyond Discipline and Punish." *Carceral*. http://www.thecarceral.org/cn4_dilts.pdf.

Foucault, Michel. 1979. *Discipline and Punish: The Birth of the Prison*. New York: Vintage.

———. 2007. *Security, Territory, Population: Lectures at the Collège de France, 1977–1978*. Translated by Michel Senellart. New York: Palgrave Macmillan.

———. 2008. *The Birth of Biopolitics: Lectures at the Collège de France, 1978–1979*. Translated by Graham Burchell. New York: Palgrave Macmillan.

Klein, Naomi. 1999. *No Logo: Taking Aim at the Brand Bullies*. New York: Picador.

Plant, Raymond. 2010. *The Neo-liberal State*. Oxford: Oxford University Press.

Reading, Bill. 1996. *The University in Ruins*. Cambridge, MA: Harvard University Press.

Van Horn, Rob. 2009. "Reinventing Monopoly and the Role of Corporations." In *The Road from Mont Pèlerin: The Making of the Neoliberal Thought Collective*, edited by Philip Mirowski and Dieter Plehwe, 204–37. Cambridge, MA: Harvard University Press.

THIRTEEN

Is There a Biopolitical Subject? Foucault and the Birth of Biopolitics

FRÉDÉRIC GROS
TRANSLATED BY SAMANTHA BANKSTON

This chapter offers some reflections on the course that Foucault gave at the Collège de France in 1979, entitled *The Birth of Biopolitics*.[1]

The course was very unique for several reasons. First, Foucault experimented with a certain number of theoretical hypotheses that would never be taken up elsewhere, not at conferences, in books, or in interviews. Thus, we find a series of studies on liberalism in this course that is completely extraordinary and rare. Accordingly, it is difficult to know whether Foucault would have kept his analyses in their original state, and what modifications they would have undergone if he had chosen to pick them up again and rework them with an eye toward eventual publication. The ideas constructed in the course should be evaluated with extreme caution since there is nothing conclusive about them. Second, it was in this course that Foucault analyzed immediately contemporary facts for the first time. This is the only course that Foucault gave at the Collège de France where he made direct reference to current political events. Third, this course was enigmatic: while its title is *The Birth of Biopolitics*, Foucault actually gave a course on liberalism and almost never spoke about biopolitics. Faced with this enigma, we can react in one of two ways: either we can consider that this is one of the hazards of research and that Foucault abandoned his initial project to exploit new theoretical paths that he was in the midst of discovering; or we can consider that by analyzing liberalism, Foucault offered a new dimension of biopolitics: a liberal biopolitics. The second critical approach is the one that I will privilege. Fourth, the 1979 course closes in on its own enigma. What I mean by that is that in 1980 Foucault made a fresh start with a course entitled *The Government of the Living*[2] (whose title might suggest a deepening of the notion of biopolitics), which actually bore on the practices of confession in ancient Christianity and the way a new Western subject is constructed through such practices. Fifth, *The Birth of Biopolitics* causes room for pause in relation to Foucault's political identity. In fact, these studies of liberalism do not include any condemnation of the principle itself, and certain commentators believe that Foucault might even have expressed a kind of sympathy for liberal ideology. Thus, this course from 1979 is truly unique and raises many questions. However, for me, painting a complete picture is not what is at stake; instead, I am trying to understand, through Foucault, how liberalism can constitute a biopolitics.

The First Meaning of Biopolitics in Foucault (1973–76)

It seems to me that understanding this notion of a liberal biopolitics first requires slightly fixing the meaning that Foucault might have given to the concept of biopolitics prior to 1979. To construct this first sense of biopolitics in Foucault, one must refer to works such as *Discipline and Punish*[3] and *The Will to Knowledge*,[4] but also the courses from the Collège de France, like the one given in 1976 entitled *"Society Must Be Defended."*[5]

During this period, Foucault distinguishes between two forms of biopower: the discipline of bodies and the regulation of populations. These two forms complement each other on one hand, and on the other hand they both oppose what Foucault calls "sovereign power."

As such, it is necessary to begin by understanding what he calls "sovereign power." For Foucault, "sovereign power" characterizes the old monarchies, but the power of the familial father can also be thought in this sense: it is the power of indisputable authority. Their main means are violence and the law. Of course, these two terms can seem antithetical for a republican tradition, but what Foucault retains from the term "law" is not the expression of the general will, but the notion of an imposed prohibition. Sovereign power is violent, but not yet in the sense of a destructive outburst. This power is violent because it proceeds by requisition (the king's officials requisition lands and property, but also time for fatigue duty and men for battle), by deduction (the king's stewards take a portion of crops and resources in the form of taxes), and finally by physical branding, as in the cases of torture where the body is quartered and burned. We can talk about violence in these examples because power is either a matter of authoritatively taking something from someone by breaking all forms of resistance, or it is a question of showing itself in the explosive brutality of spectacle. In each case, this power expresses itself in a discontinuous way: it bursts into individuals' lives to brutally take something from them or violently forbid them from engaging in certain acts. But sovereign power is also the power that states the law. It states the law in the sense that it delivers prohibitions, sketches dividing lines, and delimits what is permitted and forbidden. The law in question is an absolute, indisputable, authoritarian decree. To characterize the power of sovereignty, Foucault uses the following formula: it is a power that "makes die and lets live." It is a power that is not interested at all in people's lives in a positive sense. Of course, it is interested in their wealth and their property in order to withdraw a portion from them, but that is a relationship to things and not to life. And when this power is interested in the lives of people, it is in an immediately inverted form of the right of death over whomever. Here, it can easily be seen that in every case, this "sovereignty" that Foucault speaks of regarding "sovereign power," is very distant from the concept of sovereignty developed in modern political thought.

For Foucault, a new form of power develops after the age of reason, a form of power that is irreducible to sovereign power. This is disciplinary power. Like sovereign power, disciplinary power is also, to a certain extent, a power of extraction. But whereas sovereign power takes things, takes possession of material wealth, disciplinary power extracts use. It extracts use from individuals' living bodies, and that is how it affirms its biopolitical dimension.

Foucault's analyses of disciplinary power are well known; they are developed the most in *Discipline and Punish*. Disciplinary power is not concerned with taking things but in obtaining a certain behavior of the living person. It is a matter of obtaining good behavior, a positive attitude, or a kind gesture on the part of the soldier, worker, and student. It is an apprenticeship in docility: the body must correctly respond to demands, it must execute the right gestures at the right moment, it must maintain a positive attitude, and it must adopt good behavior. This apprenticeship will assume permanent techniques for correction and training: a regular and progressive activity program, repetitive exercises, continuous surveillance, frequent examinations, and perpetual threats of punishment. Sovereign power functions through the law: a law that forbids certain acts and fixes authoritarian limits, but remains indifferent to everything else. Disciplinary power functions through norms: it is a matter of controlling the totality of the subject's life so as to obtain specific behavior and complete docility from him or her.

I would now like to invoke some of Foucault's lesser-known analyses, which will allow me to deepen the biopolitical dimension of disciplinary power and its relation to capitalism. In 1973, Foucault gave a course entitled *The Punitive Society*.[6] It is a very important course since Foucault puts an entire series of analyses into place, which reappear in *Discipline and Punish*. The course, like *Discipline and Punish*, is concerned with elucidating the enigma of the prison. What remains enigmatic about the prison, as Foucault explains, is that it was imposed as an obvious fact on the modern West as the primary and exclusive punitive means, against all prior expectation. What that means is that the great reformers of the Enlightenment proclaimed that the criminal must no longer be confused with the sinner. For the thinkers of the Enlightenment, this was about providing a purely immanent definition of crime or offense: a crime is a social infraction, a problem concerning public order, but not a blasphemous transgression of a divine prohibition or a rupture of a sacred taboo. The criminal is a social enemy: he or she injures the common interest, as opposed to insulting some divine majesty. Yet the reformers deduce a series of possible punishments from this new definition of crime: the punishment of infamy, forced labor, the application of "an eye for an eye," or deportation, corresponding to the kind of public threat represented by the crime. But the prison was never a question. Yet, by the nineteenth century the prison becomes the rule as the primary, if not exclusive, means of punishment.

To explain the success of the prison, Foucault theorizes four great determining factors: religious genealogy, political relevance, social resonance, and economic functionality. Religious genealogy rests on the concept of the "penitentiary." Foucault claims that the protestant communities of the eighteenth century in England had established structures of confinement very early on, which must have punished irregularities in behavior of some of its members. Before, it was a matter of locating dangerous tendencies or harmful inclinations, and later it was about relying on time spent in a cell to obtain an internal transformation. The public prison's roots are in the religious and private structures of control and reeducation. That is where its penitentiary dimension is located, if by "penitentiary" we understand a notion of confinement that constitutes the punishment of the moral deviance of a vicious nature (and not an illegal act) and that must produce an internal regeneration.

At the same time, the prison presents political advantages because it permits a redistribution of illegalities. This concept of "illegalism" is important for Foucault. This amounts to saying that all of society presupposes public order and social equilibrium. Public order depends on respect for laws and a strong police force. Social equilibrium depends on alliances and tacit agreements between certain classes to bypass laws. However, from the end of the eighteenth century to the beginning of the nineteenth century, the Western world underwent several large political transformations (working-class revolutions) and economic transformations (the industrial revolution, involving the stockpiling of merchandise and the disappearance of communal land, leading to the very important development of small farm property). This new configuration of the socioeconomic world led to a new illegalism: that of *predation* rather than *fraud* as in the past, which the bourgeoisie and the working class leveled against ancestral rights or binding laws that seemed to them like an abuse of power. This new illegalism (the theft of agricultural products or stockpiled merchandise and the defacement of machines) suddenly seemed very expensive to the bourgeoisie. The problem posed to them was whether breaking with the old tradition of working-class illegalism by creating a new kind of illegalism would end the previous forms of complicity: that of crime, which, with respect to the working class, would have to both represent an alternative model while simultaneously infiltrating the populous in order to be able to thwart political plots. The "criminal" was the dangerous and immoral villain from the unwashed masses. This is the counterexample of the honest worker who above all does not want to resemble the criminal: illegalism is thus gradually devalued and rejected by the working class because it is hijacked by a contemptible form of criminality. However, the prison enables a steady environment for crime through its own logic (a logic of recidivism and carceral complicity), which proves functionally useful for the dominant classes. For crime born from prison not only serves to discourage a form of working-class illegalism by depriving it of its former cachet, but, as we know, it can also be used to prevent political revolts, where criminals are used as spies. The *carceral* is the theme of prison functionality, as the production of crime that is useful to the dominant class.

The third determining factor in understanding the prison is what Foucault called the "coercive" in 1973, which he later called the "disciplinary." This feature was already evoked above through several principles: the principle of continuous surveillance and punishment to obtain a particular attitude; the principle of regular examinations, making it possible to establish knowledge about individual behavior; the principle of exhaustive, temporal organization that establishes what each person should be doing at every moment of the day; and so on. These principles are put to work in the form of regulations in diverse institutions: in the school, army, factory, boarding school, workshop, and hospital. At this point, the prison appears as both the concentration and outcome of these disciplinary centers—concentration because the intensity of surveillance and the constraints of organized time are more developed than ever before. But it is the outcome of these centers in the sense that recorded deviations in behavior at schools, in the army, or in the factory are always at risk of being penalized by imprisonment. Even if the prison system seemed like an obvious fact beginning in the nineteenth century, despite not being a part of the penal catalog and despite the Enlightenment thinkers not having taken it into account in their reform project, that is because our societies had widely become disciplinary and punitive societies, according to Foucault.

I would like to insist on a final formulation, which is, perhaps, the most important for the problem at hand. In his punitive society course Foucault studied the economic functionality of the appearance of disciplinary power. Why did Western societies invent this mode of power? The response from the 1973 course was quite clear: discipline serves to transform the time of life into labor power. At base, Foucault says that all of Marx's work in *Capital* consisted in the denunciation of the commodification of individuals in capitalism, the alienation of existence, and the exploitation of labor power by transforming it into forces of production. But Foucault adds that in the past it would have been necessary to describe the way in which the time of life (which includes celebration, laziness, fantasy, and the caprice of desire) could have already been transformed into labor power. At base, the disciplinary power at work in different institutions (from the school to the factory, passing through barracks and the hospital) seeks this transformation. It is a matter of reducing time dedicated to distraction, pleasure, happiness, and laziness to an absolute minimum. Capitalism presupposes a chronopolitics: the transformation of the time of life into useful and productive time.

At this point, the second major form of biopower that Foucault studied in the mid-1970s can be introduced: the regulation of populations. Thus, we move from the discipline of bodies to the regulation of populations. But whereas the discipline of bodies (from the school to the factory, from the army to the boarding school) concerns vital individual forces and seeks to extract a maximal utility from them, the regulation of populations concerns a humane globalism and takes place in a public political context led by the state. When talking about the regulation of populations, it is necessary to understand the immense politics of nationality, public hygiene, or even the fight taken up by states against epidemics. Like Foucault said: in the past, the state devoted itself to the safety of its subjects. Now, its main preoccupation is their health. The questions become: how many children per family, what ways of reorganizing towns limit the spread of illnesses, what fundamental hygienic norms to inculcate in the public? This new form of concern is not free: by ensuring the health of the population, clearly, it is also a question of strengthening the state. A strong state is one whose population has astoundingly good health. Biopolitics is thus a term that designates the historical moment when the state devotes itself to governing the biological dispositions of the human race. Foucault discovers a single formula to characterize biopower. While the sovereign power "makes die and lets live," as we saw, biopower makes live and lets die. In 1976 Foucault studies several important consequences of this inversion,[7] which I will only briefly outline here: the privatization of death, which becomes an increasingly masked and hidden phenomenon; state racism: in the context of biopolitics, the state can only ground its legitimacy in killing as a means of strengthening its own vitality.

Biopolitics and Capitalism

Given these preliminary definitions, it is possible to introduce the way in which Foucault situates the problem of biopolitics in the 1979 course *The Birth of Biopolitics*. A certain number of useful concepts can be sketched here, distinguishing among three terms: economy, capitalism, and liberalism. Quite simply, "economy" can be called the art of producing, increasing, distributing, and marketing means of subsistence, resources, and material wealth, as well as the general knowledge that corresponds to these techniques. It relates to the most general form of the concept, without a doubt, and in this sense the economy has always existed: we can study the "economy" of the oldest populations, in other words, their material level.

It seems to me that by "capitalism" something other than simply the science and art of wealth is meant. Capitalism is a determinate and specific historical process: a process put to work by the modern West and which from then until now has become widespread, and even globalized. By capitalism, we mean a process of production and the massive, rational, and systematic increase of wealth. So, obviously, ever since its invention at the beginning of the seventeenth century, capitalism has seen a number of important historical evolutions and a multiplication of its forms that are impossible to summarize here. I will rest content introducing four dimensions of capitalism. Foucault recognized three of them in his 1979 course. The fourth was not mentioned, but since it has only shown itself fully within the last two decades, it is not a surprise that Foucault did not study it closely.

The first form of capitalism is commercial capitalism. The production of wealth is thus made possible through commerce, the construction of a market, a space of systematic connection between supply and demand: markets get rich by sales and make artisans who stimulate production rich at the same time. Two concepts are at the heart of this first dimension of capitalism: that of "exchange" and "division of labor." Exchange is a transaction through which two partners attempt to obtain what they do not have by offering what they do have. It rests on the calculation of equivalence and must be able to make a profit for each partner since it involves providing the surplus of what one partner has in exchange for what the other partner lacks. Division of labor constitutes the second central concept of commercial capitalism: each person must specialize in a task to become more efficient and productive since it is the product of one's specialized labor that constitutes its rate of exchange on the market. The textual reference for this dimension of capitalism is obviously Adam Smith's *The Wealth of Nations*.

Second is industrial capitalism, in which wealth is produced by factories by way of tremendous machines endowed with superhuman force, fueled by large amounts of energy. For their labor power, the workers operate these machines and thus become the conditions of mass production, whose wealth largely exceeds the necessary resources for reproducing the workers' labor power. The central concepts of this form of capitalism are those of "labor," "property," and "commodity," as we read in Marx, who was the great thinker of this form of capitalism and who theorized the effects of alienation, as well as structural injustice bound, on one hand, by the fact that only a few owned the means of production while everyone else owned only their labor power, and on the other, by the commodification of human activity and social relations.

Third, there is entrepreneurial or managerial capitalism. For this form of capitalism, it is important to highlight that the production of wealth also largely depends on the internal organization of corporations and factors such as the transmission of information, the cohesion of teams, the articulation between services, but also the adaptability toward new forms of production, the search for innovation, motivation, and the search for good investments. Far beyond the famous scientific organization of labor (characteristic of the Fordist system), it was established after World War II through a complete understanding of organizations that sought the secret of wealth production, not so much from the material power of machines than from the human dynamism of corporations.

Finally, we can talk about financial capitalism, or even shareholder capitalism, in order, this time, to highlight the revolution of wealth production that has taken place over the last few decades, which depends on stock market fluctuations. This time, the central concepts are "speculation," but also "debt," "profitability," and "investment funds." What is at stake is the creation of wealth by anticipating values, while circulating capital from one end of the earth to the other, searching for the most profitable investment.

These four forms of capitalism that I have briefly introduced (commercial, industrial, corporate, and financial capitalism) must not be thought to constitute historical figures, because, after all, all of these dimensions have been present since the beginning of the history of capitalism. But it is the predominance of one dimension over another that historically transforms capitalism. One can imagine other classifications between international market capitalism, multinational capitalism, and global finance capitalism that, to be frank, are more historical. But the forms that I am proposing have the advantage of being immediately identifiable through four major figures that constitute the four large incarnations of capitalism: the merchant, the captain of industry, the manager, and the trader.

Before posing the problem of biopolitics in this context, we must come back to the definition that I gave for capitalism. I spoke of capitalism as a process of massive, rational, and systematic creation of wealth. But perhaps it would be necessary to add an additional specification, which would be the massive, rational, and systematic creation of wealth *while ideally profiting everyone* (even if in fact it only really ends up profiting a few people).

Adding this specification is not simply ideologically discrediting capitalism, but it is an attempt to elucidate the determining factor of liberalism. For, at first glance, it seems to me that what we call "liberalism," understood as an economic doctrine in this instance, could, on one hand, be none other than the idea of wealth creation—whether in the form of commerce, industrial production, internal corporate organization, or financial speculation—that is unable to withstand being commanded by a political authority, in the form of legislative restrictions or systematic planning; and also, on the other hand, the idea that this wealth creation, in a competitive form, always ends up creating general prosperity: the public good is thus not the deliberate result of a political will, but the product derived from a multitude of egoistic and private calculations. If capitalism is in fact a historical process of the simultaneous creation and confiscation of wealth, liberalism is the demonstration according to which these structural inequalities must be the condition of general wealth production. As for the first part of the delineation of liberalism, the aforementioned thesis—according to which the state does not have to directly intervene to dictate what, how, and how much must be produced—will be studied by Foucault in the form of what he calls an injunction to "not govern too much." But liberalism is also the idea that to create the conditions of collective wealth, each person must egoistically and freely pursue her own interests, and that healthy competition reigns amongst all corporations.

But now we must return to biopower. Earlier, we distinguished between two forms of biopower conceived by Foucault between 1973 and 1976: the biopolitics of the regulation of populations and the biopolitics of the discipline of bodies. What these two forms had in common was an opposition to sovereign power. Quite clearly, the biopolitics of the discipline of bodies had been introduced in 1973 as the *ethical* condition of capitalism, but obviously in a very different sense from what Max Weber proposed when he said that capitalism presupposed an ascetic search for indefinite wealth in its development, a joyless capitalism, and this search was explained by a certain doctrine of safety.

As we saw, Foucault also tries to determine the ethical conditions of capitalism, but not by investigating the morality of the captains of industry who invested all of their profits in perfecting their means of production to become even richer, but the resignation of the exploited worker who sells her labor power to an owner, so that the latter transforms it into forces of production. For the worker to be able to sell objective, quantifiable, assessable labor power, this presupposes a previous apprenticeship in regularity and constancy: no longer being allowed to be led by natural rhythms (laziness, lassitude), fighting with oneself against desires for relaxation or change, and so on. As we saw, discipline carries out the transformation of vital, anarchist, expansive, unpredictable potentials into monotonous and gray labor power. Discipline, like biopower, is thus a long process in which the individual's vital powers are directed and transformed into labor power that fuels factories and machines. A new definition of biopower could be established from this analysis: biopolitics is an ensemble of appeals to which the individual is subjected, at the level of vital potential, in determinate directions, to intensify both the production of wealth and the power of the dominant classes.

It seems to me that one important feature from the 1979 course is that it enables thinking about forms of biopower (from the perspective of vital potential) for two other dimensions of capitalism: commercial capitalism, on one hand, tied to the classical liberalism of a theorist such as Adam Smith, and managerial capitalism, on the other hand, tied more to German, or ultimately, American neoliberalism.

In his analysis of classical liberalism, Foucault conducts a surprising analysis of the famous image of the invisible hand in Adam Smith.[8] He claims that we often emphasize this hand, but without fully questioning its invisible nature. What does he mean by that? The context of this image must be reconstructed here. Adam Smith calls to mind each person's egoistic search for her own particular interest, and the way in which a general profit ensues from the individual quest for one's small benefit. The metaphor of the "hand" refers to a providential logic: it is as if some divine hand orchestrated the intersection of these egoisms in such a way that they are harmoniously arranged, complementing each other to produce the most efficient universal interest, as if it had been sought for its own sake. From his perspective, Foucault emphasizes the invisible nature of this hand, an invisibility that he makes even more radical: if the hand is invisible, he says, it is because the subject is especially blind. The economic subject is a blind subject in the sense that she is blinded by the stubborn search for her own personal profit and wants nothing to do with the other, nothing that could resemble a collective logic, mechanisms of solidarity, public good, or a common interest. The subject only sees and seeks her own interest: all that exceeds this quest is happily invisible to her. Furthermore, Foucault emphasizes another feature of the economic subject: her irreducibility to the legal subject. This subject that blindly seeks her own interest cannot be superimposed on the citizen who defends and values rights: they are incompatible and it is impossible to hope for a harmonious synthesis of the two. In the course he gave in 1979, Foucault simply situates himself at the level of doctrines to establish the heterogeneity between, on one hand, the legal subject—whose political existence is founded on a social contract, and which, through the state, attempts to construct a public good and pays respect to their fundamental rights—and, on the other hand, the economic subject belonging to civil society who egoistically calculates her usefulness at the moment of undertaking this or that action. But we can also try to provide a biopolitical translation of this dissociation. We will start by saying that the living person is cut through by a multitude of desires: egoistic needs, but also social passions like sympathy, or even political passions like justice. The biopolitical operation consists in depoliticizing the subject and only addressing the demand for personal satisfaction within her. By stimulating her egoist appetite as a priority, by only appealing to the level of her private desires, we effectively end up extracting a pure consumer subject from polymorphous, vital potentialities, who will calculate her usefulness and will pursue her egoist satisfaction, but will remain blind to all other appeals. Biopolitics is that by which the subject is rendered blind and deaf to anything other than a need for consumption and personal satisfaction.

We already studied how, from 1973, Foucault had described discipline as a biopower allowing the time of life to be adapted to the machine of production. Thus, I will now move to the third form of capitalism. In his 1979 course Foucault is specifically interested in the third dimension of capitalism: the organizational dynamic of corporations. According to him, American neoliberalism prolongs and radicalizes a part of German neoliberalism that consisted in making the corporation a paradigm for reconfiguring its life and that of others, specifically in the name of *Vitalpolitik* (politics of life).[9] The neoliberalism of the Chicago School is essentially about constructing a self-relation that is based on the capitalist model of corporation. For example, one considers herself a holder of certain capital that is both innate (this is our genetic heritage) and acquired (these are the fruits of education), and our life is constructed as a maximal increase of the value of this capital. Out of which a series of formulations arise, like: "living is increasing the value of one's human capital," or even, "each person must become the entrepreneur of himself." This means that each person should construct a self-relation by means of efficient management and profitable investment. The ethical problem is no longer about mastering one's passions or disclosing an authentic identity, but becoming the best manager of one's natural talents and skills.[10]

This entrepreneurial relation will have to structure the totality of activities, even those that may seem the most distant, like education, for example. What is education? Ever since antiquity and the Renaissance, we have been accustomed to thinking about education as learning civic values, the development and flourishing of natural faculties, and also a way of fighting against poverty and ignorance. Meanwhile, neoliberals teach us that educating has nothing to do with forming a citizen. Educating is about making an investment; it is about increasing the value of capital. And that can apply to even more relationships. For example, friendship must be constructed like a profitable investment. The couple, too, becomes a small business. But relations with others can be shown to be more aggressive than this simple search for profitable investments. After all, these other people are themselves are constructed as corporations, which signifies the reign of indefinite competition between everyone. From there, I will have to interpret every instance of the other's success as a danger to me. Sadly, we see what space is left for aesthetic pleasures and happiness in this context since the two presuppose a fundamental gratuitousness. But we also clearly understand that the biopolitics of the "manager" of one's own existence (which is distinct from the egoist consumer and disciplined worker) makes all collective political projects impossible, in the sense of the construction of a common good, since from the start we are given over to an irreducible and contradictory multiplicity of microcorporations that pursue their own increase in value. Finally, we understand the ease with which each person accepts the alienation of the real capitalist enterprise (the primacy of efficiency and profitability, permanent competition, and the cold rationality of calculation): each person thinks of him- or herself as a corporation, and accepts the idea that if she ruins herself, or is excluded, that she did not make good investments at the right time. Because, in a world where one does nothing but increase the value of her human capital, there would never be any known victims of injustice or bad luck, but only winners and losers.

I will only make mention of the theoretical possibility of envisioning a fourth biopolitics that would bear on the last dimension of capitalism: financial globalism. Foucault did not study this form of capitalism, as such, but it seems to me that we could demonstrate how the biopolitics relevant to this last form of capitalism would entail an intensification of flows: each person must be constituted as a pure network, a point of passage for flows, rather than a stable identity.

We can now pick up our definition of biopower again. Biopolitics would thus be a strategy seeking to transform certain vital tendencies or fundamental biological traits of individuals or the human race with the intent of using them to strengthen economico-political forces. In fact, this transformational schema is found in each form of capitalism. The first feature of life is that it is *slanted*: life is animated by tendencies, desires, and tensions. Commercial capitalism tries to polarize the vital passions around the sole desire of consumption. Second, life is *dynamism*: it is activity, work, and the expenditure of creative energy. It is not the repetition of the same or simple reproduction, but the invention of forms. Industrial capitalism exploits this power for its own profit by disciplining it, and by systematically rendering it useful. Third, life is a process of flourishing: it *actualizes potential*. Managerial capitalism forces us to rationalize and maximize our potential through efficient choices, wise investments that transform existence into a process of indefinite capitalization of our innate talents. Last, the living person is permeable: she is traversed by *flows* that she retains, transforms, or rejects. Financial capitalism asks us to construct ourselves as pure points of exchange of flows of images, information, commodities, and so on.

Conclusion: Governed's Rights ("Droit des Gouvernés")

To conclude, we might pose the problem of governed's rights ("droit des gouvernés"). Up to this point we have evoked three large dimensions of biopolitics: biopolitics as the state's consideration of the population as a biological given; biopolitics as the vectorization of vital tendencies with the aim of increasing economico-political forces.

Each time what is ultimately at stake is a process of domination, exploitation, and capture. But Foucault does not convene by denouncing these processes or focusing on human rights, but on what he calls "governed's rights." We should reflect on the politico-theoretical import of this expression. Foucault does not condemn biopolitics in the name of a sacred respect for the individual. More broadly, Foucault rejects the idea of fundamental human rights, which would be grounded in a metaphysical reason, rooted in eternal nature, and guaranteed by a divine transcendence. These sacred rights would have to be protected against vicious and cynical powers. But for Foucault, is not a diabolical entity that would enact violence on an innocent humanity. Power is a relation, strategy, and a battle. Thus, there is not vicious power on one side and fragile individuals on the other. There are processes of capture and resistance. When Foucault talks about "governed's rights," it is to point out that being caught in biopolitical process gives us rights: the right to not accept everything, the right to reject this or that. It is through governed's rights that Foucault tries to conceive biopolitical resistance. It is about saying that certain social situations, certain political decisions, create the intolerable. For Foucault, denouncing the intolerable is not an appeal to transcendent justice or eternal principles that would be external to power, but it is inside a power play, asserting its governed's rights: as a citizen, I have grounds for denouncing certain abuses, resisting certain politics, and so on *since it is me to whom they apply*. For Foucault, these governed's rights belong to biopolitics as long as they do not involve reviving the sovereign subject of natural rights. Governed's rights are not inscribed in any superior normativity. They are not the expression of an eternal nature, but the manifestation of vital energy faced with reductive operations. Faced with an existing power play, governed's rights point out the will to existing otherwise. Yet, Foucault explains that, historically, this activation of governed's rights was made possible by "liberalism." Prior to being a doctrine of sacred rights of individuals or the ideological justification of capitalism, liberalism is the form of thought that, in the West, asked the question about "too much government." Are we too governed, and in what sense? This is the historical importance of liberalism, of being allowed to ask the state the question about "too much government" ("the question of liberalism, understood as the question of 'too much government'"[11]). This liberalism is not a political liberalism or an economic liberalism; it is what we can call a critical liberalism. And through it, Foucault explains, something like political life, in the sense of debates, struggles, contradictions, is made possible in the West. Thus, if economic liberalism justifies the biopolitical exploitation of vital forces for the profit of a small number of politico-economic powers, critical liberalism fuels biopolitical resistance.

Notes

1. Michel Foucault, *The Birth of Biopolitics: Lectures at the Collège de France, 1978–1979*, trans. Graham Burchell (New York: Palgrave Macmillan, 2008).
2. Michel Foucault, *Le gouvernement des vivants*, ed. M. Senellart (Paris: Gallimard/Le Seuil, 2012).
3. Michel Foucault, *Discipline and Punish: The Birth of the Prison*, trans. Alan Sheridan (New York: Random House, 1977).
4. Michel Foucault, *The History of Sexuality*, vol. 1, *The Will to Knowledge*, trans. Robert Hurley (London: Penguin Books, 1998).
5. Michel Foucault, *"Society Must Be Defended": Lectures at the Collège de France, 1975–1976*, trans. David Macey (New York: Picador, 2003).
6. Michel Foucault, *La société punitive*, ed. B. Harcourt (Paris: Gallimard/Le Seuil, 2013).
7. They are the last chapters of *The Will to Knowledge*.
8. The March 28, 1979, lecture, *Birth of Biopolitics*.
9. March 21, 1979, lecture, *Birth of Biopolitics*.
10. March 14, 1979, lecture, *Birth of Biopolitics*.
11. Summary of the course, *Birth of Biopolitics*, 327.

FOURTEEN

Discordant Practices of Freedom and Power of/over Lives: Three Snapshots on the Bank Effects of the Arab Uprisings

MARTINA TAZZIOLI

The Arab uprisings and the others in connection. This could be a title for describing the general context in which the Arab revolutions have been read on the northern shore of the Mediterranean and in the United States, presenting the Arab turmoil at once as the forerunner of new political practices—for instance, the occupation of the Kasbah in Tunis and of Tahrir Square in Cairo that staged new ways of dealing with public space—and the "late" followers of the conquest of democracy. And the simultaneity of Occupy movements and the *Indignados* with the events of the Arab uprisings has clearly fostered the easy connections posited by many critical analysts, activists, and scholars: "From Tahrir Square to Oakland," "From Cairo to Wall Street. Voices from the Global Spring," "Egypt supports Wisconsin," "Turn Wall Street into Tahrir Square." An inverse and double contamination seemed to be envisaged in all these analyses: the "not yet" of democracy going south and new fresh practices of political participation migrating from the southern Mediterranean countries northward. It is beyond the scope of this work to make a comparative analysis of the European and American Occupy protest movements and the revolutionary uprisings in the Arab region, as well as to investigate their mutual influences. Rather, without denying these nexus, I focus here on the spatial upheavals triggered by the Arab uprisings on the northern shore of the Mediterranean on the geography of the European space. And at the same, I draw the attention on the politics of/over life that migration governmentality put into place to tame and regulate migrants' turmoil.

The goal is not to bring out possible connections with events, practices, and movements currently at stake on the northern shore but, on the contrary, to illuminate how migrants' spatial upheavals have multiplied—

triggering a sort of domino effect—and, most important, engendered political and spatial transformations unsettling the tempos and the conditions of the politics of mobility.

In these three snapshots are at stake the instabilities produced at different levels by migrants' spatial upheavals "discordant"[1] and practices of freedom that do not fit into the script of the insurgent democracy or of majority politics: they trouble the "isomorphism of the surface" that makes different political experiences easily connect and influence each other. Through the technique of snapshots, I turn the attention to the production of new spatialities as an outcome of these practices of migration. Spatialities framed and enacted by political technology through economic or juridical measures, to respond, to catch, to (re)frame, to neutralize, and to profit from revolutionary events; but also spatialities subverted and produced by the crossing and the burdening of borders that migrants have done. The advantage of such a perspective is, as William Walters observes, that "the zonal allows us to map irregular spaces that are neither national nor global and it avoids the teleological assumption that developments in migration control have an inevitable direction or end point" (Walters 2011, 53–54). The interplay between these two movements of spatial production—and practices in space—allows us to account for the complex and conflicting regime of power over/of migrants' lives. Such a spatial gaze could work as a litmus paper to scrutinize the subtle dynamics and the instabilities between power over migrants' lives and power of migrants. In fact, if political technology continuously plays in space and by making spaces to govern people's movements, at the same time migrants' practices at times not only trouble those channeled spaces of mobility but also put into place different modalities of enacting and persisting in space. The reorganization of the border regime in the face of migrants' spatial upheavals basically takes place through two main strategies: "strategies of detection" and "strategies of b/ordering"—that is, techniques and knowledges through which migrants' conducts are regulated in their spatial persistence and in their mobility.

The three snapshots that I take into account here concern new border spatialities as a result of power over/of migrants' lives.[2] I look at them as the emergent outcomes of struggles, conflicts, and negotiations between techniques of governmentality and migrants' movements, both relying not only on biological life but mostly, and in a broader way, on the wide materiality of existence, conceived as an historical construction and as an hetero-normed condition (Butler 2006). Nevertheless, it does not follow that escapes, practices of migration, and strategies of resistance are coextensive and isomorphic to power: rather, I would focus mainly on the unsettling unexpected nature of resistances that try at times to dodge and times to reverse the political technology aimed at capturing and capitalizing people's movements. In this regard, by drawing attention to acts of refusal and flights, we should account for the asymmetrical dimension of resistances (Revel 2008a, 2008b) conceiving it as the capacity of transgressing and redefining meanings and uses of spaces, along with the invention of forms of struggles and strategies of survival.

The three snapshots that I figure out are (1) spaces involved in the North Africa emergency, (2) struggles/escapes/riots that happened in 2011 in the Italian detention centers, and (3) two highly symbolic urban occupations made by migrants in Brescia and Massa. The aim is to make play the coexistence of different spatialities with their nature of "emergent-emergency spaces," meaning by that the production of new political practices as correlative of, at the same time, a temporal and a temporary dimension. "Emergent" alludes to a temporal aspect, namely to spaces that arise suddenly: anomalous spaces of governmentality and of spatial overturns that challenge representability and rights as the main axes of the political domain (Papadopoulos, Stephenson, and Tsianos 2008). "Emergency" refers to the temporary dimension vested in these spaces as political responses to a disruptive and troubling array of events, subjects, and transformations framed as threats to social and spatial order: "emergency," in the context we consider here, is nothing but the name given to a space carved out by heterogeneous political technologies. In this regard, this article argues that we should bypass the form of the "camp" as the exceptional space *par excellence* (Agamben 1998) to understand what is at stake in the power over/of migrants' lives: rather, the spatialities I deal with are characterized at once by a specific economy of spatial practices and by a movement that constantly encroaches on the outside.

North Africa Emergency and "Libyan" Migrants Scattered in Italy

One of the most visible and politically immediate bank effects generated on the northern shore of the Mediterranean was the mobilization of actors, structures, and spaces related to the so-called North Africa Emergency declared by the Italian government in February 2011 and that ended in March 2013. The North Africa emergency was officially designed as a program of aid and hosting addressed to people coming from the Libyan conflict: the program established that all asylum seekers arriving in Italy would be hosted in the Hosting Centres for Asylum Seekers (CARA) until their demand of asylum was processed. Due to the sizable number of arrivals in the first half of 2011, asylum seekers were hosted in alternative structures managed by cooperatives that had never dealt with migration, each of them receiving €43 per day per refugee. However, before entering into details I have to clarify from what point of view I take into examination the North Africa emergency: in fact, the aim here is to sidestep any approach grounded on the logic of exception, focusing rather on the spatialization of the power's functioning and on the ways in which political technologies or administrative measures shape migrants' lives. Thus, I bring the attention to the spatial effect of this mechanism of discharge and how it effectively impacted on asylum seekers. Then, of particular concern here is the dispersion on the territory that the mechanism of the "modular hosting"[3] meant for migrants.

Mineo/Bari, August 2011: First of all, what should be noticed is the patchy arrangement of the system of hosting put into place by ENA: a variegated geography of centers and structures was activated, giving rise to highly nonhomogeneous procedures and conditions of hosting. But most of all it is important to underscore the ambivalence of this mechanism of capture and strand of migrants, drawing the attention to the strategy of chasing away and leaving undetected migrants' presence. "Parked" in improvised structures or instead in new durable centers, the "Libyan" asylum seekers were discharged regarding their pattern to protection and kept substantially unaware of their future possible destinations. On the one hand, migrants were de facto left to their own fate despite being controlled in their mobility, and on the other hand they have been stranded in remote zones of Italy (in the Alps or in the middle of the countryside, in overcrowded centers, or in small houses) without their demands being processed. In this regard, one can see the manifold riots that took place in detention centers as exasperated reactions to such a situation, and this was obviously an important factor. However, riots and protests were also markers of the threshold of acceptability that people coming from Libya and Tunisia could tolerate.

Bari, August 1, 2011: Asylum seekers coming from Libya started a riot in the Cara of Bari Palese protesting the delays in the processing of their demands, and then they exited from the center blocking the motorway and the railway, paralyzing the traffic of the city. After violence by the police against the migrants, twenty-nine of them were arrested for the first time. But despite the "cost" they paid for the insurrection, they launched an "ultimatum" to the territorial commission of asylum arguing that if their demands were not processed in a few days, they would rise up again. Thus their strong reaction was a way to put the Italian authorities over a barrel, demanding to be really taken into account as asylum seekers and not as negligible presences on the territory. But again, the "costs" of the protest in the long run could not be overlooked: as in the case of the riots triggered by the Tunisian migrants, especially those in the detention centers of Gradisca d'Isonzo, Brindisi, and Modena, the Italian press talked about an "external and orchestrated organization."[4] And some months later, forty-five asylum seekers involved in the struggle were processed and then deported.

Mineo, Sicily: The biggest Cara of Italy detaining more than two thousand people is situated in the middle of a desert of oranges, 10 kilometers from Mineo, the closest village on the top of a hill. Every day asylum seekers receive an allowance of €3.5 that ultimately they cannot spend outside but only in the internal shop of the center, where they can buy only cigarettes. Thus, this has generated a circuit of informal economy in the village of Mineo, where asylum seekers go to resell the cigarettes in exchange for money to buy everything else other than cigarettes—for instance, for women it was their only possibility to buy tampons since the managing institution of the camp did not provide them.

However, the point here is not to show up the illegalities that institutional actors perpetrated within detention centers since if one contests the very existence of migration detention institutions, the criticism against the opacity of administrative procedures does not hit it on the nail and at the same time risks shifting the battleground over detention centers to claims for more human conditions or for legality. Instead, on the one hand it could be argued that the violence and the power exercised in detention centers cannot be reduced to a question of legality/illegality (Benjamin 1986): as migrant detention centres "criminalize and illegalize some people for their unauthorized presence in a certain space, we cannot narrow the critique to the more or less non-human treatment in the centres, rather we should point at their substantial illegitimacy.

The protests in August were not the first at the Mineo camp: on May 10, 2011, two hundred asylum seekers living in the camp organized a huge block of traffic demanding that their demands be processed; in that case, they obtained that the territorial commission start its work after a few days. And, after August, many other unrests took place since the starting of the processing of the demands by the territorial commission was finally only a partial victory: first, the work of the territorial commission was going ahead very slowly, and most of the time people were examined only by one member instead of five, often without appropriate language translation; and second, in autumn 2011 the first rejections of protection started to come. "This is not an admissible decision, we all fled from the same war, the Libyan conflict, and we are here because of that war," I have been told by all the rejected refugees I met out of the camp. In fact, through the statement "all of us come from Libya and escaped a war" that circulated in many border zones and sites of the Mediterranean, "Libyan" migrants cracked the tenability of the Geneva Convention's principles, which are grounded on a nation-based logic—primacy of the country of origin—a settled normativity—overlooking, for instance, African internal migrations and so who could find themselves in dangerous situations or contexts of persecutions that are not their country of origin.

All these protests seem to unpack the logic of stranding and discharging, affirming the duty of humanitarian and governmental forces to finish their jobs. At the same time, more than practices of resistance, they strongly posited the unacceptability of such a treatment, challenging and trying to interrupt the meshes of governmentality that exercise a specific government over their lives.

December 2013: The end of the North Africa emergency was postponed to March 2013, due to the actual impossibility for the cooperatives that host refugees to drop them and to make as if they suddenly do not exist anymore. "What to do with them?": this was the main unresolved question for the governmental actors and that migrants themselves reiterated through their persisting in those centers, demanding that a sustainable solution be found for them. This political impasse related to the difficulty of making people disappear into the void and not letting them circulate all over the Italian territory as irregular presences, made the government give temporary humanitarian protection to almost all the asylum seekers from Libya and grant them €500 as a way of winding refugees up from the humanitarian regime. That said, once they were supposed to have become autonomous subjects, entitled to the rights related to the temporary protection and to the permit of freely circulating in the Schengen space, "not-to-chase" migrants found themselves in an existential impasse: the growing impact of the economic crisis made it impossible for migrants both to get any social protection and welfare assistance and to find even an informal job. Therefore, Cara and reception centers continued to be their main dwellings, and finally the possibility to have food and shelter became inevitably preferable to the perspective of being at the mercy of economic uncertainties and anti-immigrant raids made to defend citizens against migrant "job thieves." From their part, the corporations that managed the centers were interest in keeping the Cara open to continue to get the €43 per day per person. Thus, the Cara of Mineo was more overcrowded than at the time of the North Africa emergency, with 3,000 people living in the center while in December 2011 there were 1,800 asylum seekers: a full speed machine that nevertheless shifted from a humanitarian rationale to a dispositive to "park" people in an indefinite wait, when also the channels of informal economy were clogged up. In the meanwhile, asylum seekers still continued to arrive in Mineo, transferred there as soon as they arrived in Lampedusa, as the two Eritrean guys I met out of the center confirmed to me and then argued: "So, finally this is Italy, right? I didn't imagine it like this; but we know that also here there is the economic crisis. We don't want to stay here, maybe Sweden or Switzerland . . . and what about London? If you could just tell us what train we should take to get there . . . where there are not too many controls onboard."

The working of the categories in the government of migrations and refugees basically generates a fragmentation effect, just due to the embedding by migrants of the "mobility profiles" they are labeled with. Indeed, the juridical status or, in any case, the juridical nonstatus, that is the legal limbo, through which migrants are defined materialize for migrants into forced temporal geographies: who waits for the result of the asylum process cannot but stay stranded in place; who is rejected moves as an irregular presence on the territory; who gets the temporary protection but does not have the permit to circulate in Europe, yet has to wait for an indefinite time before going. More broadly, migration governmentality works according to a "dividing through categories" rationale or, to put it differently, a mechanism of "categorizing for partitioning": the multiplication of mobility profiles and the uneven array of juridical conditions in which migrants situate and that make them very often be at the limits of a defined legal profile, enhancing the dividing-mechanism among migrants themselves. In fact, beyond the full spectrum of migration categories and mobility profiles, we should consider how in practice the complexity and singularities of migrants' stories blur any clear distinction between the different juridical categories; and in this way, the multiple partitioning set becomes a de facto and also more complex and blurred producer of different temporal geographies. Second, these differences are further sharpened due to the effective migrant experiences that never fully fit into the existing legal bordered identities. This consideration raises a questioning about the drive to pluralize migration that underlies many critical analyses: indeed, if on the one side a political epistemology of migrations needs to unpack the migration catchword, highlighting the heterogeneous conditions and practices of mobility that the name "migration" embraces and eclipses, on the other side we should be cautious of the underway multiplicative factor of migration governmentality and its crumbling effects. And if one could object that it is not the same kind of multiplicity that is at stake, it remains that it is not enough to praise for the multiplication of migration and the recognition of singularities: in this regard, the important criticism against the monolithic understanding of migration by governmental agencies and its objectification into an overwhelming category should take into account at the same time the "partitioning through differentiation" on which the asylum politics is predicated.

May 2013, Paris, Hamburg, Berlin: If the so-called North Africa emergency was one of the most visible bank effects of the Arab uprisings on the northern shore of the Mediterranean, we should consider also how the North Africa program has resonated beyond Italy through the logic of the "spread hosting." Despite the changed political situation and the deadening of organized struggles by asylum seekers in Italy, the scattering across of "Libyan" asylum seekers and refugees coming from Italy has produced "knots of struggle" in many European towns, in which they protested against any possible deportation or return to Italy and fought to get accommodation: migrants from Libya entitled to temporary protection named themselves the "Lampedusa in Hamburg" group,[5] arguing "we did not survive the NATO war in Libya to die in the streets of Hamburg" and in this way staging preeminently, as the marker of their present migrant condition, the geographies they enacted and the border-site in which they were blocked, identified, or detained. "We are here and we will stay. No European country can evade the responsibility. We will not be played with anymore by the European policy. We think that the social and economic situation in Italy and other southern European countries is well known and that there is no possibility of livelihood for us. . . . If we get the humanitarian protection but at the same time every possibility to survive is denied, are we effectively subjects of concern of the humanitarian?" Ultimately, the very formula of "measures to favor exit patterns" for the "Libyan" migrants is quite ambiguous: in fact, more than programs of integration or of labor training, these measures have resulted up to now in ways of discharging and encouraging people to move away.

Organized Collective Escapes and Extreme Resistances: The Struggle for "the Oxygen of Freedom"

April 2011, Lampedusa and beyond: The second snapshot I take could start in the island of Lampedusa, where 26,000 Tunisian migrants arrived in 2011, just in the aftermath of the Tunisian revolution: through their presence, they produced an outstanding turmoil at the level of both European politics and Italian law, and of migrants' struggles, radicalizing claims, and forms of protest. Moreover, gazing at the unexpected arrival of Tunisian migrants all coming in the span of a few months makes us interrogate the juridical and political transformations and the troubles that their presence generated, triggering deep spatial redefinitions. As Federica Sossi persuasively argues, Tunisian migrants had immediately to confront practices of border enforcement that were activated to respond to that "sudden and effective upheaval of the space" (Sossi 2012, 23). In this regard, instead of lingering on the island of Lampedusa, I move across the Italian territory, taking snapshots of the multiplicity of struggles that happened in the Italian detention centers at the time of the Arab uprisings. The will and the struggle for freedom that triggered and made effective the Tunisian upheaval, I suggest, has been pushed forward by Tunisian migrants as people who took part in the revolution and then enacted that freedom as freedom of movement, in part breaking with the norms of their country (Mezzadra 2006) and in part presenting themselves as the "sons of the revolution" (Sossi 2012).

The first freeze-frame of this snapshot refers to March 27, 2011—so, in the midst of the "Tunisian revolution in Italy": it shows a group of Tunisian migrants clinging to the wire netting of a detention center in Sicily, protesting about being detained like criminals arrived from the other shore of the Mediterranean Sea. One of them screams, "The world is not mine or yours, it neither belongs to Obama nor to Berlusconi, it belongs to everyone. So, if I want to breathe the oxygen of Italy, I can do it; if I want to breathe the oxygen of Canada, I can do it. No wire exists for me. I'm here not to steal or to rob, I'm here for the oxygen of freedom."[6] What is stunning in these words is, on the one hand, the insistence on an unconditioned freedom and, on the other hand, the inflection of freedom itself ultimately as freedom of movement and at the same time as an unconditional right to stay everywhere, to trample the soil and to move on.

Moreover, this discourse displaces the spread discourse that sees migrations as the result of economic individual needs or of global market dynamics. Indeed, the sentences of the oxygen of freedom quoted highlight a very different point: the economic factor is only one among others that induced people to leave Tunisia, and most of the time it was not primary; rather, what they strived for is the effective possibility of moving, of living elsewhere. The claim about "the oxygen of freedom," apart from the unconditional yearning to freedom that it reveals, also makes glaringly clear the persistence of a partitioning line that, at present, works as one of the main political technologies for the production of social inequality: in fact, it points at the crosscutting geopolitical gap between those people on earth who can move freely with their passport and those who are not allowed to move without a permit. Thus it shows how this massive divider is the source of tangible effects: the oxygen of freedom to which the Tunisian migrant makes reference is precisely what he has been denied, first in Tunisia under the regime of Ben Alì—where the *crime of emigration* has existed since 1975—and then by European governments. If it is fundamental to sidestep the paradigm of forced migration and opening to the idea of migratory projects (Mezzadra 2006), Tunisians' migrations taking place in 2011 radically displaced any economic reductionism for understanding the political troubling force of migrant struggles: "I don't want to be given any sandwich, I want to be left free to move away from Lampedusa,"[7] a Tunisian migrant firmly stated during the protests that took place in Lampedusa at the end of March 2011, when migrants shouting "*hurrya*!" ("freedom" in Arabic) asserted, as neither negotiable nor deferrable, their will to move. To the contrary, both the Tunisian uprisings and migrations across the Mediterranean were narrated by mainstream media respectively as struggles for bread and escapes for finding a job. These two issues were both effectively at stake, and it is also crucial to stress the existence and the legitimacy of these aspects in order not to fall into the error of detaching practices of freedom from any concrete concern and reiterating the gesture of seeing in events and practices what we are accustomed or willing to see. In this regard, Sandro Mezzadra draws attention to the connection between revolutionary uprisings and practices of migration: "Why," Mezzadra asks, "should that scream [*hurrya*] be bordered within institutionally defined spaces? Rather, practices and ways of bordering which gave rise to that scream bear the marks of continuous trespasses of existences taking place in other spaces" (Mezzadra 2011, 86).

If we consider the power over/of migrants' lives, the fact of keeping the two spatial upheavals disconnected—viewing the revolution as a conquest of democracy and, inversely, the arrival of Tunisian migrants in Lampedusa as the marker of the risks and the crisis of such ungoverned democracy—led to framing practices of migration exclusively in terms of "power over migrants lives," reducing de facto migrants' existences to lives governed by the regime of control over people's mobility. Moreover, the image of Tunisian migrants crammed in Lampedusa or in the detention centers, as waiting for a response from the European Union (EU) or from the Italian government, portrayed migrants as people lacking in life's projects or "worthy" political demands.

Thus, to grasp the complex entanglement between the spatial troubles and transformations generated by migrants and the new geographies traced by power, I take stock of the riots, escapes, and resistance and struggles that took place in the Italian detention centers in 2011. The long series of struggles I report was "inaugurated" at the beginning of February, when about sixty Tunisian migrants exited the courtyard of the detention center in Modena, went up onto the roofs, and protested, screaming "Freedom, freedom!!" while also showing signs with the same words. In the meantime, thirty-two Tunisian migrants succeeded in escaping from a detention center in Brindisi. From then on, it was never more than seven days in between riots, and I just recall here in passing the escape of hundreds of Tunisians that happened in Manduria on April 2. It goes without saying that the high number of riots clearly depends also on the considerable number of migrants who arrived at once on the Italian coast. In particular, the Italian government put into place new sites of detention and renamed some old ones, inventing an anomalous spread-out space of detention in which people were not formally detained.

Taking on Foucault's considerations on rising up by people, what is noticeable here is that the unintelligibility of those movements through the lens of traditional democratic processes of transformation colludes with the eradication of political fear that Tunisians experienced—"we do not have fear anymore"—and that they translated into the unacceptability of obedience to their confinement: "The impulse through which a single individual, a group, a minority or an entire people says: *I will no longer obey,* and throws the risk of their life in the face of an authority they consider unjust, seems to me to be something irreducible" (Foucault 2000, 449). The sentence that they passed along sounded like "we, who actively took part to the revolution; now we are here and we continue it" stressing the hypocritical position of European governments, addressing especially to France—"you, who acclaimed our democratic uprising and promoted the human rights rhetoric; you are chasing us away, the sons of the revolution." Ultimately, the question is not if they effectively took part in the upheavals of January 14; rather, what is significant in our perspective is that they found and asserted that relationship between the two spatial upheavals. They did not ask to be settled and to become integrated in the Italian society, and their claim never referred to citizenship; instead, they want to exercise their freedom gained in Tunisia that made it possible to materially burn the borders, crossing then the Italian border and moving on to France and to other European countries. In the meantime, something had changed also in relation to the intolerable nature of power, namely to the threshold of tolerance as regards the confinement of people's movements into wired places and, more widely, against their hampered mobility (Balibar and Brossat 2011).

Actually, as the chronology of riots displays, not only did migrant struggles take place also as organized escapes, but these latter proved to be the most successful strategies: the vast majority of flights succeeded because of the cunning strategies adopted, combined with promptness of action and coordination among people. Along with escapes, during the first half of the year we witnessed symbolic protests consisting of staging one's will for freedom and bringing the echo of the demonstration beyond the wall of the center: so, despite their juridical status of "irregular," in the first period Tunisian migrants put into action a strategy of visibility, claiming their will to move on. From this perspective, the most remarkable gesture was the protest that happened in Lampedusa on March 31, when migrants took to the streets in a disorganized and peaceful way, crying: "We want to move elsewhere."

These riots and strategies of resistance highlight an overall inclination by migrants to continue their practices of movement, not backing down. In such a way, I would argue, they also convey a specific content to the idea of resistance to power over migrants' lives: these migrant struggles contributed to crack the very distinction between "legitimate" and "ordered" migration and disordered mobility, placing their will of living differently at the top of it. In this sense, gazing closely at the dynamics here at stake, we observe that the unconditioned right to trample on the soil that Tunisian migrants ultimately seem to claim is actually an odd form of right, at most a non-juridical right—that is, a non-sense—since as Nicholas De Genova points out it is not a question of migrants' rights but rather of the expression of migrant mobilization: "that is to say they erupt from mobilities which cannot be fixed into place, categorized and regimented. They refer us to practices and processes of open-ended becoming, that actively produce and transform space" (De Genova 2010b, 115). Beyond any "social contract" or representative model, Tunisian migrants acted their freedom regardless of any condition or restriction to accept for making their claims legitimate. In a nutshell, they "placed" their freedom without seeking to be in accord with and recognized by powers and rules that fix both the contents and the borders of "legitimate" political practices: they did not look for any correspondence between their practices and the political order; they did not seek to fit into that. In this sense, I would suggest, freedom here is claimed as an "unfitting practice." After all, it is important to remember that "the sheer autonomy of migrations, especially that of unauthorized migration, remains a permanent and incorrigible affront to state sovereignty" (De Genova 2010a, 39).

The Church and the Crane: Migrants' Unsettling Spatial Struggles

Massa, May 2011/Brescia, November 2010: Through this last snapshot I step aside from the unexpected turmoil produced by the events of the Arab Spring and from Tunisian migrants coming from the Tunisian revolution. However, such distance is taken on purpose, to interrogate the political resonance that the Tunisian migrants' upheavals eventually produced, supporting other struggles at a distance, through their revolutionary impetus. More exactly, the idea here at stake is that migrants' strategies of resistance on the one hand and political technologies on the other hand gave rise to a spatial redefinition as well as a change in the ways of enacting movements and presence in spaces. In a certain way, such a resonance could be figured out as a kind of accelerator of existing social conflicts, leading to a radicalization of the ways of acting. Radicalization here designates not necessarily an increased use of violence but rather the exposure and the direct involvement of subjects—basically, they bring to the fore contradictions and paradoxical elements of the government of migrations. This short-circuit of power (ir)rationality produced by migrants' struggles has occurred above all in the symbolic dimension they are hinged on.

In this regard, the snapshot I take concerns two urban protests, whose force lay precisely in the way in which symbolic places are used and made to play in the struggle. Moreover, both protests are characterized by a "troubling of space," namely a persistence in space, despite the "illegitimacy" to stay there, that at the same time entails a redefinition of that space itself. This troubling persistence was put into practice also through the occupation of symbolic places: a church and a crane. In both cases, it is noticeable how a juridical claim has turned into a protest that, although starting from a punctual demand addressed to governmental actors, spilled over that perimeter, encroaching on a broader range of political issues.

The occupation of the cathedral by migrants happened in the Italian town of Massa in May 2011. It started on May 1 and lasted for more than twenty days. What is very significant to our purpose is the considerations that led to the decision to occupy the church and the way in which migrants really acted within the existing political conditions and framework, evaluating how to interrupt the functioning of the "governmental machine"—composed of different political agents, like local institutions, political parties, the Catholic church, employers and trade unions—making visible the absurdity that underlies the governance of migrants' lives. Every step of the occupation was adopted according to a strategic evaluation, grounded on the idea that the political keystone of the struggle should be to make crash on to one another the different social and political forces/actors.

Instead of occupying the trade union office, migrants opted for the church for two reasons. First, because they considered that traditional forms of occupation are by now empty of all political meanings since what refers to factory work no longer represents the modalities and the extension of migrants' workforce exploitation: the precariousness in which they constantly live is not reduced or bordered to the place of the factory; and besides, salaried work does not correspond any more to the forms of work they have to lend themselves to. Second, they were aware that the challenge was much broader than the question of legal permit, and thus they translated the juridical demand on a political level; or, to put it bluntly, they claimed the political dimension of that protest, refusing to be seen as struggling only for being regularized. Quite to the contrary, the place of the church would have put to the test the hypocrisy of the Catholic church—which appeals to an unconditioned welcome for migrants—forcing it to make a stand on that issue. Hinging on the centrality played by the Catholic church in Italy, migrants looked at the church as a place from which to address all political forces. Nevertheless, they did not ask for hospitality within the church; rather they overturned the space of the church as a sanctuary space: "we don't want to be tolerated or hosted: we know that this space is not a property of the Italian State, but we argue that it does not belong to the catholic Church too, since it is named—by the Church itself—as a universal space. So, we simply decided to stay here, without asking anybody to be hosted."[8] In this way, they strategically played on the ambivalence of discourse of the church, putting offside the regime of tolerance. If we think about the Tunisian oxygen of freedom—the episode that I mentioned in the second snapshot—it seems to me that we find a quite similar logic at stake: in fact, both migrants are clearly careless of terms and conditions fixed by the democratic regime of discursivity, as, for instance, the insistence on rights' claim. In other words, the occupation of the church in Massa resounds the "discordant freedom" I mentioned about Tunisian migrants. To qualify this expression in a slightly different way, I would say that it addresses freedom as a practice—which does not seek any correspondence or alliance with the conditions of democratic society as claimed by "liberal States": "migrants are not the lacking subjects of freedom; quite to the contrary, just because they are not subjects of right of any pre-established citizenship . . . they deeply put into question its structural limits," and therefore, compared to democracy as the normative order of our societies, "migrants represent a contradiction on the move" (Mometti and Ricciardi 2011, 7). The occupation of the church in Massa suggests to us also a creative dimension that is more and more at stake in migrants' struggles: in fact, as I argued before, what is really interesting is not the form of struggle they put into place—the occupation—but rather the way in which they replayed and reworked it, asking people and institutions to come there to negotiate and redefining the space of the church as a common space. As L.D., one of the Senegalese migrants I met, pointed out, "we made visible, as in the back-light, the lack of inventiveness characterizing Western political tradition and the weakness in producing practices which effectively short-circuit power regimes of discursivity" in order to open spaces for political agency refusing the well-known dance of political representation as the master ground for political action. In the same perspective, we should read the gesture they performed in front of the main institutional buildings, showing at the same time a box of tomatoes and a residence permit: why, they asked, do we have fewer rights than a box of tomatoes? In fact, if the certified box of vegetables could freely circulate all over Europe, this is not the case for migrants who, even though they get a permit for staying and working in Italy, cannot search for a job in France or live in Spain. That is to say, the freedom of movement, very rhetorically promoted by the EU and effectively asserted in international treaties, is allowed and enacted by governments through a deep asymmetry between people and goods. This is also the reason they refused to accept a temporary permit from the local government: indeed, what they wanted to sidestep was the logic of precarization underlying the concession of permit; and second, they refuse to narrow their struggle to a battle for getting the resident permit basically accepting and playing a logic that says " we demand to be a subject of rights and, ultimately, to be a subject as such": indeed, "we use this targeted struggle for spilling over the issue of residence permit, making clear that what we want is freedom of movement, and not only juridical regularization. In this sense, it's clear that we have never claimed rights, what you call 'rights' is eventually what we are stripped of."[9] In some way, it could be argued that the symbolic level of these struggles pushes to redefine the self-evident function of right, shifting it from a claiming question to a pick-lock to unhinge that regime of political epistemology traditionally legitimized by right. But despite the fact that a huge literature asserts the strength of outcast people to challenge and to reshape the borders of citizenship and political belonging (Balibar

2002; Isin 2002; McNevin 2011; Nyers 2011), it cannot be overlooked that what is at stake here is not a demand of belonging or "recognition as integral rights-bearing members of the polity despite lacking legal status."[10] It is rather the proposition "to belong to" that is actually challenged by migrants' practice of insistence on the space, thus effectively shaking the discursive framework of Western political thought (Arendt 2003). The idea of belonging necessarily entails a political space of citizenship given in advance, into which one is supposed to fit, much as you can force the borders of that frame—"to belong to what?": it is precisely the object of belonging that is taken for granted, while what is asserted is eventually the possibility to enlarge the range of subjects entitled to it. In this regard, I would say that the shift of focus from citizenship to belonging does not push the question on: in fact, the normative ideal-type is in both cases—citizenship and belonging—the "ethical" construction of the good citizen or at least of the good civic subject: civilization and (urban) civility overlap as the underlying frame of reference.

After considering the choice of the place—the church—we should dwell on the day they decided to act: May 1. We see how symbolic element and strategic evaluation are intertwined. In fact, strategically speaking May 1 was a day in which a lot of people would have supported their struggles—as a nonworking day and as a bank holiday, which conveys a specific political meaning, so that people are more inclined than usual to take part in workers' struggles. Then, on the other hand, hinging on a very symbolic day, migrants ultimately aimed at unsettling and reversing the very meaning of May 1: they stress that far from being a holiday to celebrate, it should be understood as a moment for recalling the precariousness of workers' lives as much as the absence of work as quite a constant that people have to confront. In other words, the first of May sounded at once like a way of gathering as many people as possible, far beyond the communities of migrants—playing just on the political meaning of that day—and as a stage for rethinking its function.

Now I take a step back, turning to the occupation of a crane that took place in Brescia in November 2010, when nine migrants decided to go up the crane and stay there to protest against the 2009 fraud act of regularization, as in the case of migrants in Massa. At this point, it is relevant to see how certain strategies of resistance could produce a sedimentation of knowledges and practices—I would say of ways of acting—that then are remobilized and replayed by other people, resounding in other contexts. What the occupation of the crane clearly staged is above all the overcoming of the fear to openly question the regime of clandestinization set by the

Italian government and, more broadly, by the politics of migration control. Second, migrants actually broke with the condition of "being-spoken-by"—trade unions, political parties, and so on—since they chose a highly symbolic workplace, the crane—thus stressing the exploitation of migrants' labor force—but refusing to be represented by anybody.

The hypothesis that I put forward is that migrants struggling in Brescia put into place a way of facing political forces, and at the same time a way of exposing their own precariousness, leaving a sort of practical-political legacy that was then very soon reused by others. Such a legacy has been reshaped and mostly radicalized in the aftermath of the political uprisings that happened in North Africa. Paving the way to a series of migrants' struggles playing on building occupations and symbolic acts, something happened between the occupation of the crane in Brescia and the other struggles. The snapshot on Massa highlights that migrants reactivated the practical *savoir-faire* put in place in the struggle in Brescia, readapting it in light of a by-then acquired fearlessness to face power. Finally, it is important to stress that when the struggle in Massa happened, the political movement in Brescia had been relaunched and spread to other Italian towns like Padova.[11]

Rethinking Migrant Struggles and Freedom of Movement

"Migrant struggles" is the name for identifying the multiple concrete struggles in which migrants are engaged: more or less organized struggles that defeat, escape, or trouble the order of mobility, the regime of labor, the politics of detention and control, or the space of citizenship. These struggles take place at the border or before and beyond the borderline; they that gain the scene of the public space or remain invisible. Assumed in this first empirical meaning, "migrant struggles" pluralize and unpack the catchword of "migration" highlighting the heterogeneity of migrant conditions and the different ways in which migrants are confronted with powers. However, we should resist seeing any practice of migration as deliberate agency or as challenges to the border regime. Rather, it is important to keep in mind the ambivalences that cross practices of migration: they could play as resistances against some mechanisms' containment or against the social norms of the country of origin, but at the same time being the other side of the selected mobility that partitions channels of allowed and fast mobility, on the one hand, and unskilled migrant labor force that is supposed to move illegally, on the other. From this point of view, the stake becomes to bring out what also in these irregular-but-wanted practices of migration exceeds the economic frame in which their presence is required and expected. But "migrant struggles" could be framed also in a more structural way, starting from the consideration that every migration is situated in and grapples with a certain struggle field, and in this sense is always crisscrossed by and involved in multiple struggles. It basically depends on the fact that practices of mobility that are labeled as migrations are captured, filtered, and managed by migration policies and techniques of bordering. Starting from these considerations, I contend that practices of freedom played by Tunisian migrants make visible that there are multiple and conflicting ways of conceiving and enacting freedom of movement: the adjective "discordant" stresses precisely the uneasiness both of governmental forces and of European citizens in grappling with practices of movement that do not fit into the established frame of regulated mobility. In this way, Tunisian migrants have troubled the presumed universality of the paradigm of freedom of movement presented by governmental discourses aiming at managing and selecting practices of mobility. "The free mobility of whom and at what costs?" should be perhaps the question to pose every time one engages in a critical and political analysis on the struggles over freedom of movement. Such an analytical posture should redouble the gesture of discordant freedoms enacted by migrants, resisting the assumption that the freedom of mobility proposed by "liberal democracies" is the normative yardstick to judge the legitimacy of migration. Moreover, it means to challenge the conflation of freedom of movement with free circulation or, better, to reduce the former to a question of easily circulating in space (Bigo 2010): the freedom of movement I talk about and that Tunisian migrants glaringly staged does not concern the mere possibility of movement but the effective and equal possibility to move and stay without playing along with the pace of the politics of mobility

The second point that I raise in this regard concerns the relationship between "discordant" practices of freedom and the will of "not being governed like that and at that cost" that Foucault pointed out (1997, 45). This means to interrogate to what extent the practices of migration crack the going-without-saying of governmental rationality, its indisputable reality and necessity. Obviously such a question opens a field of problems that goes beyond the scope of this research. To grapple with such a question, we should first distinguish two different meanings of government—and government over lives. In fact, these struggles undo the assumption that mobility is something that should be governed or managed. And the ways in which Tunisian migrants discredited the governmental rationale was less for their acts of border crossing in itself (the practice of the "harraga," namely the act of "burning frontiers") than on the basis of what I called "discordant" practices of freedom: namely, migrant struggles that, despite sometimes concerning specific demands, exceeded any specific address toward governmental actors of "liberal democracies." Indeed, if we look at Tunisian migrants arriving in Europe, they did not claim any "reasonable" political solution—for instance, international protection or Italian citizenship—and finally, they did not address power at all since mostly during the first months of their presence on the Italian territory, they simply tried to move toward France or northern Europe but were blocked by many state controversies. However, another level of government is also at stake, I believe, when we refer to migrants' strategies of resistance. This second meaning draws on the definition of power given by Foucault in 1978: "what defines relationships of power is that it's a mode of action that does not act directly and immediately on others; instead, it acts upon their actions. . . . Power is less a confrontation between two adversaries or their mutual engagement than a question of government. To govern is to structure the possible field of action of others" (Foucault 1982, 789). Now from this point of view, it should be stressed that, as I previously showed, practices of migrations always take place within a struggle field of governmentality and that migrants are continuously "captured" in the meshes of power. Moreover, drawing on Foucault's definition of government, by challenging the techniques of bordering that constitute the migration regime, migrants put into place and enact different ways of exercising their freedom. In other words, insofar as there is a strategy of resistance at stake, however "intentional but nonsubjective" (Foucault 1990, 94) it could be, the goal is to slacken the grasp of power, to interrupt some mechanisms of capture, or to not be governed by those laws of mobility, and in this sense it is always a relational practice confronting and situated within power relations.

Ultimately, our main focus on emergent spatialities produced by practices of movements highlighted that the spatial upheavals that migrants generated forced power to reassess its strategies and targets; and migrants' *struggles for movements* opened new spaces that do not fit within existing maps.

Notes

1. Many thanks to Arnold Davidson for suggesting this expression to me.
2. In this regard, I argue that the notion of "life" is conceived here not in biological terms but as the material and complete existence of individuals as historical subjects. As Revel (2008a, 149) contends: "We have to account for life as an historical construction. Biopowers should not be conceived only as biological powers but

also as dispositives of subjection, exploitation, capture and regulation of existence. Historical and epistemological determinations which produce life. It's necessary to think at the same time the determinations that make us being what effectively we are and the possibility to detach from ourselves."

3. "Emergenza umanitaria Nord Africa: l'accoglienza dei," Protezione Civile, accessed March 3, 2015, http://www.protezionecivile.gov.it/jcms/it/view_dossier.wp;jsessionid=2AFFBDF4FAE9E6A47A503ACB356059C0?contentId=DOS24091.
4. "A Restinco appena pacificato ora arrivano decine di tunisini. Cie in emergenza," accessed March 4, 2015, http://www.brindisireport.it/cronaca/a-restinco-appena-pacificato-ora-arrivano-decine-di-tunisini-cie-in-emergenza.html.
5. *Lampedusa in Hamburg*, accessed March 4, 2015, http://lampedusa-in-hamburg.org/.
6. "Dalla Tunisia al C.A.R.A. di Mineo—'L'oxygène de la liberté,'" accessed March 4, 2015, http://www.storiemigranti.org/spip.php?article922.
7. "Lampedusa Next Stop," accessed March 4, 2015, http://www.storiemigranti.org/spip.php?article872.
8. Interview with one of the migrants who occupied the church, Massa, December 24, 2011.
9. Interview with migrants who occupied the church in Massa, December 24, 2011.
10. McNevin (2011, 131).
11. After many sit-ins in early May, on May 22 migrants occupied the churchyard of the Dome in Brescia, while on June 16 a group of migrants went up a crane in the city of Padova and in Milan, after the occupation of the tower in November 2010, they repeated the same action in September 2011.

References

Agamben, Giorgio. 1998. *Homo Sacer: Sovereign Power and Bare Life*. Stanford, CA: Stanford University Press.

Arendt, Hannah. 2003. *The Origins of Totalitarianism*. New York: Harcourt.

Balibar, Étienne. 2002. *Politics and the Other Scene*. London: Verso.

Balibar, Étienne, and Alain Brossat. 2011. "Les voix des deux rives: A propos du Printemps Arab et des migrations." *Outis* 1:143–62.

Benjamin, Walter. 1986. "Critique of Violence." In *Reflections: Essays, Aphorisms, Autobiographical Writings*, edited by Peter Demetz, 277–300. New York: Schocken.

Bigo, Didier. 2010. "Freedom and Speed in Enlarged Borderzones." In *The Contested Politics of Mobility: Borderzones and Irregularity*, edited by Vicki Squire. London: Routledge.

Butler, Judith. 2006. *Precarious Life: The Powers of Mourning and Violence*. London: Verso.

De Genova, Nicholas. 2010a. "The Deportation Regime: Sovereignty, Space and the Freedom of Movement." In *The Deportation Regime: Sovereignty, Space, and the Freedom of Movement*, edited by Nicholas De Genova and Natalie Peutz, 33–65. Durham, NC: Duke University Press.

———. 2010b. "The Queer Politics of Migration: Reflections on 'Illegality' and Incorrigibility." *Studies in Social Justice* 4(2): 101–26.

Foucault, Michel. 1982. "The Subject and Power." *Critical Inquiry* 8(4): 777–95.

———. 1990. *The History of Sexuality*. Vol. 1, *The Will to Knowledge*. London: Penguin.

———. 1997. "What Is Critique?" In *The Politics of Truth*, edited by Lotringer Sylvère and John Rajchman, 41–82. New York: Semiotext(e).

———. 2000. "Useless to Revolt?" *In Power: The Essential Works of Michel Foucault 1954–1984*. Vol. 3. London: Penguin.

Isin, Engin. 2002. *Being Political: Genealogies of Citizenship*. Minneapolis: University of Minnesota Press.

McNevin, Anne. 2011. *Contesting Citizenship: Irregular Migrants and New Frontiers of the Political*. New York: Columbia University Press.

Mezzadra, Sandro. 2006. *Diritto di fuga: Migrazioni, cittadinanza, globalizzazione*. Verona: Ombre Corte.

———. 2011. "Le avventure mediterraneee della libertà." In *Libeccio d'oltremare: Il ventodelle rivoluzioni del Nord Africa si estende all'Occidente*, edited by Pirri Ambra. Rome: Ediesse Edizioni.

Mometti, Felice, and Maurizio Ricciardi. 2011. *La normale eccezione. Lotte migranti in Italia*. Rome: Alegre.

Nyers, Peter. 2011. "Forms of Irregular Citizenship." In *The Contested Politics of Mobility: Borderzones and Irregularity*, edited by Vicki Squire. London: Routledge.

Papadopoulos, Dimitris, Niamh Stephenson, and Vassilis Tsianos. 2008. *Escape Routes: Control and Subversion in the 21st Century*. London: Pluto Press.

Revel, Judith. 2008a. "Identità, natura, vita. Tre decostruzioni biopolitiche." In *Foucault oggi*, edited by M. Galzigna. Milan: Feltrinelli.

———. 2008b. "Resistances, Subjectivities, Common." Translated by Arianna Gove. *Generation Online*. http://www.generation-online.org/p/fprevel4.htm.

Sossi, Federica. 2012. "Qui e lì sono la Stessa Cosa. Sommovimenti di spazi e narrazioni." In *Spazi in migrazione: Cartoline di una rivoluzione*, edited by Sossi Federica. Verona: Ombre Corte.

Walters, William. 2011. "Rezoning the Global: Technological Zones, Technological Work and the (Un)making of Biometric Borders." In *The Contested Politics of Mobility: Borderzones and Irregularity*, edited by Vicki Squire, 51–73. London: Routledge.

PART FIVE

Biopower Today

FIFTEEN

Biopower Today

PAUL RABINOW AND NIKOLAS ROSE

> Q: Isn't it logical, given these concerns, that you should be writing a genealogy of biopower?
>
> MF: I have no time for that now, but it could be done. In fact, I have to do it.
>
> —Foucault (1984, *344)*

What is "biopower"? In a book ostensibly devoted to the history of sexuality, *La Volonté de savoir*, published in 1976 (English trans. 1978), Michel Foucault includes six highly provocative pages on this theme in a chapter entitled "Right of Death and Power over Life." For a long time, he argues, one of the privileges of sovereign power was the right to decide life and death, a right that, by the classical age, had been constrained to occasions when the sovereign himself was threatened from enemies without and within. This was the juridical form of sovereign power—the right of a ruler to seize things, time, bodies, and ultimately the lives of subjects. It was the model of power that was codified and generalized in classical political philosophy—a model that remained essentially unaltered when the "king's head" was displaced from sovereign to state. But, Foucault argues, since the classical age, deduction has become merely one element in a range of mechanisms working to generate, incite, reinforce, control, monitor, optimize, and organize the forces under it. While external wars are bloodier than ever, and regimes visit holocausts on their own populations, these wars are waged not in the name of the sovereign, argues Foucault, but in the name of the existence of everyone:

> Entire populations are mobilized for the purpose of the wholesale slaughter in the name of life necessity. . . . It is as managers of life and survival, of bod-

> ies and the race, that so many regimes have been able to wage so many wars, causing so many men to be killed. (Foucault 1978, 137)

Power, Foucault argues, is now situated and exercised at the level of life.

Foucault promised to flesh out his sweeping generalizations in one of the six proposed volumes of the *History of Sexuality* whose titles appear on the book's back jacket. That promise was not fulfilled, although he devoted a number of his 1976 lectures to this theme. But he did propose a rather simple and now familiar bipolar diagram of power over life. In this diagram, one pole of biopower focuses on an anatomo-politics of the human body, seeking to maximize its forces and integrate it into efficient systems. The second pole is one of regulatory controls, a biopolitics of the population, focusing on the species body, the body imbued with the mechanisms of life: birth, morbidity, mortality, longevity. He claims that this bipolar technology, which begins to be set up in the seventeenth century, seeks "to invest life through and through" (Foucault 1978, 139). And, by the nineteenth century, he argues, these two poles were conjoined within a series of "great technologies of power" of which sexuality was only one. In so establishing themselves, new kinds of political struggle could emerge, in which "life as a political object" was turned back against the controls exercised over it, in the name of claims to a "right" to life, to one's body, to health, to the satisfaction of one's needs.

At its most general, then, the concept of biopower brings into view a field comprised of more or less rationalized attempts to intervene on the vital characteristics of human existence. The vital characteristics of human beings, as living creatures who are born, mature, inhabit a body that can be trained and augmented, and then sicken and die. And the vital characteristics of collectivities or populations composed of such living beings. And, while Foucault is somewhat imprecise in his use of the terms, within the field of biopower, we can use the term "biopolitics" to embrace all the specific strategies and contestations over problematizations of collective human vitality, morbidity, and mortality; over the forms of knowledge, regimes of authority, and practices of intervention that are desirable, legitimate, and efficacious.

More than quarter of a century after the introduction of this concept, at the threshold of our current "biological century," this contested field of problems and strategies is more crucial and enigmatic than ever.[1] Yet surprisingly little work has been done to develop Foucault's own sketchy suggestions into a set of operational tools for critical inquiry. The term "biopower" is more likely to be taken to refer to the generation of energy from renewable biological material. The term "biopolitics" has been taken up by advocates of a range of environmental and ecological causes.[2] However, we believe that Foucault's concepts of biopower and biopolitics retain considerable analytical utility. As a first step toward some conceptual clarification, and drawing on our previous work in this area, we propose that the concept of biopower designates a plane of actuality that must include, at a minimum, the following elements (Rabinow 1994, 1996, 1999; Rose 2001, 2006):

- One or more truth discourses about the "vital" character of living human beings and an array of authorities considered competent to speak that truth. These truth discourses may not themselves be "biological" in the contemporary sense of the discipline, for instance, they may hybridize biological and demographic or even sociological styles of thought, as in the contemporary relations of genomics and risk, merged in the new language of susceptibility.
- Strategies for intervention on collective existence in the name of life and health, initially addressed to populations that may or may not be territorialized on the nation, society, or pre-given communities, but may also be specified in terms of emergent biosocial collectivities, sometimes specified in terms of categories of race, ethnicity, gender, or religion, as in the emerging forms of genetic or biological citizenship.
- Modes of subjectification, through which individuals are brought to work on themselves, under certain forms of authority, in relation to truth discourses, by means of practices of the self, in the name of their own lives or health, that of their family or some other collectivity, or indeed in the name of the life or health of the population as a whole: Rabinow has examined the formation of new collectivities in terms of "biosociality," and Rose has examined the formation of kinds of human subject in terms of "somatic individuality."

The Limits of Biopower

We frame our initial specification in these limited terms partly in response to the ways in which the terms "biopower" and "biopolitics" have been used by two of our leading contemporary philosophers—Giorgio Agamben and Antonio Negri. Agamben and Negri have each made these terms central to their critical analyses of the politics of the present. Their work has had a very significant impact on social and political thought, especially in the United States. What appears to attract many is the generality of their claims to characterize the nature and essence of the present epoch. These authors suggest that contemporary biopower takes the form of a politics that fundamentally depends on the domination, exploitation, expropriation, and, in some cases, elimination of the vital existence of some or all subjects over whom it is exercised. Contemporary biopower, they imply, is a form of power that ultimately rests on the power of some to threaten the death of others. Yet we consider the epochal philosophical deployments of the terms to be misleading; it is relevant to consider them in a little more detail to counterpose them to our own approach.[3]

Empire

For Hardt and Negri, biopower is an encompassing, totalizing term—biopower serves to secure the dominion of a global form of domination that they term "Empire" (Hardt and Negri 2000). Theirs is a neo-Marxist reading: their first premise is that the work of power should be understood as the extraction of some kind of "surplus value" from human life on which Empire depends. This is what they mean when they assert that all contemporary politics is biopolitics: it is a "form of power that regulates social life from its interior" (2000, 23). Further, they conflate this omnipotent and all-pervasive biopower with an idea loosely derived from a short and speculative essay by the French philosopher Gilles Deleuze, in which he argues that we have moved from "societies of discipline" to "societies of control" (Deleuze 1995). Michel Foucault (1977) had characterized "disciplinary societies" as those in which the management of inclusion and exclusion was accomplished by an archipelago of disciplinary institutions dotted across the social field—asylums, factories, schools, hospitals, universities—each seeking to implant a mode of conduct into body and its correlate soul. Today, argued Deleuze, writing in the closing decades of last century, control was not confined within such institutions but was immanent in the flexible, fluid, and fluctuating networks of existence itself. Hardt and Negri take up this idea when they suggest that biopolitics is a form of power "expressed as a control that extends throughout the depths of the consciousnesses and bodies of the population" (2000, 24). They claim that this biopolitical power is exercised in the name of multinational and transnational corporations that, since the second half of the twentieth century, have chosen "to structure global territories bio-politically" (2000, 31).

Biopower, here, is enrolled in an attempt to resurrect a revolutionary view of world history, which ends with a twist of Christianity to inspire resistance to Empire: Hardt and Negri cite a legend about St. Francis of Assisi who "refused every instrumental discipline, and . . . posed a joyous life . . . against the will of power and corruption. Once again in post modernity we find ourselves in Francis's situation, posting against the misery of power the joy of being" (2000, 413). Now we would certainly agree that it is necessary to extend the scope of traditional analyses of economic exploitation and geopolitics to grasp the ways in which the living character of human beings, and indeed of other living beings, is being harnessed by "biocapital."[4] But it is difficult to see what analytical work can be done by such an expanded concept of biopower: in the end, Hardt and Negri merely provide a superficial description of certain aspects of our present, framed within the kind of grand narrative of history that other theorists of postmodernity had proclaimed a thing of the past. But political evaluation of the forms of biopolitics is evaded, submerged under their simplistic Manichean opposition of a mysterious global Empire to an even more phantom "multitude." This version of the concept of biopower is quite antithetical to that proposed by Foucault: the concept is emptied of its critical force—it can describe everything but analyze nothing.

It might be useful here to remind ourselves that when Foucault introduced the term in the last of his Collège de France lectures of 1975–76, *Society Must Be Defended* (2002), he is precise about the historical phenomena that he is seeking to grasp. He enumerates them: issues of the birthrate and the beginnings of policies to intervene on it; issues of morbidity, not so much epidemics but the illnesses that are routinely prevalent in a particular population and sap its strength, requiring interventions in the name of public hygiene and new measures to coordinate medical care; the problems of old age and accidents to be addressed through insurantial mechanisms; the problem of the race and the impact on it of geographic, climatic, and environmental conditions, notably in the town. The concept of biopower is proposed after ten years of collective and individual research on the genealogy of power over life in the eighteenth and nineteenth centuries.[5] Foucault himself had lectured on the politics of health in the eighteenth century in Japan and in Brazil; his seminar members were producing detailed historical studies of the role of medicine, town planning, royal shipyards, and a host of other sites in which experiments about how to produce and regulate ways of maximizing the capacities of both the population and the individual as a target of power are being carried out. The concept of biopower—like that of discipline—was not transhistorical or metaphoric, but precisely grounded in historical, or genealogical, analysis.

We should also note that biopower, for Foucault, does not emerge from or serve to support a single power bloc, dominant group, or set of interests. While initially linking biopolitics to the regulatory endeavors of developing states, Foucault recognized that "the great overall regulations that proliferated throughout the nineteenth century . . . are also found at the sub-State level, in a whole series of sub-State institutes such as medical institutions, welfare funds, insurance, and so on" (2002, 250). This is the point at which Foucault began to develop his concept of "governmentality," a concept whose whole rationale was to grasp the birth and characteristics of a whole variety of ways of problematizing and acting on individual and collective conduct in the name of certain objectives that do not have the state as their origin or point of reference. As he develops this line of thought concerning the multiplicity of forms and sources of authority, Foucault also distances himself from the view that such power over life is unambiguously nefarious.[6] This is also the turning point that leads Foucault to a fascination with ancient modes of subjectification and the possibilities of freedom. In this context, it is worth remembering that medicine is perhaps the oldest site where one can observe the play of truth, power, and ethics in relation to the subject, and to the possibilities of a good or, as the Greeks would have it, a flourishing life.

Homo Sacer

Giorgio Agamben, in a series of haunting books, identifies the Holocaust as the ultimate exemplar of biopower and biopower as the hidden meaning of all forms of power from the ancient world to the present. In particular he explores the moments that he terms, after Carl Schmitt, "states of exception," when a sovereign state declares a time or a place where the rule of law can be suspended in the name of self-defense or national security (Agamben 1995, 1996, 1998, 2000a, 2000b, 2005). Much can be learned from these studies of the profound traumas that mark European histories: we agree that Holocaust is not an exceptional moment of throwback to a singular barbarianism, but an enduring possibility intrinsic to the very project of civilization and the law. However, Agamben grounds his analysis in a particular way that we find problematic. He argues that all power rests ultimately on the ability of one to take the life of another—it is a power over life grounded in the possibility of enforcing death. He characterizes this power by reference to the obscure metaphor of *homo sacer*—the enigmatic figure in Roman law whose crimes made his sacrifice impossible but who could be killed with impunity. Like this figure, who is reduced from *bios*—crudely, the way of life proper to an individual or a group in a polity—to *zöe*—"bare life"—he suggests that the birth of biopower in modernity marks the point at which the biological life of subjects enters politics and belongs entirely to the state. The ultimate grasp of the sovereign or the state over the lives of subjects is exemplified, for him, in the concentration camps, labor camps, and death camps of the Nazis: sovereign states depend on their capacity to create states of exception. Such states may be exceptional but are nonetheless immanent in modernity itself—a fourth space added to that of state, nation, and land, in which inhabitants are stripped of everything but their bare life, which is placed without recourse in the hands of power. Indeed they are the "nomos" of modernity: "This is why the camp is the very paradigm of political space at the point at which politics becomes biopolitics and *homo sacer* is virtually confused with the citizen" (Agamben 1998, 171).

Agamben takes seriously Adorno's challenge—how is it possible to think after Auschwitz (Mesnard and Kahan 2001)? But, for that very reason, it is to trivialize Auschwitz to see it as the hidden possibility in every instance where living beings enter the scope of regulation, control, and government. The power to command under threat of death is exercised by states and their surrogates in multiple instances, in micro forms, and in geopolitical relations. But this does not demonstrate that this form of power—commands backed up by the ultimate threat of death—is the guarantee or underpinning principle of all forms of biopower in contemporary liberal societies. Nor is it useful to use this single diagram to analyze every contemporary instance of thanatopolitics—from Rwanda to the epidemic of AIDS deaths across Africa. Surely the essence of critical thought must be its capacity to make distinctions that can facilitate judgment and action.[7]

Holocaust is undoubtedly one configuration that modern biopower can take. Racism allows power to subdivide a population into subspecies, to designate these in terms of a biological substrate, and to initiate and sustain an array of dynamic relations in which the exclusion, incarceration, or death of those who are inferior can be seen as something that will make life in general healthier and purer. As Foucault put it in 1976, "racism justifies the death-function in the economy of biopower by appealing to the principle that the death of others makes one biologically stronger insofar as one is a member of a race or a population" (2002, 258). It is true that in this lecture he suggests that it is "the emergence of biopower that inscribes [racism] in the mechanisms of the State . . . as the basic mechanism of power, as it is exercised in modern States" (2002, 254). But the Nazi regime was, in his view, exceptional—"a paroxysmal development":

> We have, then, in Nazi society something that is really quite extraordinary: this is a society which has generalized biopower in an absolute sense, but

> which has also generalized the sovereign right to kill . . . to kill anyone, meaning not only other people but also its own people . . . a coincidence between a generalized biopower and a dictatorship that was at once absolute and retransmitted throughout the entire social body. (2002, 260)

Biopower, in the form it took under National Socialism, was a complex mix of the politics of life and the politics of death—as Robert Proctor (1999) points out, Nazi doctors and health activists waged war on tobacco, sought to curb exposure to asbestos, worried about the overuse of medication and X-rays, stressed the importance of a diet free of petrochemical dyes and preservatives, campaigned for whole-grain bread and foods high in vitamins and fiber, and practiced vegetarianism. But, within this complex, the path to the death camps depended on a host of other historical, moral, political, and technical conditions. Holocaust is neither exemplary of thanatopolitics nor the hidden dark truth of biopower.

Sovereignty

Our criticism here is linked to a disagreement about "sovereignty." While Hardt and Negri differentiate "Empire" from the forms of sovereignty that emerged in the nation state, the diagram remains more or less unaltered: although "imperial sovereignty . . . is organized not around one central conflict but rather through a flexible network of microconflicts," Empire nonetheless gathers unto itself the power relations that traverse all those "elusive, proliferating and non-localizable contradictions" (2000, 201). The monolithic image of Empire thus tries to condense and unify all those forms and relations into a single sovereign power, to which can only be opposed some force that is radically other, gestured to in the name "multitude": the multitude, then, is the contemporary incarnation of the regicide, who, in eliminating the sovereign, will inaugurate an epoch in which sovereign power is reappropriated by subjects themselves. Despite its apparent radicalism, anticapitalists would do well to be wary of the religious underpinnings of this fable of resistance as deliverance to a promised land.

For Agamben, sovereignty also has something of a sacred form—the ancient ritual declaration of *homo sacer* remains present today in the capacity of the sovereign state to establish the state of exception, to commit those stripped of the rights of bios to those zones, and to torture or kill those reduced to the status of *zöe*—bare life—without legal restraint. But this sovereign power is no longer confined to those who are explicitly agents of the state—it apparently extends to all those who have authority over aspects of human vital existence. Hence Agamben argues that the power over life exercised today by "the jurist . . . the doctor, the scientist, the expert, the priest" arises from the alliance with the sovereign into which they have entered; like those who populated an earlier image of power, Althusser's ideological state apparatuses, wittingly or unwittingly they do the sovereign's will (Agamben 1998, 122). *Homo sacer*, for Agamben, is thus not a historically marginal phenomenon: it demands our attention as critical thinkers precisely because it is the ordering principle of contemporary societies. Against such a "growing bio-political nightmare" the only solution seems equally sacred: no wonder Agamben invokes the figure of a messianic "end of time," taken from Walter Benjamin, as one possible way out.

The interpretation of contemporary biopolitics as the politics of a state modeled on the figure of the sovereign, and of all forms of biopolitical authority as agents of that sovereign, suits the twentieth-century absolutisms of the Nazis and Stalin. But we need a more nuanced account of power, and of sovereign power, to analyze contemporary rationalities and technologies of biopolitics. Sovereignty did, of course, entail the right to take life, but the essence of premodern sovereign power was its sporadic and discontinuous nature—that indeed was the rationale for its excesses. The totalization of sovereign power as a mode of ordering daily life at all times and places across a territory would be too costly; indeed, as many historians have argued, the excessive form in which this power is exercised, for example, in spectacular public executions and the elaborate rituals of the courtroom, seeks to compensate for its sporadic nature. Sovereignty, in this sense, is precisely a diagram of a totalized and singular form of power not a description of its implementation. Certainly some forms of colonial power sought to operationalize it, but, in the face of its economic and governmental costs, colonial statecraft was largely to take a different form. The two megalomaniac state forms of the twentieth century—Stalin's Russia, Hitler's Germany—also sought to actualize it, as have some others in their wake: Albania under Hoxha, North Korea. But no historian of premodern forms of control could fail to notice the dependence of sovereign rule on a fine web of customary conventions, reciprocal obligations, and the like—in a word, a moral economy whose complexity and scope far exceeds the extravagant displays of the sovereign. Sovereign power is at one and the same time an element in this moral economy and an attempt to master it. A cursory glance at the work of Jacques Le Goff—whose work Foucault knew well—or Ferdinand Braudel and the whole *Annales* project, or, for English readers, the writings of E. P. Thompson should be sufficient to dispel such recent misreadings (Hay 1975; Thompson 1975; Braudel and Labrousse 1976; Le Goff 1980, 1990).

Over the eighteenth and nineteenth centuries, nation states, in addition to their theaters of power and public display, began to be key mobilizers of the internal forces of their territories so as to secure their objectives of prosperity and security. But the governmentalized states of the late nineteenth century took the shape that they did through the prior formation of ever-growing apparatuses of knowledge collection and problematization that formed alongside the state apparatus, often in conflict with it, in the emergent terrain of the "social." From this time on, states can rule only because of the ways in which they manage to connect themselves up to these apparatuses, which have their own logics and viscosity. So long as regimes aspire to liberalism, such apparatuses and authorities will exercise demands and constraints on central powers. Nonstate bodies have played a key role in biopolitical struggles and strategies since the origin of "the social"—philanthropic organizations, social investigators, pressure groups, medics, feminists, and assorted reformers have all operated on the territory of biopower.

Since the end of the Second World War, and taking here only the example of health, a range of powerful agencies within states and a range of transnational bodies have taken on a new importance. So have a host of bioethics commissions, regulatory agencies, and professional organizations: a whole *"bioethical complex,"* in which the power of medical agents to "let die" at the end of life, the start of life, or in reproduction, are simultaneously enhanced by medical technology and regulated by other authorities as never before. Further, we have seen the rise of new kinds of patients' groups and individuals, who increasingly define their *citizenship* in terms of their rights (and obligations) to life, health, and cure. And, of course, new circuits of *bioeconomics* have taken shape, a large-scale capitalization of *bioscience* and *mobilization* of its elements into new exchange relations: the new molecular knowledges of life and health are being mapped out, developed, and exploited by a range of commercial enterprises, sometimes in alliance with states, sometimes autonomous from them, establishing constitutive links between life, truth, and value. This is a far from homogeneous field of agents, tactics, strategies, and objectives. Yet, at the same time, states do retain power to designates zones of exception, even when their legality is dubious—the camp remains a grim reality from the wars in the Balkans, through Guantanamo Bay to the "detention centers" springing up across Europe to incarcerate "asylum seekers" and others who trespass on the spaces of bios but are not admitted. Do these all form part of a single configuration of biopower? This remains to be demonstrated. And even if they did, we doubt that such a biopower could be characterized solely, or even principally, in terms of its propensity for "making die": for while death is part of the picture, it takes the form of "letting die" as much as of "making die." But also, of course, central to the configuration of contemporary biopower are all those endeavors that have life, not death, as their telos—projects for "making live."

We have suggested that the concept of biopower seeks to individuate strategies and configurations that combine three dimensions or planes—a form of truth discourse about living beings and an array of authorities considered competent to speak that truth; strategies for intervention on collective existence in the name of life and health; and modes of subjectification, in which individuals can be brought to work on themselves, under certain forms of authority, in relation to truth discourses, by means of practices of the self, in the name of individual or collective life or health. Although we draw these elements from Foucault's all too brief interventions on the concept, it is worth remembering that his principal site of investigation was historical. He studied the emergence of forms of power in the eighteenth century, their transformation in the nineteenth, and, to some limited extent, an examination of the forms taking shape at the end of the nineteenth century. Rationalities, strategies, and technologies of biopower changed across the twentieth century, as the management of collective life and health became a key objective of governmentalized states, and novel configurations of truth, power, and subjectivity emerged to underpin the rationalities of welfare and security as well as those of health and hygiene (Donzelot 1979; Rose 1985, 1999; Ewald 1986; Rabinow 1989). It would certainly be misleading simply to project Foucault's analysis forward as a guide to our present and its possibilities. One key mutation concerns the relations between what one might term, clumsily, the macro and the micro, or, following Deleuze, the molar and the molecular, poles of this mode of power. That is to say, on the one hand, the emphases and relations on ways of thinking and acting at the level of population groups and collectivities, variously defined; and, on the other hand, the individualization of biopolitical strategies. Undoubtedly, in the era of the social state—and in those locales where such states still form the organizing principle of political struggle—it was the molar that was privileged. In the twentieth century, states not only developed or supported insurantial mechanisms of security, but gathered together, organized, and rationalized the loose threads of medical provision, specified and regulated standards of housing, engaged in campaigns of health education, and the like. Even liberal states also played their role in the battle against degeneracy, imposing immigration controls, sometimes legitimating compulsory or quasi-compulsory sterilization, encouraging organizations giving eugenic guidance on marriage and procreation and so forth. Of course, each of these was to have its "molecular" counterpart, for example, in the transformation of the home into a machine for health, and the education and solicitation of mothers as ancillary workers in the health care of their children.

Today, much of this configuration remains, and, indeed, some of it has been translated to a supranational level in the endeavors of the European Union, the World Bank, and the like. But, with the decline of the domain of the social as a privileged site of national objectivation and intervention in the "advanced liberal" societies of the West, we observe new collective formations emergent everywhere (Rabinow 1996; Rose and Novas 2005). At the same time, as we can observe in the politics surrounding the sequencing of the human genome, we see the birth of new modes of individualization and conceptions of autonomy with their associated rights to health, life, liberty, and the pursuit of a form of happiness that is increasingly understood in corporeal and vital terms (Rabinow 1994; Rose 2001).

Analytics of Biopower

To develop this argument further, drawing on our current research, we will focus on three topics that seem to us to condense some of the biopolitical lines of force active today: race, reproduction, and genomic medicine. Of course, to place all these diverse developments within the ambit of biopower is not to imply that there is some unity at work here, or some essence—truth or falsity—that all these forms exemplify or embody. We need to recognize dispersion, contingency, and virtuality, although not with deconstructionist intent. Before we can see whether some general political rationality is emerging, the task of analysis is to articulate some preliminary diagnoses at a smaller scale. Placing the evidence from such analyses in the framework of biopower, we think we can begin to identify and analyze elements of such a domain, though it is neither stable nor homogeneous, nor does it merely repeat patterns familiar from history. A modest empiricism, attentive to peculiarities, to small differences, to the moments when shifts in truth, authority, spatiality, or ethics make a difference for today as compared to yesterday, reveals configurations that do not conform to the images provided by our philosophers.[8] In these configurations, race, health, genealogy, reproduction, and knowledge are intertwined, continually recombining and transforming one another. By this we mean that knowledge of health transforms the idea of race, that ideas of genealogy are reframed by new conceptions of reproduction, that changing ideas of genealogy radically impact on the politics of race, races, and racism. Let us turn to explore some of these issues in some more detail.

Race

Race, together with health, and in variable relations with it, has been one of the central poles in the genealogy of biopower.[9] Conceptions of race formed a prism not just for the imagination of the nation, but also for the political management of national health and vitality, and of international competitiveness, from the so-called war of nations in the eighteenth century (the topic of several lectures by Foucault in *Society Must Be Defended*), through the massive biologization of race in the nineteenth century, linked to pre- and post-Darwinist evolutionary thinking and applied both within states and in their colonial dominations, to the later nineteenth-century obsession with degeneracy and race suicide, and the strategies of eugenics that spread from the United States to Japan and elsewhere in the first half of the twentieth century. After the Second World War, official racialist discourses were discredited: by 1963, for example, the United Nations Declaration on the Elimination of All Forms of Racial Discrimination took as one of its premises "that any doctrine of racial differentiation or superiority is scientifically false, morally condemnable, socially unjust and dangerous, and that there is no justification for racial discrimination either in theory or in practice" (United Nations General Assembly 1963, preamble).[10] Of course, racialist practices hardly subsided, but a biological understanding of racial categories was no longer "in the true" in political or policy discourse. In part due to the persistent interventions of radical critics, the link between biological understandings of distinctions among population groups and their sociopolitical implications seemed broken or at least denaturalized. Many biologists still believed they encountered such differences, not least in examining the prevalence of particular diseases in different regions or the efficacy of medicines in different national populations, but such arguments tended to remain in the technical literature. Some individuals and groups persisted in making public claims for politically pertinent correlation between human qualities and racially differentiated biological capacities in a whole number of controversies from education to criminality, but even those with scientific credentials, such as William Shockley, largely argued this from outside the truth discourses of biology. In many countries, not least the United States, race was crucial as a socioeconomic category, a mark of discrimination, and a mode of identification that remained extremely salient socially and politically, from the allocation of federal funds to the manifestations of identity politics. But, despite the fact that race functioned as a marker of belonging and the basis of a claim to disadvantage, even when groups or individuals sought to trace their "roots" they seldom related this genealogy to a biological substrate. The same is true of the murderous racist wars that spread across Europe in the wake of the demise of the Soviet empire, from Armenia to the Balkans. Appeals to racial identities to ground the elimination of other groups needed no justification in the truth discourse of biology. While, in Rwanda, Hutus referred to Tutsis as "cockroaches," such epithets were hardly elements in a political rationality drawing on biological understandings of racial difference.

At the turn of the new century, however, race is once again reentering the domain of biological truth, viewed now through a molecular gaze. At a certain moment, when it became clear that humans shared more than 98 percent of their genome with chimpanzees, and that intergroup variations in DNA sequences were greater than intragroup variations, it appeared that genomics itself would mark the terminal point of biological racism (perhaps even species-ism). But this humanitarian dream proved to be short-lived. A new molecular deployment of race has emerged seemingly almost inevitably out of genomic thinking. Critics denounced the model of a single genome that underpinned the Human Genome Project (HGP), fearing that it would establish a white male norm. The first move here was cast as ethical: as the initial proposer of this work, L. Luca Cavalli-Sforza, put it: "to explore the full range of genome diversity within the human family" and "to help combat the widespread popular fear and ignorance of human genetics and . . . make a significant contribution to the elimination of racism" (quoted in M'Charek 2005, 5–6). Despite criticism, this effort to ensure the recognition of diversity in the framing of scientific truth as an essential dimension of genomic knowledge was later adopted by the HGP, which provided US $1.2 million to set up workshops to develop the technical and organizational aspects of the project, to consider the social and ethical implications, and to conduct a pilot study (Reardon 2001, 2005; M'Charek 2005).

Genome mapping led to the conclusion that, while the DNA sequence of any two randomly selected individuals will be 99.9 percent identical, the variations at the level of single DNA base—called Single Nucleotide

Polymorphisms (SNPs)—are very significant, notably in relation to susceptibility to disease. On average, it was claimed, one letter in 1,000 differed between two individuals—which made a total of many million variations between them—estimates of the number ranged from 6 million to 15 million. The National Institutes of Health (NIH) and the Wellcome Trust have given considerable funds to research into the establishment of genomic differences at the single nucleotide level. Thus, in 1999, Wellcome announced a consortium with ten pharmaceutical companies to find and map 300,000 common DNA sequence variations. Further, it became clear that sets of nearby SNPs on the same chromosome are inherited in blocks. The pattern of SNPs on a block is termed a haplotype. While blocks may contain a large number of SNPs, a few SNPs—known as tag SNPs—are enough to uniquely identify a haplotype. Haplotype mapping promised a more economical way of identifying SNPs relevant to disease, and the NIH and Wellcome, together with labs in Japan and China, are collaborating in an international HapMap project. Such funding has been justified precisely in biopolitical terms, as leading toward and ensuring the equal health of the population in all—or some—of its diversity. For example, to create the HapMap, researchers will analyze DNA from blood samples collected in Nigeria, Japan, China, and the United States—from US residents with ancestry from northern and western Europe. While the samples will be anonymous, they will be identified by the population from which they were collected. The SNP differences that account for 0.1 percent of the 3 billion base pairs of the human genome seem to provide ample space for population differentiations as they affect differences that have significance for human health.

The science itself, and the recognition of the variability of the human genome at the level of the single nucleotide, thus immediately open up a new way of conceptualizing population differences—in terms of geography and ancestry—at the molecular level. In addition to the ethical humanism of the state projects, additional pressure to proceed in this direction came in some areas from the demands of patient groups for genomic self-knowledge, and in others from the commercial aspirations of pharmaceutical companies and the biomedical industry for a genomic strategy for diagnosis, drug development, and marketing. By 2003 multiple projects were underway to map diversity at the level of the SNP. Strikingly, Howard University in the United States has generated a database of DNA sequences to be used to exploring the genomic bases of disease among black Americans and also to trace individuals' "roots" to their preslavery origins in very specific regions of Africa. This contemporary program to identify biological differences is not undertaken in the name of population purity but of national economic development, the search for health in biosocial communities, and the growing sense of many individuals that genetics in some way holds the key to their "identity."

It would be tempting to say that this highly sophisticated genomics has produced new complexity into the figure of humanity. But, despite the heated debates in the medical literature, the core racial typology of the nineteenth century—white (Caucasian), black (African), yellow (Asian), red (Native American)—still provides a dominant mold through which this new genetic knowledge of human difference is taking shape and entering medical and lay conceptions of human variation. Medical researchers and gene mappers specify their populations and their samples in such terms, drug companies seek to target specific pharmaceuticals to groups designated, for example, as "African Americans," and individuals seek to trace their "African" roots through matching the patterns of their SNPs with those from villages pillaged by the slave trade in Niger or Cameroon. It is undoubtedly the case that SNP mapping will produce typologies of difference between "population groups" and almost inevitable that these population groups, in the name of health, will be coded in terms of broad cultural conceptions of race. New challenges for critical thinking are raised by the contemporary interplay between political and genomic classifications of race, identity politics, racism, health inequities, and their potential entry into biomedical truth, commercial logics, and the routine practices of health care. We do not think it helpful to assert in advance that such endeavors are based on faulty premises, let alone to suggest that they are implicitly racist and will exacerbate discrimination. Contemporary genomics is principally directed at illness conditions rather than gross characteristics such as intelligence or personality. It understands most of those conditions as arising out of interactions between multiple coding regions, where gene expression can be activated and inactivated by many environmental factors at levels ranging from the cellular to the familial, the social, and the environmental. It seeks not to pronounce on destiny per se but rather to render the future as probabilistic and to open it to hope and to technical intervention. It would be unhelpful and misleading to regard this configuration as a replay of the past or to submerge it within some imagined global logic of biopower: instead we need to identify the points where critical judgment, diagnosing new possibilities and dangers, might play a part in the direction it takes.

Reproduction

For Foucault, sexuality was crucial, in part because it was the hinge that linked an anatomo-politics of the human body with a biopolitics of the population. But today, perhaps for as long as the last fifty years, these issues have become decoupled. Sexuality has been disengaged, to a degree, from the symbolics and practices of reproduction, and reproduction itself has become the object of a series of forms of knowledge, technologies, and political strategies that have little to do with sexuality. From about the 1970s one can see a triple movement. The question of reproduction gets problematized, both nationally and supranationally, because of its economic, ecological, and political consequences—overpopulation, limits to growth, and so on. A new politics of abortion emerges, taking different forms in different national contexts. And, in the West at least, a related issue of "reproductive choice" begins to take shape, when a small number of couples in the West, in alliance with some doctors, strove to define infertility as a potentially remediable medical condition and, consequently, the site of legitimate interventions. All of these sites jointly, yet differently, combined in making reproduction a problem space, in which an array of connections appear between the individual and the collective, the technological and the political, the legal and the ethical. This is a biopolitical space par excellence.

The new reproductive technologies involving the micromanipulation of eggs and sperm, and preimplantation genetic diagnosis and selection, although they have attracted the most attention in the Anglo-American world, have actually been rather restricted in their impact on national populations, let alone on global population politics. Though they have been the site of a discursive explosion, the focus of regulatory attention, and political and ethical controversies in many Western countries, it is hard to discern some unified biopolitical strategy underlying these developments. The rhetoric of choice clearly resonates with the ethic of autonomy at the heart of advanced liberal modes of subjectification, and the transformation of infertility into a treatable illness exemplifies the reimagining of human capacities as open to reengineering and enhancement by medicine. However, we need to recognize the limited scope of these procedures, and the fact that they are far from routine and often unsuccessful. Biopolitically, reproductive choice in the form of embryo selection, far from being in the service of general racial improvement or even individualized "designer babies," has been almost entirely limited to the identification of fetuses with major malformations or crippling and terminal genetic disorders (Franklin 1997, 2006; Throsby 2004). Even then, the use of diagnostic techniques has not inevitably led to termination but often to providing anticipatory information in the services of the kinds of life planning that have become intrinsic to forms of life in contemporary liberal societies. Perhaps, as many feminists have argued, the principal biopolitical achievement here lies on the axis of subjectification: these strategies exhibit the characteristic formation in which apparent choices entail new forms of "responsibilization" and impose onerous obligations, especially, in this case, on women.

Less attention in the Western academy has been paid to the "molar" pole of the management of reproduction—the campaigns for population limitation that have spread across the Indian subcontinent, China and South East Asia, and many Latin American countries. These biopolitical strategies are undoubtedly underpinned by truth claims, although they are those of demography and economics, not of heredity and eugenics. Take, for example, the 1972 publication of the report from the Club of Rome entitled *Limits to Growth*. Using a model derived from system dynamics for its analysis, the report concluded that

> if the present growth trends in world population, industrialization, pollution, food production, and resource depletion continue unchanged, the limits to growth on this planet will be reached sometime within the next one hundred years. The most probable result will be a rather sudden and uncontrollable decline in both population and industrial capacity. (Meadows et al. 1972, 23)

Fundamental to their prescription to avert this problem was birth control to stabilize population, by limiting family size to two children, especially in those countries where it currently greatly exceeded that, but even this path was no guarantee of success.

> We end on a note of urgency. We have repeatedly emphasized the importance of natural delays in the population-capital system of the world. These delays mean, for example, that if Mexico's birth rate gradually declined from its present value to an exact replacement value by the year 2000, the country's population would grow from 50 million to 130 million. We cannot say with certainty how much longer mankind can postpone initiating deliberate control of its growth before it will have lost the chance for control. (Meadows et al. 1972, 182)

These dire warnings resonated with a raft of analogous concerns about the impact of population growth on economic wealth and the need for governments—especially those of less developed states—to introduce policies to curtail reproduction—especially among the poor—as a prerequisite to modernization. These varied from the coercive—China's One Child Policy or the sterilization campaigns in India are the two best-known examples—to those that gradually came to adopt principles of informed consent to what was euphemistically termed "voluntary surgical contraception"—for example, in Mexico. They were based on demographic data and algorithms linking population growth to economic performance developed by geographers and mathematicians, embedded in educational programs for development workers and others, proselytized by numerous private pressure groups and policy advisory bodies, and built into the policies of development agencies such as the "Office of Populations" of the "Bureau for Global Programs" of the United States Agency for International Development (USAID). The "population time bomb" became part of the common sense of public opinion in the West, and a major justification for aid from advanced industrial societies to poorer countries was that this would enable them to limit their population and hence the danger that their population growth posed.

By the end of the 1980s, policies for the limitation of procreation among the poor stressed the importance of voluntary assent and informed choice, and argued that the aim was to prevent the misery of maternal deaths and perinatal mortality in the Third World. Voluntary female sterilization is the most prevalent contraceptive method today, used by more than 138 million married women of reproductive age compared to 95 million in 1984 (Robey et al. 1992). There is particular controversy over the increasing use of the quinacrine pellet method developed by Dr. Jaime Zipper in 1984, distributed to nineteen countries around the world, including Bangladesh, Chile, China, Colombia, Costa Rica, Croatia, Egypt, India, Indonesia, Iran, Malaysia, Morocco, Pakistan, Philippines, Romania, the United States, Venezuela, and Vietnam, but subject to later banning in some countries. The use of quinacrine, often surreptitiously, through direct relations between NGOs and individual doctors, often aimed at particular segments of the population considered problematic or undesirable, leads critics to conclude that these repeat Nazi nonsurgical sterilization practices and are contemporary successors to the sterilization and population limitation campaigns of the 1970s and 1980s. Despite their rhetoric of informed choice, they amount to global eugenics.

From the perspective of biopower, however repugnant these policies, it is misleading to make that criticism through a rhetorical association of them with the eugenics of the mid-twentieth century. If we use the term "eugenics" to apply to any intervention on the reproduction, morbidity, and mortality of the population, it covers everything from contraception through abortion to public health, and its use becomes merely part of a general critical rhetoric. Eugenics—the improvement of the biological stock of the population—did indeed take both negative and positive forms, but in each case, it was directed to maximizing racial fitness in the service of a biological struggle between nation-states. The forms of biological knowledge that inform our ways of governing others and ourselves are no longer those of the survival of the fittest. Limiting population in the interests of national economic prosperity does not operate according to the biopolitical diagram of eugenics and is not the same as purification of the race by elimination of degenerates.

This is not to say that there are no forms of eugenics around. One visible form is linked to public health. In Cyprus, there are systematic programs of nationwide testing, with the assent of the population, the church, and the state, to identify and eliminate cystic fibrosis—not by embryo selection but by marriage counseling.[11] We can see something of the same strategy at work in practices for the control of Tay Sachs among Ashkenazi Jews in North America and in Israel—practices that have been developed by authorities arising from within those "biosocial communities" themselves.[12] By any definition, this is a strategy aimed at reducing the levels of inherited morbidity and pathology in a population considered as a whole by acting on the individual reproductive choices of each citizen, through various forms of authoritative calculation and guidance, sanctioned by a range of religious and secular authorities, including bioethicists, and approved of by the population. If, as we suggest, this is a type case of contemporary biopolitics, it would clearly be misleading to diagnose it as a form of genocide or the reawakening of the specter of the camp. Political violence between ethnic groups is certainly endemic in the two countries that we have cited, but that violence turns on a different, nonbiopolitical register.

These examples, and others that we could cite, lead us to argue that the economy of contemporary biopolitics operates according to logics of vitality, not mortality: while it has its circuits of exclusion, *letting die* is not *making die*. With the development of ever more sophisticated, cheaper, and readily available forms of genetic testing, biopolitics at both poles—the molar and the molecular—might well be changing. As endless conferences and books have argued, there is all the difference in the world between using genetic techniques to diagnose and even select against embryos with Down syndrome or fetal tube syndrome, and seeking to use those techniques to diagnose intelligence and eliminate the "feeble minded." It is not clear what configurations will take shape if current research succeeds in identifying genetic markers for susceptibilities to common complex disorders such as stroke or heart disease, or for risks of depression or schizophrenia. Our own current research pays close attention to this work, the scientific and technological techniques directed at these ends. But there is no evidence to suggest that the forms of biopolitics that are taking shape around these have, as their strategic objectives, wholesale management of population qualities. Their logic is different and notably involves attempts to develop and maximize targets for pharmaceutical markets and other health-care interventions that entail enrolling individuals, patient groups, doctors, and political actors in campaigns of disease awareness and treatment in the name of the maximization of quality of life. This is capitalism and liberalism, not eugenics, by either the front or back door, at least insofar as eugenics has acquired an inescapably negative meaning in our contemporary culture. We still need to develop the conceptual tools for the critical analysis of the ways in which biopolitics plays out in relation to biocapital and bioeconomics, in circuits in which health and vitality become key stakes in market relations and shareholder value.

The possibilities of genomic management of the population—designer babies, engineered futures, the "sorting society," and the like—have a powerful symbolic presence in contemporary biopolitics, especially in those polities where twentieth-century eugenics took its most corrosive form. However large-scale genetic management of the population has not taken place, and, indeed, it is currently technically impossible. More significant, with the exception of some minor sects, few forces embrace such a rationality. Nonetheless, currently feasible practices such as sex selection do seem to be having molar consequences outside Europe, even though this is the product of individual choices aimed at personal aspirations and shaped by specific sociocultural contexts, and is often explicitly discouraged by official policies. Over and above these alterations in the gross characteristics of the population, in which genetic technologies merely amplify existing cultural forms, we think that it is extremely unlikely that the micromanagement of population characteristics through intervention at the point of reproduction will be scientifically and technically feasible. Even if it is feasible in relation to certain specific conditions, as we see in relation to sickle cell or Tay Sachs, the forms and extent of such genomic management will be shaped by the concerns of particular biosocial communities, rather than by a state commitment to the engineering of population quality for national ends. We will neither repeat the past, nor enter the utopias or dystopias of futurology: to understand and intervene in possible futures we need an analytic that is more modest and empirical, attuned to all the small mutations where today is becoming different from yesterday.

Genomic Medicine

The first biopolitical strategies, in the eighteenth century, concerned the management of illness and health. These provided a model for many other problematizations operating in terms of the division of the normal and the pathological. They have a peculiar saliency in liberal societies because they establish links between the molecular and the molar, linking the aspiration of the individual to be cured to the management of the health status of the population as a whole. The poles of this biopolitical field extend from the management of collective health by means of pure water, through annual health check-ups, health insurance, preventive medicine that operates in large domains between collectivities and individuals, to the field of clinical interventions on the body of the sick person in the name of health. Despite the contemporary focus on the individuated body, action on the collective pole has been the main motor of increases in longevity and quality of life. Variations in the rationalities and technologies directed to this collective pole are the key factors that have led to the scandalous variations in life expectancy and life chances that we can observe today around the globe. In the vast majority of these instances, the causes and the remedies are known, and require no further scientific advance or technological innovation, but only political will. Even in apparently novel disorders, such as SARS, whose outbreak rapidly called forth the whole panoply of modern biological medicine including the rapid identification and sequencing of the pathogen, the preventive modes of intervention required were archaic. They were basically those of quarantine, first applied to epidemic outbreaks such as plague, that have been deployed at least since medieval times and have merely been updated to take account of contemporary mechanisms of mobility and communication. There strategies proved highly effective without any significant contribution from genomic medicine.

What, then, of genomic medicine? It will have become clear that a judgment as to whether a new regime of biopower will take shape—that is to say, will form a qualitative new configuration of knowledge, power, and subjectivity—depends on many factors. Some of these depend on the uncertain outcome of genomic research itself, but many others depend on contingencies external to genomics and biomedicine. As we write, it is still not clear whether the new forms of molecular and genomic knowledge are actually capable of generating the kinds of diagnostic and therapeutic tools that its advocates hope for.[13] The stakes here are high, economically, medically, and ethically. They lie in the presumed capacity of genomics to form a new "know how" that will enable medicine to transform its basic logic from one in based on restoring the organic normativity lost in illness to one engaged in the *molecular reengineering of life itself*. Genomics promises to identify the key processes that control the manufacture of proteins and, in doing so, to open these to precise intervention to produce therapeutic effect. The political economy of these knowledges is, as Carlos Novas has pointed out, one of hope: the hope of individuals, campaigners, scientists, health-care systems, science policy advisers, and the pharmaceutical companies that a new kind of "know how" of life itself will emerge that will generate cures, along with their attendant biovalue (Novas and Rose 2000). For its advocates, the genomic identification of functional pathology must inevitably open a path toward molecular intervention. But to the degree that this logic proves impossible to realize, genomics will remain only one dimension of health care and biological understanding; one that gains its intelligibility within a wider field of knowledge on the aetiology, prognosis, and treatment of disease.

How, then, might we begin to think through the implications of the nascent advances in molecular and genomic technologies? The belief that something significant is at stake here mobilizes the strategies and tactics of a whole variety of forces whose characteristics have been documented in detail in numerous empirical studies. National governments invest in genomics, set up bio-banks, and fund research into basic and applied genomic medicine. Pharmaceutical and biotech companies invest billions and employ tens of thousands of talented scientists and technicians in subtle and elegant experiments and inventions. Patient groups invest hope, political capital, their own tissue samples, and money in the search for genetic treatments. Pressure groups lobby for and against some or all of these developments on the basis of their own ethical or biopolitical concerns. So clearly a modified biopolitical rationality in relation to health is taking shape, in which knowledge, power, and subjectivity are entering into new configurations, some visible, some potential. This formation involves many elements that have played their part in previous apparatuses, and many that took more or less their current shape after the Second World War: patient groups are not new, pharmaceutical companies preexisted genomics, and governments have invested increasingly large sums in promoting and regulating basic and applied medical research in the name of population health, economic development, and international competitiveness. But alongside these previous configurations, which have by no means disappeared, we believe that something new is taking shape, something that is beginning to colonize and mutate the major apparatuses for the management of the health of each and of all, at least in the industrial democratic world.

Let us take two small examples of these new investments. Rabinow's research in 2003 was an anthropological investigation of Celera Diagnostics, in Alameda, California.[14] This company is an offshoot of Celera Genomics, the company that accelerated the race to map the human genome (and other nonhuman genomes as well). With several hundred million dollars at its disposal, it identified roughly a dozen major disease areas and adopted an approach that seeks to identify clusters of SNPs (single nucleotide polymorphisms) in functional areas of the genome. Hence Celera Diagnostics combines massive, expensive machine capacity, diverse alliances with multiple disease associations and university researchers, and a strategy that this will enable the diagnostic identification of predispositions to complex disease involving variations in numerous genes. Its model for polygenetic conditions moves beyond the search for the "gene for" model of the 1990s, a model that is obviously inadequate for understanding the most common disorders such as cancer, heart disease, and other complex disorders, and probably also inadequate for understanding the genomics of most diseases and susceptibilities. The goal is to produce diagnostic tests that would be used massively in reference laboratories in a routine fashion, to enable presymptomatic diagnosis and preventive interventions on a previously unimaginable scale within the next five years.

If this model were to succeed and to be deployed widely, not only in the developed but also in the less-developed world, the logics of medicine and the shape of the biopolitical field would be altered, and new contestations would emerge over access to such technologies and the resources necessary to follow through their implications. Further, as the forms of knowledge generated here are those of probability, new ways of calculating risk, understanding the self, and organizing health care would undoubtedly emerge. It is still not known whether this model will prove operable. If it does, while it is clear that the shape of the biopolitical field would mutate, there is no technological determinism here: multiple responses are possible. And if we remember, as we always should, that, even in the world's most prosperous nation, millions are still denied access to the basic health technologies and medical interventions that have been established for half a century, the political and social implications are evidently shaped more by the political side of the biopolitical than the medical side. If success is partial and patchy, if hopes are deflated, if venture capital and stock market investments move elsewhere, this still does not mean that nothing will emerge. Rather it means simply that, as with so many previous medical advances, the mutations that will take place in therapeutics will be smaller, more dispersed and their effects harder to see in the short term, though perhaps evident from the perspective of the future.

In a related but distinct area of the field, Rose's research in 2003 focused on the development known as pharmacogenomics and in particular on its engagement with mental disorders. The research site here was the take-up, principally in Europe, of the new generation of antidepressant medication, in the context of a belief, underscored by the World Health Organization and accepted by international health management agencies, that by 2020 depression will become the second largest cause of morbidity in both the developed and less-developed world, second only to ischaemic heart disease. There are clearly many factors that have led to this belief, which cannot be addressed here in any detail. They include the humanistic belief of doctors and others that much misery is the result of an underdiagnosed clinical condition for which safe and effective drug treatments are now available, the concern of national governments about the cost to their budgets of days lost through depression, the significance of the key indicator of suicide rates in international health comparisons, and the intensive marketing and "disease awareness" campaigns of the pharmaceutical companies.

What is the link between this and genomics? First, the new (third) generation of antidepressants claim to be fabricated at a molecular level to target the precise neuronal mechanisms that underlie depressive symptoms. Second, because there are more than a dozen of these drugs on the market, there are no clear symptomatic or other markers to enable doctors to choose between them, and yet the different drugs are variable in their effects, some having beneficial effects with some individuals, while having no therapeutic effects in others and generating adverse effects in a third group. Third, some argue that genetic testing may enable medics to choose the right drug at the right dose for the right individual, thus maximizing both therapeutic benefit and compliance, minimizing adverse effects, increasing the efficiency of the targeting of health-care resources, and hence acting not just at the individual level but also on key financial and population health indicators. If successful, driven by the wish of all concerned, including patients, to have effective drugs that have minimal side effects, genetic testing may migrate from the genetic counselor's office to the general practitioner, and become as routine as blood tests, opening up the population as a whole to a genetic understanding of their health, illness, and predispositions. If only partially successful, the routinization of genetic testing prior to treatment decisions may be slow, patchy, and limited, but the genetic rewriting of mental illness will nonetheless once more enter the field of truth, not in the name of population purification and the elimination of degeneracy, but in the name of quality of life, even happiness. In each case, the potential is there for a reshaping of the biopolitics of mental health, not only rewriting its epistemology along biological lines, but also reconfiguring the relations of knowledge, power, and expertise that govern it, perhaps engendering new strategies for minimizing mental disorder at the individual and collective levels, and reshaping the ways in which individuals themselves think about, judge, and act on themselves in the name of mental health.

Conclusion

One might well imagine what it might have been like in 1800 for an analyst attempting to grasp the transformative implications of the forerunners of the "birth of the clinic." Today we may well be in an analogous situation, where the drivers of change can be discerned, some mutations can already be detected, some consequences predicted, but where the overall directions remain obscure and their implications still in doubt. Thus it is no surprise that it is hard to tell whether we are at the early stages of a momentous shift, in the middle of a process that is well underway toward stabilizing new forms, or in a conjuncture that will prove to be a dead end or at least marginal to other changes that we cannot envisage today. But in attempting to make a diagnosis from "in the middle," we think that the concept of biopower focuses our attention on three key elements that are at stake in any transformation—knowledge of vital life processes, power relations that take humans as living beings as their object, and the modes of subjectification through which subjects work on themselves qua living beings—as well as their multiple combinations.

In the new political economy of vitality, transnational flows of knowledge, cells, tissues, and intellectual property are coupled with local intensifications and regulated by supranational institutions. Mobilizations of persons, tissues, organs, pathogens, and therapeutics operate at different speeds and encounter local obstacles and incitements. Individualizing and collectivizing subjectifications are also mobile and transnational: cystic fibrosis groups cut across national and class barriers as do their caregivers; models of patient activism spread and are taken up and reinterpreted from Bangladesh to Toronto. Who, in 1955, could have imagined depressed people as a global category, not only as targets but also as active subjects in a new biopolitics of mental health? If we are in an emergent moment of vital politics, celebration or denunciation is insufficient as an analytical approach. The concept of biopower, used in a precise fashion, related to empirical investigations, and subject to inventive development, would surely take its place as a key part in an analytical toolkit adequate to the diagnosis of what Deleuze (1989) has termed "the near future."

Notes

This essay was published originally in BioSocieties *1 (2006): 195–217. This paper was first presented at "Vital Politics: Health, Medicine, and Bioeconomics into the Twenty-first Century," London School of Economics and Political Science, September 5–7, 2003.*

1. Sydney Brenner is among those who have examined key features of our "biological century": see, e.g., Brenner (2000). Gilles Deleuze, in his *Foucault* (1988), has a provocative appendix where he talks about the future of biopower.
2. For example, the Biopolititics International Organization, based in Greece, which focuses on environmental protection, while in *Christian Biopolitics: A Credo and Strategy for the Future*, by Kenneth Cauthen, seeks to nurture "an emerging new consciousness among many potential dreamers and doers in the churches who can help provide us with the visions and the values we need to promote a movement toward an ecologically optimum world community full of justice and joy in which the human race can not only survive but embark on exciting new adventures of physical and spiritual enjoyment" (1971; available at http://www.religion-online.org/showbook.asp?title=2301).
3. In France, the reception of Negri has been minimal and that of Agamben has turned more specifically on his claims about the concentration camps; see Mesnard (2004).
4. On biocapital, see Rose (2006).
5. Note that these lectures were given in the same year that the first volume of *History of Sexuality* was published in France, and that, aside from a few passing comments, Foucault never returned to the concept of biopower again.
6. See, e.g., his interview on social security entitled "The Risks of Security" (English trans. Foucault 2000, 365–81).
7. This is a point that Agamben himself makes elsewhere.
8. We have discussed our approach to empiricism elsewhere: see our introduction to Foucault (2003) and the introduction to Rose (1999, esp. 11–15).
9. This argument is developed in more detail in *The Politics of Life Itself* (Rose 2006, chap. 6).
10. To be found online at http://www.un-documents.net/a18r1904.htm.
11. This is the subject of ongoing research by Stefan Beck.

12. Discussed in Prainsack and Siegal (2006). See also Rose (2006, chap. 6).
13. This paper was originally written in August 2003.
14. Since writing this paper, this research has been published as Rabinow and Dan-Cohen (2004, 2006).

References

Agamben, G. 1995. *Homo sacer*. Turin: G. Einaudi.

———. 1996. *Mezzi senza fine: Note sulla politica*. Turin: Bollati Borlinghieri.

———. 1998. *Homo sacer: Sovereign Power and Bare Life*. Stanford, CA: Stanford University Press.

———. 2000a. *Means without End: Notes on Politics*. Minneapolis: University of Minnesota Press.

———. 2000b. *Remnants of Auschwitz: The Witness and the Archive*. New York: Zone Books.

———. 2005. *State of Exception*. Chicago: University of Chicago Press.

Braudel, F., and E. Labrousse. 1976. *Histoire economique et sociale de la France*. Paris: PUF.

Brenner, S. 2000. "Genomics—The End of the Beginning." *Science* 287 (5461): 2173–74.

Cauthen, K. 1971. *Christian Biopolitics: A Credo and Strategy for the Future*. Nashville: Abingdon Press.

Deleuze, G. 1988. *Foucault*. Minneapolis: University of Minnesota Press.

———. 1989. *Qu'est-ce qu'un dispositif?* In *Michel Foucault, philosophe*, edited by F. Ewald. Paris: Editions du Seuil.

———. 1995. "Postscript on Control Societies." In *Negotiations*, 177–82. New York: Columbia University Press.

Donzelot, J. 1979. *The Policing of Families*. New York: Pantheon.

Ewald, F. 1986. *L'Etat providence*. Paris: Grasset.

Foucault, M. 1976. *La Volonté de savoir*. Paris: Gallimard.

———. 1977. *Discipline and Punish: The Birth of the Prison*. New York: Random House.

———. 1978. *The History of Sexuality*. Vol. 1, *The Will to Knowledge*. London: Penguin.

———. 1984. "On the Genealogy of Ethics." In *The Foucault Reader*, edited by P. Rabinow. New York: Pantheon.

———. 2000. "The Risks of Security." In *Power: The Essential Works of Michel Foucault, 1954–1984*, edited by J. D. Faubion, 365–81. New York: New Press.

———. 2002. *Society Must Be Defended: Lectures at the College de France, 1975–76*. New York: Picador.

———. 2003. *The Essential Foucault: Selections from Essential Works of Foucault, 1954–1984*. Edited by P. Rabinow and N. Rose. New York: New Press.

Franklin, S. 1997. *Embodied Progress: A Cultural Account of Assisted Conception*. London: Routledge.

———. 2006. *Born and Made: An Ethnography of Pre-implantation Genetic Diagnosis*. Princeton, NJ: Princeton University Press.

Hardt, M., and A. Negri. 2000. *Empire*. Cambridge, MA: Harvard University Press.

Hay, D. 1975. *Albion's Fatal Tree: Crime and Society in Eighteenth-Century England*. New York: Pantheon.

Le Goff, J. 1980. *Time, Work, and Culture in the Middle Ages*. Chicago: University of Chicago Press.

———. 1990. *The Medieval World*. London: Collins & Brown.

M'Charek, A. 2005. *The Human Genome Diversity Project: An Ethnography of Scientific Practice*. Cambridge: Cambridge University Press.
Meadows, D. H., et al. 1972. *The Limits to Growth: A Report for the Club of Rome's Project on the Predicament of Mankind.* New York: Universe Books.
Mesnard, P. 2004. "The Political Philosophy of Giorgio Agamben: A Critical Evaluation." *Totalitarian Movements and Political Religions* 5 (1): 137–57.
Mesnard, P., and C. Kahan. 2001. *Giorgio Agamben a l'epreuve d'Auschwitz.* Paris: Editions Kime.
Novas, C., and N. & Rose. 2000. "Genetic Risk and the Birth of the Somatic Individual." *Economy and Society* 29 (4): 485–513.
Prainsack, B., and G. Siegal. 2006. "The Rise of Genetic Couplehood? A Comparative View of Prenatal Genetic Testing." *BioSocieties* 1 (1): 17–36.
Proctor, R. 1999. *The Nazi War on Cancer*. Princeton, NJ: Princeton University Press.
Rabinow, P. 1989. *French Modern: Norms and Forms of the Social Environment*. Cambridge, MA: MIT Press.
———. 1994. "The Third Culture." *History of the Human Sciences* 7 (2): 53–64.
———. 1996. "Artificiality and Enlightenment: From Sociobiology to Biosociality." In *Essays on the Anthropology of Reason*, 91–112. Princeton, NJ: Princeton University Press.
———. 1999. *French DNA: Trouble in Purgatory*. Chicago: University of Chicago Press.
Rabinow, P., and T. Dan-Cohen. 2004. *A Machine to Make a Future: Biotech Chronicles*. Princeton, NJ: Princeton University Press.
———. 2006. *A Machine to Make a Future: Biotech Chronicles*. 2nd ed. Princeton, NJ: Princeton University Press.
Reardon, J. 2001. "The Human Genome Diversity Project: A Case Study in Coproduction." *Social Studies of Science* 31 (3): 357–88.
———. 2005. *Race to the Finish: Identity and Governance in an Age of Genomics*. Princeton, NJ: Princeton University Press.
Robey, B., S. O. Rutstein, L. Morris, and R. Blackburn. 1992. *The Reproductive Revolution: New Survey Findings.* Baltimore: Population Information Program.
Rose, N. 1985. *The Psychological Complex: Psychology, Politics and Society in England, 1869–1939*. London: Routledge & Kegan Paul.
———. 1999. *Powers of Freedom: Reframing Political Thought*. Cambridge: Cambridge University Press.
———. 2001. "The Politics of Life Itself." *Theory, Culture & Society* 18 (6): 1–30.
———. 2006. *The Politics of Life Itself: Biomedicine, Power and Subjectivity in the Twenty-first Century*. Princeton, NJ: Princeton University Press.
Rose, N., and C. Novas. 2005. "Biological Citizenship." In *Global Assemblages: Technology, Politics and Ethics as Anthropological Problems*, edited by A. Ong & S. Collier, 439–63. Malden, MA: Blackwell.
Thompson, E. P. 1975. *Whigs and Hunters: The Origin of the Black Act*. London: Allen Lane.
Throsby, K. 2004. *When IVF Fails*. London: Palgrave Macmillan.
United Nations General Assembly. 1963. *Declaration on the Elimination of All Forms of Racial Discrimination*. New York: United Nations.

SIXTEEN

A Colonial Reading of Foucault: Bourgeois Bodies and Racial Selves

ANN LAURA STOLER

Foucault has long figured as a shadow presence in my work, first as a source of inspiration and then as a source of reflection and critique. My initial thinking about knowledge and power, like so many others in colonial studies, was transformed by him. For decades and through different venues, I have worked off of and reworked his insistence on sexuality as "an especially dense transfer point for relations of power."[1] How colonial authorities chose to speak about the carnal and iterate its dangers did more than substantiate his general claim. In broaching how colonial agents positioned sex as "the truth" of the racialized self, it was to Foucault that I turned again.

But as a blueprint for understanding what joined carnality to power in the nineteenth-century Netherlands Indies, Foucault's contentions made only partial sense. His assertion that sexuality was "originally, historically bourgeois" seemed at once more reasonable and more questionable the more time I spent in the colonial archives. What was absent, obviously so in retrospect, was the issue of race.

For a student of colonial history and presence, reading Foucault incites and constrains. Volume 1 of the *History of Sexuality* is an uncomfortable reminder of how much empire and its colonial landscapes have remained in the peripheral vision of even critical European history—much less than more conventional historiographies of race. It is this reflection that frames the challenge of this chapter. In it I revisit a question that informed *Race and the Education of Desire* to ask how the making of a European bourgeois self might look different when that history is rendered less self-referential, when imperial politics is placed center stage.

My colonial reading of Foucault draws on cumulative contributions to colonial studies to ask what implications such a rereading might have for how we think about the intimacies of empire, European history, and the colonial etymologies of race. I return to some arguments of my book, but that return is selective. Here I am interested in why Foucault's elusive and suggestive treatment of race remains so marginal to what colonial historians take from him today.

Writing *Race and the Education of Desire* led me to pursue some issues that were more predictable than others. Engaging the interface of Foucault's work on the nature of racial discourse and on the history of sexuality prompted rethinking the technologies of colonial rule and their sites of production. Questions about the "education of desire" turned me less to sexual desires per se than to the wider array of sentiments that carnal knowledge may express.[2] Working back through Foucault's Nietzschean bent in a colonial context not least raised some hard questions and left them unanswered: How does one guard against history writing in "the comfort zone"?[3] What would make up a discomforting colonial history of the present?[4]

Foucault's insistent return to the political entailments of knowledge production has been usefully disruptive at a number of registers. Take, for example, the critique it might invite of "comparative colonialisms." What could be more reassuring than the argument on which comparative studies of colonialism has thrived; namely, that differences in colonial policies derive from European distinctions of national character? In such a model, some country's legacy was always more benevolent, another's violences were truly atrocious, and yet another's integrative efforts were more effective or more benign. The positioning may shift with perspective, but the narrative structure stays the same. Such comparisons signal more about the national underpinnings of comparison than they do about significant differences in strategies of rule. These are not "wrong" answers so much as unproductive questions.

Attention to the perceptions and practices of colonial regimes in different times and places could start with other kinds of questions. What are the premises and "ready-made syntheses" that have made comparative studies of colonialism possible? Can one compare colonialisms without defining the protean criteria for assessing race? What sorts of comparisons are invidious, and which are not? What assumptions allow a comparison between "mixed bloods" in the Indies and "coloreds" in South Africa and métis in Indochina? What sorts of colonial racial grammars made architects of imperial policy think they could—and us today think we should with a different political agenda—do so? Such questions place comparison itself as a political project and incommensurable, nonviable comparisons as a political choice.[5]

The task of writing new colonial genealogies, then, is an opportunity of many sorts. It presses us to learn from Foucault as we push past his insights and push our own further to think harder about how the making of race has figured in placing sexuality at the center of imperial politics. I ask about the colonial state's investments in managing the assessment of what was normal and what was not; how the management of sexuality in part framed what sentiments could be expressed and to whom they could be directed. Colonial states had a strong interest in affective knowledge and a sophisticated understanding of affective politics. While the politics of sentiment figures more centrally in my own work than in Foucault's, I ask what we might glean from his insights and where we have not yet taken them.

Where Is Race in Foucault?

Race and the Education of Desire was based on some basic questions. At a time when Foucault's work has had an enormous impact across the disciplines and on the discursive and historic turn within them, why have contemporary scholars dealt in such an oblique way with the slimmest—some might argue the most accessible—of his major works, *The History of Sexuality*? More precisely, why has colonial studies, where issues of sexuality and power are so prominent on the intellectual and political agenda, had so little to say about that work? In a field in which reading that volume has been de rigueur and reference to it conferred intellectual authority, what accounts for this absence of an engagement that is analytically critical and historically grounded?

Race and the Education of Desire was intended initially to be a colonial reading of *The History of Sexuality*, one that questioned both Foucault's account of the making of European bourgeois subjects and the centrality of empire and race to that process. But what seemed an exciting but straightforward task got abruptly halted in midprocess. This was when I first heard about and then sat riveted in Paris's Saulchoir library listening to the scratched tapes of his then unpublished lectures from 1976 given at the Collège de France. These were the now published eleven full lectures (*Il faut défendre la société*) devoted to theorizing the history of racisms, racial discourses, and racisms of the modern state. More striking still was the fact that they were delivered the very winter that the first volume of *The History of Sexuality* was in press. Here was Foucault, in just two years, engaged in two parallel histories and sequential pursuits: a history of sexuality and a treatise on power through a genealogy of race.

The lectures at once confirmed Foucault's Europe-bound vision but as quickly unsettled a quick dismissal of him for sidestepping issues of race. They placed racism more centrally in his thinking than any of his then-published work would suggest. But they also prompted a rush of questions. Did this work on race really represent a "turning point" in Foucault's intellectual itinerary?[6] What should we do with his often-repeated statement that all his writings were autobiographical? Was it significant that he had spent 1966 through spring 1968 in the former French protectorate of Tunisia where he wrote *The Archaeology of Knowledge,* wrote friends about anti-Semitism, and had his work repeatedly interrupted by student strikes against the governing policies of a newly empowered state?[7] In *Race and the Education of Desire,* I chose to avoid the obvious but still opaque discrepancy between his knowledge of postcolonial conditions in Tunis and the absence of attention to France's colonial history in his work. I have begun to rethink that position today.[8]

Indeed, what has colonial studies at any particular moment imagined is a "usable" Foucault? Why does *Discipline and Punish* more often provide a model than the nuanced methodological insights of *The Archaeology of Knowledge*? Why is there so much recent work on colonial "governmentality" but so little on the complex ways in which he understood the movements of subjugated knowledges and their "resurrections"? And why, until very recently, is there even less on race?[9]

This is *not* a prelude to the argument that we have all missed the "real" Foucault. Identifying the tensions between his lectures and his written work is only part of a more general effort—to ask what Foucault offers for our understanding of the bourgeois underpinnings of colonial regimes and in turn how attention to colonial sites refigures his parameters and challenges what gets counted as part of European historiography.

Two basic arguments of *Race and the Education of Desire* are relevant here. First, the proliferating discourses of sexuality that Foucault registers in Europe in the eighteenth and nineteenth centuries simply cannot be chartered in Europe alone but through a more *circuitous* imperial route than he was to suggest. These were refracted by men and women whose affirmation of a bourgeois self was contingent on imperial products, perceptions, and racialized others that they produced. I argued that you could not get from eighteenth- to nineteenth-century technologies of sex in Europe without tracking them across colonial ground. I thus approached *The History of Sexuality* through several avenues by comparing its chronologies and strategic ruptures to those in the colonies. But, as important, I argued that a "comparison" of these two seemingly dispersed technologies of sex in colony and in metropole misses the extent to which they were so tightly bound.

Second, I argued that racial entailments were not relevant in the colonies alone. By bringing anxieties and struggles over citizenship and nation back within our frame (as Foucault did not), bourgeois identities in metropole and colony emerge as tacitly and emphatically coded by race. In rerouting the history of sexuality through the history of empire, modern racism appears less "anchored" in European technologies of sex than Foucault claimed. Both racial and sexual classifications appear as ordering mechanisms that *shared* their emergence with the bourgeois order of the early nineteenth century. Racial thinking was not subsequent to the bourgeois order but constitutive of it.

The implications are several. First, racism was not a colonial reflex, fashioned to deal with the distant other, but part of the very making of Europeans themselves. "Internal colonialism," in this perspective, was not a peculiar form of empire building (a variant frequently used to describe the expansion of the American West) but, according to Foucault, its earliest and enduring incarnation. Second, racisms have rarely adhered to the visual and social clarity of difference among widely disparate groups. They have flourished in similar, contiguous populations. Racisms have riveted on ambiguous identities—racial, sexual, and otherwise—on anxieties produced precisely because such crafted differences were not clear at all. Racisms gain their strategic force, not from the fixity of their essentialisms but from the internal malleability assigned to the changing features of racial essence.[10] In the nineteenth-century Indies, cultivation of a European self was affirmed in proliferating discourses on pedagogy, parenting, and servants. These were microsites in which bourgeois identity was rooted in notions of European civility, in which designations of racial membership were subject to gendered appraisals, and in which "character," "good breeding," dispassionate reason, and proper rearing were part of the changing cultural and epistemic indexing of race.

For a long time, the story goes, we supported a Victorian regime, and we continue to be dominated by it even today. Thus "the image of the imperial prude is emblazoned on our sexuality, restrained, mute and hypocritical."[11] Colonial readings of Foucault, for the most part, have applied the *general principles* of a Foucauldian frame to a specific ethnographic time and place, engaging the outlines of his analytic apparatus more than the historical content of his analysis. This sort of passion for Foucault's general strategies is apparent in readings of each of his texts—nowhere more than in treatments of *La volonté du savoir*, or *The Will to Knowledge*, rendered in English as volume 1 of *The History of Sexuality*.

That book engages a disarmingly simple thesis: why, if in nineteenth-century Europe sexuality was indeed something to be silenced, hidden, and repressed, was there such a proliferating discourse about it? Foucault tells us in the first line of the first chapter that we have gotten the story wrong: the "image of the imperial prude . . . emblazoned on our restrained, mute and hypocritical sexuality" misses what that regime of sexuality was all about. It did not seek to restrict a biological instinct, to overcome a "stubborn drive," nor was it an "exterior domain to which power is applied."[12] For Foucault, sexual discourse is not opposed to and subversive of power but a "dense transfer point" of it, charged with "instrumentality."

The gist of the message in much recent work, including my own, has turned with varied inflection on a similar premise—that the management of the sexual practices of colonizer and colonized was fundamental to the colonial order of things and that discourses on sexuality at once classified colonial subjects into distinct human kinds while policing the domestic recesses of imperial rule.[13] Such readings take seriously the fact of a relationship between colonial power and the discourses of sexuality but neither confirm nor confront the specific chronologies Foucault offered or the selective genealogical maps his works suggests.

Students of empire have shown little interest in Foucault's most basic rejection of Freud's repressive hypothesis. On the contrary, we have exhibited strong allegiance both to a Foucauldian perspective on power and to implicit Freudian assumptions about the psychodynamics of empire and the sexual energies "released." We have applied an ejaculation theory of history to show how such regimes extend and work.

Some of the problems are Foucault's, some our own. *The History of Sexuality* seems to impede an alternate venture. Tracing the deployment of sexuality within an analytic field confined to Europe—to "modern Western sexuality"—Foucault presents a familiarly pat binary world of *ars erotica* (the Orient) and *scientia sexualis* (the West).[14] The opening image of the "imperial prude" is the first and only reference to empire. For Foucault, that image of the prude is a mainstay of our misguided reading of nineteenth-century sexuality that is dismissed, replaced, and not further discussed. Overseas empire disappears along with its caricature.

Such origin myths of the making of European cultural practices are less credible today as the bracketed domain of European history has been unsettled, its sources reassessed, its boundaries blurred. More than four decades after *The History of Sexuality* first appeared, as colonial studies has increasingly turned to tensions that cut across metropolitan and colonial sites of imperial rule, we are still prompted to ask whether the shaping of bourgeois subjects can be located outside those force fields in which imperial knowledge was promoted and desiring subjects were made. Foucault's impulse to write a history of Western desire that rejects desire as a biological instinct or as a response to repressive prohibitions pushes colonial studies in a direction that feminism has long urged, to question how shifts in the imperial *distributions* of desiring male subjects and desired female objects might shape that story as well.

Moreover, in re-viewing the colonies as more than sites of exploitation and as more than "laboratories of modernity," what constitutes metropolitan versus colonial inventions and importations has precipitously shifted course. Timothy Mitchell's study of colonial Egypt placed the Panopticon, that supreme model institution of disciplinary power, as a colonial invention that appeared first in the Ottoman Empire, not northern Europe.[15] Gwendolyn Wright and Paul Rabinow argued that the modern was played out in colonial settings, that French policies on urban planning were experimented with in Paris and Toulouse but probably in Rabat and Haiphong first.[16] Mary Louise Pratt stretched back further to argue that those modes of social discipline taken to be quintessentially European may have been inspired by seventeenth-century imperial ventures and only then refashioned for a later bourgeois order.[17] Such reconfigured histories have pushed a rethinking of European cultural genealogies across the board. They prompt us to ask whether those most treasured icons of modern Western culture—liberalism, nationalism, state welfare, citizenship, culture, and "Europeanness" itself—were not clarified among Europe's colonial exiles in Asia, Africa, and Latin America and only then brought "home."

But the point here is not simply to turn the tables and thus argue that "modernity" or "capitalism" was invented in the colonies rather than in Europe, as some studies of the colonial so eagerly contend. It is rather to imagine new ways of subverting deeply statist historiographies by tracing people's transnational itineraries and circuits of knowledge production through movements of global breadth. Ironically, the crucial element of *The History of Sexuality* that does speak to the imperial world of the nineteenth century has been largely ignored. This is Foucault's strategic linking of the history of sexuality to the construction of race. Contrary to the received account—that the book is about sexuality and biopower—references to racism appear in virtually every chapter. That few of Foucault's interlocutors have noticed or noted them is surprising, given that the final two sections of the book deal directly with the "instrumentality" of sexuality in the rise of racisms and the convergence of both with the biopolitical state. One could argue, as Etienne Balibar does, that racism is what the concept of biopower sets out to explain.[18] The 1976 lectures, and the one published in *Temps modernes* in 1991 with the subtitle "La naissance du racisme" (The Birth of Racism), give weight to Balibar's argument while rendering the silence more profound.

This silence may reflect the constraints of a political and intellectual field in which Foucault was situated at the time. The concept of class and the social transformations to which capitalism gave rise remained foundational in critical social and political theory. Race and racial theory were not. In US scholarship, histories of racism occupied a different space, bracketed as a subtheme in the history of slavery, in Britain as a politically anesthetized, ahistorical field of "race relations," in France as a history of Jewish genocide, and in Germany as a history of Teutonic particularism that produced histories of horrors to dispose of that were over and past.

What is more, after this short-lived turn to the history of racisms, Foucault precipitously abandoned the project. The "war of races," the explicit discussion of "state racism," disappears from his subsequent lectures and his written work. Whether he was "deadlocked" on theorizing racism, as one of his close colleagues suggests, is hard to know. Whatever the case, in his lectures racism is situated squarely at the core of state and societal processes of normalization and regularization, processes that were to occupy him for many years before and after. In the lectures, racism was a normalizing feature of a range of state formations, not an aberrant feature of them.

But the refusal in France to engage him on racism may have little to do with Foucault. It may reflect a long-standing and more widespread refusal to consider racism as fundamental to France's contemporary history. The surge in the 1990s in publications and public discourse on the force of the Far Right contrasted sharply with its confined and limited treatment earlier.[19] It was really only in that period—following the National Front's electoral victories—that a wider intellectual constituency sought to account for the Front's diverse appeal and acknowledge French racism as more than an uncharacteristic blip on its political horizon.[20] If some commentators sought to emphasize the foreign and singularly marginal status of the National Front's adherents, more convincing analyses described it as a movement spawned on French soil with a deep history and solidly "made in France."[21] Even among the latter, none explore or entertain a relationship between French racism and the making of a French modern state. Still, Foucault remained decidedly off the radar of those in France writing on race, despite the fact that the discourses he described—those of a "war of the races" and of a "defense of society" against itself—have been familiar refrains in the National Front's claims to defend the nation and to protect a French society at "civil war."

But Foucault's contention in the 1976 lectures that racisms are basic to the way biopower develops in all modern states was not well received, nor was it twenty years later when they were published. A 1997 review in *Le Monde* refers to this aspect of Foucault's argument as a surprising and disturbing "*sursaut*" (a leap with the sense of an abrupt, almost involuntary start), making the point that Foucault had gone "too fast" ("Il va soudain trop vite"), and too far. Philosophers too had little to say about the lectures (in contrast to the sustained surge of interest in them today).[22]

If some surprises in the lectures were ill received, as an exemplar of Foucault's thinking, other surprises are in store that concern his methodology. First, for those who understand his notion of power as always capillary and micro rather than concerned with the state's macromonopolies, the lectures should give pause. Here his focus is on modern biopolitical states and the conditions of possibility that produce and condone state-sanctioned disenfranchisement and murder within them.

Second, rather than a focus on the discontinuous epistemic rupturing in history—for which he has been so well known—here Foucault alerts us to a more complex process. This is one that includes the simultaneous "re-inscription," "encasement," and "recovery"—terms he uses repeatedly in the lectures—of older racial discourses as they are reshaped into new ones, understood as a layering of sedimented hierarchical forms. It is through this tension between recuperation and rupture that Foucault explores racisms' tactile and polyvalent mobility. Attention to Foucault's thinking on racism, then, is not a presentist reading of his work. Racism is a complex subtext of how he understood biopower and thought about states. The lectures show his strained efforts to grapple with racism and elide it in a historical frame so locked in Europe that "colonial genocide" (a term that appears only once in passing) could by his account derive from Europe's internal politics, subsumed and unexplained.

References to racism in *The History of Sexuality* are neither incidental nor perfunctory, but carefully signposted in each part of the book—which makes sense since that volume was the blueprint for what was never realized, the six volumes he had set out to write, with volume six to be titled *Population and the Races*. Although references to racism are sparing in *The History of Sexuality*, the fact of modern racism is fundamental to it. In Foucault's genealogy, racial discourse was a part of the technologies of sex that arose in the eighteenth century to regulate sexual conduct and by which populations could be expanded and controlled. These would become "the anchorage points for the different varieties of racism of the nineteenth and twentieth centuries." It was the scientific arbitrators of sex who authorized the "hygienic necessity" of cleansing and invigorating the social body in forms, he writes, that "justified the racisms of the state, which at the time were on the horizon." Note here that racism is a potential waiting to be born, not yet on the terra firma that produced the rigid racial taxonomies of the late nineteenth century.[23]

A colonial perspective could offer a different chronology with other prefigurings—of which Foucault was peripherally aware. Colonial technologies of rule bear witness to explicit racially based policies of earlier date and widespread use. Why then did he categorically reject the standard story of nineteenth-century sexuality but embrace this version of the history of race? *The History of Sexuality* hints at some reasons, but the lectures provide more. Colonialisms were outside his purview, as were their racialized systems of social classification. But, as important, as the lectures make clear, Foucault was concerned with "state racism," and it was to this form alone that he thought the term "racial discourse" should be applied.

Not least, Foucault set out to explain the Nazi state and the final solution it embraced. (In the lectures it is Stalin's Soviet state as well.) Neither apartheid Africa nor the segregationist United States was mentioned. The teleology of his argument was to account for a state's right and obligation to kill not only its external enemies but its internal enemies as well. His concern was with those discursive forms and those categories that made common sense of purgings within, how a state establishes that right and then turns it into a moral obligation. It is a discourse, he argued, that confers on a citizenry the right to kill its own designated members as an act of beneficent purification. The 1976 lectures, published as *Society Must Be Defended*, renders this discourse of defense a key impetus for expanding the domains of security at the intimate level and at large.

These issues are not spelled out in *The History of Sexuality*, but in the 1976 lectures they frame his project. There he traces the transformation and reversal of a discourse on the unjust state that will reappear in the nineteenth century. "The state," he argues in his fourth lecture in January 1976, "is and must be the protector of the integrity, the superiority, the purity of the race."[24] Modern racism is born out of this conversion from a discourse on *races* to a discourse on *race*, from a discourse directed against the state to one organized by it. In *The History of Sexuality*, racism is embedded in early discourses on sexuality but not yet in explicit form. It is only in the late nineteenth century that the "series composed of perversion-heredity-degeneracy" comes to make up the "solid nucleus" of a new technology of sex that "took the exasperated but coherent form of state-directed racism."[25]

Such references may seem to suggest a progressive story of racism emerging out of earlier technologies of sex, but Foucault's story is more complicated. In the final chapter of *The History of Sexuality,* the concept of biopolitics and its ties to racism are pivotal to his argument. "Biopower" is identified as having two distinct forms: one concerned with the life of the individual, the other with that of species. It is the micromanagement of the individual body and the macrosurveillance of the body politic—and the circuits of control between them—that linked the fate of the two. Note here the crucial link: the "encasement" of a disciplinary power targeting the individual within a state power targeting the social body. It is this convergence that allows for racism in its contemporary form.

If we turn to colonial terrain, these formulations are compelling but make what Foucault does *not* say about the gendered coordinates of colonialism more pronounced. In the Indies, in South Asia, and in the North and South American colonies, the sexual arrangements of company officials, subaltern military, and settlers were monitored if not successfully regulated early on in ways that repeatedly positioned women of different hue as desired objects and more obliquely as unruly desiring subjects as well. Connections among the making of racial categories, the prescribing of women's reproductive functions, and the managing of sexuality are hard to miss.

By the mid-nineteenth century, as we have seen, Dutch children in the Indies—abandoned, illegitimate, and "mixed blood"—had become the sign and embodiment of what needed fixing in colonial society. More clearly defined bourgeois prescriptions that encouraged white endogamy, attentive parenting, Dutch-language training, and surveillance of servants made up the web of directives designed to shore up the state's priorities. Together these discourses of sexual and racial transgression provided social reformers with ready evidence that colonial policies had to distinguish between the "real" Dutch and those assimilated natives, Indos, and poor whites considered of "fabricated" European status, between citizens and subjects, and between colonizer and colonized.

From a Symbolics of Blood to an Analytics of Sexuality

In *The History of Sexuality,* Foucault writes:

> Beginning in the second half of the nineteenth century, the thematics of blood was sometimes called on to lend its entire historical weight toward

> revitalizing the type of political power that was exercised through the devices of sexuality. Racism took shape at this point (racism in its modern, "biologizing," statist form).[26]

Describing the reappearance of a "symbolics of blood" in nineteenth-century science as a technology that consolidated racism makes for a not unfamiliar story. What is dissonant is the selective Europe-bound genealogy Foucault derives for it. He traces it solely through an aristocratic symbolics of legitimacy and descent, with no sign of an imperial politics of exclusion worked out earlier and reworked later on colonial ground. Science and medicine may have fueled the reemergence of the beliefs in blood, but so did a folk theory of cultural contagions, as threatening as those of bodily ones.

The revival of a symbolics of blood derived from an imperial logic that cultural hybridities were subversive, that subversion was contagious and that native sensibilities and affiliations were the invisible bonds that differentiated "mixed-blood" against "full-blooded" Europeans who claimed the right to rule. For Foucault, modern racism appears as a consequence of that class body in the making. In my perspective, race was constitutive of it. We may have so few colonial readings of *The History of Sexuality* because questions of what constituted European identities in the colonies, who counted as European and could claim to be white, are only now foregrounded in critical cultural theory and analysis.

Colonial Cultivations of the Bourgeois Self

Colonialism was not a secure hegemonic bourgeois project.[27] It was only partly an effort to import cultured sensibilities to the colonies but as much about the *making* of them. Indeed, most of the European population in the Indies never enjoyed the privileges of what Benedict Anderson once labeled a "bourgeois aristocracy."[28] That ill-defined population included poor whites, subaltern soldiers, minor clerks, mixed-blood children, and creole Europeans whose economic and social circumstances made their ties to metropolitan bourgeois civilities often tenuous at best.

These were neither players absent from the colonial stage (as some official histories would have it) nor the rebel vanguard against European rule (as some authorities feared and claimed). Rather they were people precariously poised, economically vulnerable, and socially askew. Effaced from view and then pushed to center stage at strategic moments, these were people about whom colonial officials virulently disagreed. Where they were placed in the state's racial taxonomy affected the extent of the state's financial responsibility for the impoverished and the reach of its moral authority. Note here that colonial comparative statistics on indigent Europeans in South Africa, Australia, India, and the Indies were never simply intended to count which Europeans were poor but which of the poor were "real" Europeans and should be included as such. Statistics was a moral science, underwritten by moral judgments that distinguished the deserving from the undeserving poor. They were also assessments that depended on a definition of race. Only then could officials construct social policy for European welfare and a definition of poverty.

Our blind spots are one issue, Foucault's another. His account of what sexuality meant to the eighteenth-century bourgeoisie refused what he saw as a facile economic reductionism—emphasis on the exploitation of the other. Thus he wrote:

> The deployment of sexuality *must be seen as the self-affirmation of one class rather than the enslavement of another*; a political ordering of life[,] . . . it provided itself with a body to be cared for, protected, cultivated, and preserved from the many dangers and contacts, to be isolated from others so that it would retain its differential value.[29]

Note how even the syntax absents key actors. Foucault makes no room for the fact that these bourgeois bodies were produced in practices never contingent on the will to self-affirmation alone. This "body to be cared for, protected, cultivated, and preserved from the many dangers and contacts" required other bodies that would perform those nurturing services and provide the leisure for such self-absorbed administerings and self-bolstering acts. It was a gendered body, *dependent* on the intimate set of sexual and service relations between French men and Vietnamese women, Dutch women and Indo men, and shaped by the politics of race. Native women who served as concubines, servants, nursemaids, and wives in European households at once threatened the "differential value" of adults' and children's bodies that they were there to protect and affirm.

Here was a fundamental imperial tension: between a culture of whiteness that cordoned itself off from the native world and a set of domestic arrangements and class distinctions among Europeans that produced cultural proximities, intimacies, and sympathies that transgressed them. The family, as Foucault warned, was not a haven from the sexualities of a dangerous outside world but the site of its production. Colonial authorities knew it only too well. They were obsessed with moral, sexual, and racial affronts to European identity in prisons, in schools, and in hospitals but most definitively where they had equivocal control—in the home.

As evinced in so much of the Dutch colonial archives, medical manuals and pedagogic journals insisted that mixed-blood children in poor white households needed to be salvaged from their domestic surroundings and severed from their native mothers. European children of the well-to-do too were at risk if the proper habitus was not assured, if socializing with poorer children of mixed parentage was not monitored and certain social protocols were not met. The risk for both was that their sense of "belonging" and their longings allowed in too much that was locally acquired and Javanese.

Foucault wrote as if one could easily assume that the middle class was sure of what it was affirming. I think it was not. These strategies of identity making and self-affirmation were labile, affirmed by a cultural repertoire of competencies and sexual prescriptions that altered as states weighed profit-making strategies against the stability of rule. Self-discipline, sexual morality, and self-control were visual signs of middle-class rearing, indexing what was invisible and harder to test—namely, what defined the essence of being European and whether creole and "Indo" affinities for things Javanese were a threat to it.

One might argue that such racialized notions of the bourgeois self were colonial idiosyncrasies and applicable there alone. But racial and imperial metaphors were applied to class distinctions in Europe at a very early date. While social historians generally have assumed that racial logic drew on the ready-made cultural disparagements honed to distinguish between middle-class virtues and the immorality of the poor and between the "undeserving" and the "respectable" poor, it may very well be that such social etymologies were sometimes reversed.

The lexicon of empire and its sexualized images may have provided for a European language of class as often as the other way around. From Montaigne to Mayhew, in Britain, the Netherlands, and France, imperial images of the heightened erotics of the colonized saturated the discourses of class. Nor was it first in the mid-nineteenth century that those parallels were made between the immoral lives of the British underclass, Dutch dirt farmers, Irish factory workers, and primitive Africa or Southeast Asia. To argue this case is not to embrace unwittingly the comparative categories used by colonial states. It should rather encourage a tracking of the histories of these comparisons themselves. It should prompt us to ask what kinds of equivalencies would have to be made and what truth claims would allow or disallow such comparisons to be drawn.

Empire figured too in the bourgeois politics of liberalism and nationalism in ways in ways students of colonialisms increasingly have sought to explore. As Uday Mehta has argued, eighteenth-century bourgeois liberalism fundamentally embraced a politics of exclusion based on race.[30] The most basic universalistic notions of "human nature" and "individual liberty," so dear to John Locke and John Stuart Mill, rested on breeding and the learning of "naturalized" habits that set apart those who exhibited such a "nature" and were endowed with the sensibilities that would allow them to exercise such liberty from the racially inferior—and in Mehta's case, the South Asian colonized world. Discourses of sexuality, racial thinking, and rhetorics of nationalism used visual markers to (poorly) index the cultural and affective attributes on which these folk theories of difference were based.

The quest to define moral predicates and invisible essences tied the bourgeois discourses of sexuality, racism, and nationalism in fundamental ways. Nationalist discourse staked out those sexual practices that were nation building and race affirming, marking, as Doris Sommer notes, "unproductive eroticism[,] . . . not only [as] immoral, [but also as] unpatriotic."[31] In such a frame, European women were cast as the custodians of their morally vulnerable men and of national character. Cultivation of the bourgeois self and its gendered assignments appear rooted in Europe and *inside* the nation rather than as part of the making of them.

Children's Sexuality and the Alienation of Affections

It was not only the child of the people, the future worker who had to be taught the disciplines of the body, but also the schoolboy, the child surrounded by domestic servants, tutors, and governesses, who was in danger of compromising not so much his physical strength as his intellectual capacity, his moral fiber, and the obligation to preserve a healthy line of descent for his family and his social class.[32]

Foucault's attention in *The History of Sexuality* to what he calls the "pedagogization of children's sexuality" is schematic, uneven, and only broadly framed. But the subject was central to his conception of a biopolitical state. His planned volume three of the projected six-volume *History of Sexuality* was to be titled *The Crusade for Children* (Croisade des enfants). He never finished that project, but what he did have to say about the discourse on children and masturbation makes sense of colonial fears in sharpened ways. As a principal discursive site where bourgeois culture defined and defended its interest, in colonial perspective it was also one of the key sites where racial lines were transgressed and national identities were formed.

It was a discourse in which the distribution and education of desire was lodged in the home, as Foucault put it, in that "tiny, sexually saturated, familial space."[33]

In the Indies, these discourses were animated not by fears of children touching their own bodies but by fears of their affections for those bodies that should not touch them. Foucault rightly observed that the profusion of child-centered prescriptions and protocols affirmed a bourgeois self in the making. But it was that "cast" of stereotypical others shadowed in these narratives against which the quotidian boundaries of the cultivated self were drawn. As he had put it in his *Resumé des cours* for his 1974–75 lectures on "Les anormaux,"[34] authorities saw children's transgressions as the fault of parents who entrusted them to "wet nurses, domestics, tutors, all these intermediaries regularly denounced as the initiators of debauchery."[35]

Liberal philosophers, colonial policymakers, and nationalist thinkers could not agree more. They shared a political preoccupation with the dispositions of very small children, the malleabilities of their minds, and the training of habituated practices that would "appear natural" to them as adults later on. Strict surveillance of domestic servants was one way to protect children. Removal of children from the home was another, as the proliferation and state endorsement of kindergartens and nurseries in the nineteenth century attest. Both moral philosophers and policymakers were spurred by the conviction that bourgeois households were providing poor child management and that toddlers and even infants were better off in nurseries than in a servant's care.

In the Indies, virtually all of the debates about masturbating and sexually precocious European children were about whether these children would be able to acquire the sensibilities that would allow them to grow up European. Schooling was proffered as a protection against the risks to which girls were exposed outside the family—but more dangerously within it.

These discourses did not target the sexuality of children so much as the dangers posed by alien cultural longings and affective estrangements, disrupting the culturally contained milieus in which European children rightfully belonged. Children required an environment cordoned off from those conduits of sentiment that would disincline them to "feel at home" in a European setting and, as one colonial official put it, "to think and feel" not in Dutch but in Malay or Javanese. Servants could steal more than the sexual innocence of European children. They could redirect their cultural longings, the smells they preferred, the tastes they craved, and their sexual desires.

Judith Butler once called *The History of Sexuality* a history of Western desire, but I am still not convinced this is the case. From volume one we learn little about what sorts of passions are produced and what people actually did with them. We learn even less about how pleasure was distributed, how desire was motivated and power displayed. But those of us who study sexuality and empire do not do much better. All might agree that carnality has underwritten European folk theories of race for more than two centuries. But in colonial studies the carnal often is suspended as a precultural instinct, given and unexplained. Such analyses often proceed *not* from a Foucauldian premise that sexual cravings are a social construct and sex a nineteenth-century invention, but from an implicitly Freudian (and imperial) one. In this narrative, colonialism expressed the sublimated sexual outlet of virile and homoerotic men. White women, by and large, display affective pains and pleasures but no sexuality at all.

The sexual stories that European colonials and their metropolitan observers told about their own desires and what distinguished themselves turned on instantiating class-specific sensibilities. Thus the sexual susceptibility of European children in the tropics demanded vigilant education of their desires and care that native contacts were controlled. European colonial women whose conjugal choices were deemed unfit could be stripped of the European community's protection of their womanhood and disavowed as true Europeans.

Similarly, colonial discourses of desire contrasted lower-class men of passion and bourgeois men of character. They did not silence talk of sex but knowingly detailed its social etymologies. These discourses attributed sexual excesses to those of creole, lower class, and mixed-blood background—"fictive" and not properly embourgeoised Europeans. Such representations hinged on the presence of other actors, on a marking of their sexuality as the truth of the self, as an index of the social category to which they truly belonged. British, French, and Dutch moralizing missions produced discourses contrasting desire and reason, native instinct and white self-discipline, subversive unproductive sexuality and productive patriotic sex, but these lines could never be drawn—nor were they meant to be drawn—with racial clarity.

These assessments of the sexuality of masturbating children, of promiscuous servants, and of degenerate white men and intruders in the bourgeois home congeal around what constituted the threat in these transgressive moments. Sexual intimacy and precocity were at issue, but not them alone. Evidence of emotional ties, confusions of blood and milk, impudence, disrespect of indifference were as dangerous as carnal knowledge and tainted blood. Nor was reference to particular sentiments just another way of talking about what "really mattered"—sex. Subversions of the bourgeois order were those that threatened that repertoire of sensibilities glossed as "personal character" and that marked who was eligible to be classed as white. These contexts of "cultural contagion" might disrupt dichotomies of ruler and ruled as they clarified and confused what being respectable and properly colonial was supposed to be about. Desire may have been reduced to the sexual, but the desires at issue embodied other powerful sensibilities. Cultivation of a self that was self-reliant, without pretensions, and morally pure was thought to define the interior landscapes of "true" Europeans and the interior frontiers of the superior polities to which they were constantly reminded they rightfully belonged. Thinking about "the education of desire" more broadly may help to avoid a quandary, that is, reproducing the very terms of high imperial discourse that reduced and read all desires as sexual ones. It offers another option—looking to a wider range of affective dispositions and cultural transgressions that informed what was unspeakable and what had to be said.

Further Reflections

Race and the Education of Desire turned out to be more useful than I had hoped for some, less accommodating to others. Scholars in a range of fields also have been smitten with Foucault and troubled by him. Some have shared my concern with how few people have pushed his chronologies up against those in the colonies and enthusiastically joined me in asking why he did not. Some students of the colonial have joined to query the articulation between European biopower and colonial racisms in both contiguous empires and those "overseas." Fewer have sought to ask whether a language of race developed out of the language of class (as nineteenth-century racial discourse suggests), rather than the other way around. Fewer still have attended to Foucault's efforts to think through a specific genealogy of discourses on race or the emphasis he placed in the 1976 lectures on modern racism as an inherent feature of contemporary states. Foucault's focus was decidedly not on some generic notion of racism but on what he alternately referred to as "state racism or "racism in its statist form," racism as part of the normalizing apparatus of capitalist, fascist, and socialist states. Such a shift in analytic emphasis does two things: it reconceives racism not as an aberrant, pathological development of state authority in crises but as a fundamental "indispensable" technology of rule—as biopower's operating mechanism. This demands a further question: How could this be so?

How could racism serve such a wide spectrum of political agendas at the same time?

That question is in many ways basic to the progression of the 1976 lectures. For where Foucault began his exegesis was not with a discussion of racism, as one might expect, but with a more general set of observations about the nature of power, erudite knowledge, and discursive formations. As he did in the methodological project of *The Archaeology of Knowledge*, he once again refutes—and refuses—the notion that a discursive formation can be identified by its unity and coherence but by "the different possibilities that it opens of reanimating already existing themes, of arousing opposed strategies, of giving way to irreconcilable interest, of making it possible with a particular set of concepts, to play different games."[36]

It is different manifestations of this "tactical mobility" that Foucault tracks in the race seminar. Beginning in the first two lectures with the assertion that discourses are composed of both erudite and subjugated knowledge, he makes a striking point: seemingly unified discourses are historically layered with oppositional discourses, permeated with resurrected subjugated knowledges that may resurface with them. In short, racial discourse is neither always a tool of the state nor always mobilized against it. Racial discourses diffuse over a broad field. Their genealogical histories should track their "spaces of dissension" and unique sites of dispersion.[37]

The first two lectures have been canonized (indeed fetishized) as Foucault's work that speaks most directly to the study of subaltern history. But they do so in a very specific way—by setting out a counterintuitive proposition. Disqualified knowledge can be of different sorts. Like the field of phrenology, it can be valorized at one historical moment and dismissed at another, without undermining its basic principle: namely, an alignment between the physiology and the inner states that determine race. But disqualified knowledge can work against itself and serve erudite knowledge instead. Racial discourse, for example, accrues its force not because it is a scientifically validated discourse but just the opposite. It is saturated with sentimentalisms that increase its appeal. Disqualified knowledge can do something else. It can usurp the place of erudite knowledge and in its "resurrection" turn the world upside down. Thus the first two lectures that address this resurrection introduce a specific kind of history, a genealogy of race.

A second important point worth more attention: Foucault draws on the notion of polyvalent mobility to account for why racial discourse is rarely consistent in its political affiliation or strategic claims. It may serve a reactionary political agenda or a reformist one. It may be mobilized against the state at one historical moment and usurped by the state at another. Recognition of this quality may help to account for its resilience and enduring relevance over time. I have tried to argue elsewhere that this very fact—that racial discourses contain, comply with, and coexist with a range of political agenda—is not a contradiction but a fundamental historical feature of their *nonlinear* political genealogies.[38]

This analytic focus on how counternarratives and disqualified knowledges may resurface within official discourses links to a third theme, one critical to Foucault's understanding of the resilience of racial discourses and the force of discursive formations. This is the tension he posits between processes of rupture *and* recuperation. It remains striking how much Foucault's philosophical and historical project continues to be characterized as one concerned with historical discontinuity, abrupt fissures, and dispersions in unexpected time and place. Many scholars take this as his defining contribution and the hallmark of his work.

But from the lectures it is clear that discursive formations are never built on epistemic rupture alone. Discourses about race and how race is known—that is, the popular and erudite theories that inform knowledge of it—are plural not singular, sedimentary not linear. It is within these sedimented folds that new planes and surfaces emerge. These preserved possibilities account for why racial discourses so often appear as new and renewed at the same time. But what accounts for those specific features that are available for recuperation? What qualifies some for recodification and not others? What makes the contemporary discourse of the European extreme Right on citizenship, foreigners, or national education look so similar to the wider consensual discourse on these issues *and* not dissimilar to that of racist demagogues a hundred years earlier?

No one would argue that the 1976 lectures offer a comprehensive analysis either of racial discourses or of racisms of the state. On the other hand, few others have asked such discomforting questions about modern state formations or explored the reversibilities of racial discourses and the process of reversal. If Foucault pressed on some questions more than others, it is for us twenty years later to take on the ones he could have and did not.

It is for us to understand the conditions of possibility that give racial thinking its continuing and refurbished currency, to dissect the forms in which racist visions pronounce themselves as relevant to the twenty-first century, to understand what charges them with populist and oppositional appeal. Foucault's notion that state policies are directed at "the defense of society" against itself has chilling resonance in extreme Right discourse in Europe today. Understanding more fully what joins racisms, biopolitics, and modern states may be one way of doing what Foucault encouraged—the writing of histories that nourish reversals, recuperations, and insurrections within them. It is a direction that much of my own subsequent efforts to recraft the scope and scale of imperial histories has sought to take.[39]

Notes

This essay was previously published as chapter 6 in Ann Laura Stoler, Carnal Knowledge and Imperial Power: Race and the Intimate in Colonial Rule.

1. Michel Foucault, *The History of Sexuality*, vol. 1, *An Introduction* (New York: Vintage, 1978), 103 [hereafter *HS*].
2. I discuss this relationship between sentiment and the state through the venue of moral philosophy in *Along the Archival Grain: Epistemic Anxieties and Colonial Common Sense* (Princeton, NJ: Princeton University Press, 2010).
3. On Nietzsche's notion of history in the comfort zone, see "On the Uses and Disadvantages of History for Life," in *Untimely Meditations*, trans. R.J. Hollingdale (Cambridge: Cambridge University Press, 1996), 57–124.
4. Michel Foucault, "Nietzsche, Genealogy, History," in *The Foucault Reader*, ed. Paul Rabinow (New York: Pantheon, 1972), 76.
5. See Ann Laura Stoler, "Intimidation of Empire: Predicaments of the Tactile and Unseen," in *Haunted by Empire*, ed. Ann Laura Stoler (Durham: Duke University Press, 2006), 1–22; Ann Laura Stoler, "Tense and Tender Ties: The Politics of Comparison in North American History," in Stoler, *Haunted*, 23–67.
6. See Christian Delacampagne's review titled "Foucault, généalogie du biopouvoir," *Le Monde des Livres*, February 21, 1997, 1.
7. On Foucault's productive and political period in Tunisia, see David Macey, *The Lives of Foucault* (New York: Pantheon, 1993), 183–208.
8. Ann Laura Stoler, "Colonial Aphasia: Race and Disabled Histories in France," *Public Culture* 23, no. 1 (2011): 121–56.
9. Foucault's thinking on subjugated knowledge has been widely available for some time under the title "Two Lectures" in *Power/Knowledge: Selected Interviews and Other Writings, 1972–1977*, ed. Colin Gordon (Brighton Sussex: Harvester Press, 1977); and in Nicholas B. Dirks, Geoff Eley, and Sherry B. Ortner, eds., *Culture/Power/History: A Reader in Contemporary Social Theory* (Princeton, NJ: Princeton University Press, 1994). There is no mention in either volume that these "two lectures" are the first two from the 1976 series on race.
10. See my "Racial Histories and Their Regimes of Truth," *Political Power and Social Theory* 11 (1997): 183–255
11. *HS*, 3.
12. Ibid.
13. See *Carnal Knowledge and Imperial Power: Race and the Intimate in Colonial Rule* (Berkeley: University of California Press, 2010), chap. 3; Anne McClintock, *Imperial Leather: Race, Gender, and Sexuality in the Colonial Conquest* (New York: Routledge, 1995); and Robert Young, *Colonial Desire: Hybridity in Theory, Culture and Race* (London: Routledge, 1995).
14. *HS*, 57–58

15. Timothy Mitchell, *Colonizing Egypt* (Cambridge: Cambridge University Press, 1991).
16. Paul Rabinow, *French Modern: Norms and Forms of the Social Environment* (Cambridge, MA: MIT Press, 1989); Gwendolyn Wright, *The Politics of Design and French Colonial Urbanism* (Chicago: University of Chicago Press, 1991).
17. Mary Louise Pratt, *Imperial Eyes: Travel Writing and Transculturation* (New York: Routledge, 1991).
18. Etienne Balibar, "Foucault et Marx: L'enjeu du nominalisme," in *Michel Foucault, Philosophe: Rencontre Internationale, Paris, 9, 10, 11 Janvier 1988* (Paris: Seuil, 1989).
19. See my "Racist Visions for the Twenty-first Century: On the Cultural Politics of the Radical Right in France," in *Relocating Postcolonialism*, ed. David Goldberg and Ato Quayson (Oxford: Blackwell, 2002).
20. Despite the recent rash of books on racism and the radical Right documenting the French National Front's rise, split, and fall, texts used in middle schools (college) and lycées categorically avoid mention of the history of racism in France.
21. More sophisticated analyses can be found in Jean-Yves Camus, *Le front national: Histoire et analyses* (Paris: O. Laurens, 1997), and on the National Front as an integral product of French history and politics, see the journalist Hubert Huertas's *FN: Made in France* (Paris: Editions Autrestemps, 1997).
22. The special issue of a journal entitled "Michel Foucault: The War of the Races to Biopower") that appeared in 2000 had virtually nothing to say about race. See also Yves Charles Zarka, "Michel Foucault: De la guerre des races au biopouvoir," in *Cités: Philosophe, politique, histoire* (Paris: Presses Universitaires de France, 2000).
23. *HS*, 54.
24. Michel Foucault, *Difendere la società* (Florence: Pont alle Gracie, 1990), 66.
25. *HS*, 118–19.
26. *HS*, 123.
27. See Nicholas Thomas, *Colonialism's Culture; Anthropology, Travel, and Government* (Princeton, NJ: Princeton University Press, 1994).
28. Benedict Andersen, *Imagined Communities* (New York: Verso, 1983), 137.
29. *HS*, 123; emphasis added.
30. Uday Mehta, "Liberal Strategies of Exclusion," in *Tensions of Empire: Colonial Cultures in a Bourgeois World*, ed. Frederick Cooper and Ann Laura Stoler (Berkeley: University of California Press, 1997), 59–86.
31. Doris Sommer, "Irresistible Romance: The Foundational Fictions of Latin America," in *Nation and Narration*, ed. Homi K. Bhabha (Berkeley: University of California Press, 1990), 87.
32. *HS*, 121.
33. *HS*, 47.
34. Michel Foucault, *"Les anormaux": Cours au Collège de France (1974–1975)* (Paris: Seuil/Gallimard, 1999); *Abnormal*, trans. Graham Burchell (New York: Picador, 2003).
35. Michel Foucault, *Résumé des cours, 1970–1982* (Paris: Julliard, 1989), 78.
36. Michel Foucault, *The Archaeology of Knowledge*, trans. A. M. Sheridan Smith (New York: Pantheon, 1972), 36–37.
37. Ibid., 152.
38. See my "Racial Histories," n5.
39. See *Along the Archival Grain; Imperial Debris: On Ruins and Ruination* (Durham, NC: Duke University Press, 2013); and *Duress: Concept-Work for Our Times* (Durham, NC: Duke University Press, forthcoming).

SEVENTEEN

Totalitarianism or Biopolitics? Concerning a Philosophical Interpretation of the Twentieth Century

ROBERTO ESPOSITO
TRANSLATED BY TIMOTHY CAMPBELL

A philosophical interpretation of the twentieth century: what does such an expression refer to, and what kind of weight do we want to give it? We could, of course, provide two different and in certain ways even opposing responses. The first is the one offered by the classic philosophical tradition of the twentieth century, which is to say the one supplied by Edmund Husserl, Martin Heidegger, and Jean-Paul Sartre (if we limit ourselves to only its most illustrious figures). For them the events of contemporary history are interpreted with a key supplied by philosophy itself, the only one available that might express what is essential for history. Whether the key is found in the crises of European sciences, in the unfolding of nihilism, or in the liberation of oppressed peoples—if we stick to the authors I have cited—in each case the twentieth century is understood according to the demands of a given philosophy whose task it is to make meaningful the events of the last century and to organize historical phenomena so that they move forward in an orderly fashion. A relation, therefore, is established between philosophy and history that is, so to speak, impositive. Only philosophy can impart an overarching sense to a series of facts that would otherwise be meaningless.

The first response, which produced analyses of monumental importance, is answered by another, one that overturns its logic. Rather than subordinating the movement of history to the logic of a given philosophy, it sees events as consisting of elements that are themselves philosophical. Meaning is no longer stamped on events from the outside, that is, from a point that coincides with the philosophical perspective of the person looking at the event. Instead this response focuses on how meaning originates and is constituted by the facts themselves—by their novelty, their scope, and their effects. Perhaps this change in perspective can be felt as well in what the grand tradition of twentieth-century philosophy—from Heidegger to Wittgenstein, but also including Kojève—defined on the one hand as "the end of philosophy" and on the other as the "end of history." In reality, what came to an end was a way of seeing history as the object of philosophical reflection. From that point on, history was, so to speak, no longer the object but, if anything, the subject of philosophy. No longer the form of history, philosophy becomes its content.

If contemporary events enjoy a philosophical depth, then our task is no longer to supply a proper meaning to how history is composed but rather to attend to the meaning that is originally present in the events under examination. And this is not because history has a unique, preconceived meaning—which was precisely the pretext for all philosophies of history, were they progressive or regressive, ascending or descending—but rather because history is constituted by the intersection of a number of different vectors of meaning that compete with each other. Events charged with significance, such as the attack on the Twin Towers, are precisely those that invert previous meanings and instantly open up new horizons of sense. When we say that contemporary history is above all philosophical, we are not saying that contemporary history can be understood by using philosophy in contrast to the more reductive perspectives of economics, sociology, or political science (as suggested by Augusto Del Noce some years ago).[1] We are saying rather in a stronger sense that the decisive events of contemporary history—the world wars, the emergence of technology, globalization, and terrorism—are in themselves philosophical powers that struggle to control and dominate the world—or the predominant interpretation of the world and therefore of its ultimate meaning. This is why—even before oil, weapons, and democracy—the metaphysical stakes of conflict were fundamentally concerned in how we were going to define contemporary history.

I would now like to relate these two modalities for understanding contemporary history—that of the more traditional philosophy *of* history and that of history *as* philosophy—to two hermeneutic paradigms that are often confused and superimposed but that, in my view, emerge radically as mutually exclusive in their presuppositions and effects on meaning. I am speaking about the paradigms of totalitarianism and biopolitics. Despite attempts to bring them together in a framework that makes one the continuation or the confirmation of the other (in the sense of a biopolitical totalitarianism or a totalitarian biopolitics), we are really dealing with interpretive models that not only logically diverge but that are certain to exclude each other. Their separation has less to do with their respective contents and more to do with a difference in approach that concerns precisely the relation between philosophy and history and the mode by which history is thought *by* or *within* philosophy.

In the totalitarian model, history is read and interpreted chronologically. It is traversed by an underlying break between two choices—between democracy and totalitarianism—that alternate. Thus the long phase of liberal democracy's general development is followed in the middle decades of the last century with a totalitarian period in both the West and the East; this period in turn is supplanted by two victories of the liberal-democratic model, a model that became hegemonic in the West in 1945 and 1989. What emerges, therefore, is that liberal democracy has both historical and philosophical traits. At the same time, however, modern history is situated along a vertical axis; first it ascends and advances and then, beginning in the 1920s, regresses until finally in the second half of the century it ascends again, despite the presence of new risks, some of which are linked to current developments in the Middle East. To these breaks on the vertical axis there corresponds on the horizontal axis a significant uniformity among forms, contents, and languages that are in fact profoundly dissimilar. I'm referring not only to Nazism and communism, fused in a single conceptual block, but also to liberalism and democracy, made homologous by the demands of a philosophy of history more inclined to assimilate them to each other than to differentiate between them.

That the totalitarian paradigm is born out of a traditional philosophy of history is demonstrated by the ongoing (and contradictory) recourse it has to the category of the origin. That this term appears in the title of two of the most important political works of the last century—Hannah Arendt's *The Origins of Totalitarianism* and Jacob Talmon's *The Origins of Totalitarian Democracy*—is not an accident; it is the clearest sign we have that totalitarianism may be a novel category but that its philosophical framework is absolutely classical.[2] In all the philosophical essays on totalitarianism, the interpretive gaze employed points to the origin but then struggles to find it. Where exactly is the origin of totalitarianism to be found, what produced it, and what is the principal basis for the absolute originality of twentieth-century totalitarianism? It is precisely here, in these demands made on the origin, that the first antinomy of the entire paradigm becomes apparent. How does one trace the genesis of the totalitarian phenomenon, which is also said to be, as Arendt herself argues, incapable of being assimilated to any other form of government and that therefore avoids any kind of causal genetic sequence? Why bother to find the origin of what does not seem to have an origin? And then how are we to keep together what in principle is discontinuous, that is, what is absolutely new and at the same time continuous—which is to ask again: what derives from an origin?

Two possibilities present themselves, both of which are typical of philosophical historicism. The first, which is the one adopted by Arendt, traces the entire Western philosophical tradition back to an originary loss of the Greek polis. Here, then, all of subsequent history is condemned to a process of depoliticization, which is certain to merge with the antipolitical drift of totalitarianism. Thus, twentieth-century totalitarianism, understood as a dynamic and, what's more, as a logic, winds up appearing as the outcome—not necessarily decided beforehand but virtually so, given certain conditions—of a homogeneous logic found in modernity. It is true—we are still speaking of Arendt—that in her analysis an unexpected quickening is established between depoliticization and totalitarianism, which sets off their respective meanings, though they form part of the same trajectory. This quickening begins with Thomas Hobbes, whom Arendt awkwardly interprets as the philosopher who supplied political thought with all the assumptions for its racial theories, assumptions that would eventually lead to the abyss of Auschwitz and Kolyma. The other possibility, taken by Talmon (as by François Furet, though somewhat differently), consists in a search for the origin of totalitarianism within the same democratic tradition that it should be opposed to.[3] Here the meaning of totalitarianism is located in an originary disease *[malattia]* situated in the past, which when no longer found in Hobbes is traced to Jean-Jacques Rousseau, and when no longer placed at the origin of modernity moves to the decisive event—the French Revolution. But here, too, the paradigm of totalitarianism remains imprisoned by a second antinomy it cannot escape: if the reference to the French Revolution—which is to say the most extreme form of democratic despotism, as Furet puts it—also holds true for communism, then how do we explain Nazism with reference to it?

Even Arendt's monumental essay is not spared these logical and historical difficulties. *The Origins of Totalitarianism* is divided in two. In the first section, we find the magisterial genealogical reconstruction of Nazi anti-Semitism, which dates back to the years of the war; the subsequent section is much weaker and is linked to the appearance of Stalinism (evidently conditioned by the Cold War). The reason for this imbalance, which can be traced empirically to the closing of the Soviet archives, really concerns, however, the critical point of the entire interpretive model, namely, the difficulty of locating the roots of Soviet communism in the same degenerative drift—from the crises of the nation-state, to colonial imperialism, to the explosion of biological racism—that brought us Nazism. The question then is: how are we to hold together in a single categorical horizon a hypernaturalistic conception such as that of Nazism with the historicist paroxysm of communism? From a philosophical point of view, what does a theory of absolute equality—which is what communism at least in its principles purports to be—have to do with a theory and indeed a practice of absolute difference such as found in Nazism? A one-shaded drawing, based on the vertical opposition between the temporality of democracy and the temporality of totalitarianism, seems to carry the day over great logical, categorical, and linguistic caesurae, which in a complicated fashion extend across modern history in the paradigm of totalitarianism.

If Arendt's *Origins of Totalitarianism* remains a monumental work on Nazism, it is not surprising to find that on account of the logical and historical difficulties noted above, Raymond Aron, Talmon, and Furet will concern themselves instead with communism and only communism.[4] Aron in *Democracy and Totalitarianism* makes clear the reason for choosing to leave out (or having to choose to leave out) from his analysis the other pole of the totalitarian paradigm. What really interests him are only those regimes that profess to be democratic but that instead are practically derived from a perverse deviation from democracy. Both Talmon and Furet (but also Marcel Gauchet and Claude Lefort) agree with Aron; totalitarianism (of the Left, of course) is born from an infected rib of democracy and hence is a part of it.[5] Furthermore, the totalitarian regime does not arise out of a defect but rather from an excess, a surplus, of democracy, from a democracy so radical, so extreme and absolute, and so full of egalitarianism as to break down its own formal limits and so to collapse on itself, turning into its opposite. Communism—this is Gauchet's thesis—is instituted through a perverse inversion of the democratic model that distorts its features but is always based on the same suppositions. Communism is both democracy's dream and its nightmare. At this point, the chain of aporias of the entire totalitarian paradigm clearly emerges. In the first instance, if communism is not only situated on the conceptual horizon of democracy that emerges from the French Revolution but in a certain sense brings it to fulfillment in its excess of egalitarianism, how then can it withstand the distinction, fundamental to the entire discourse, between totalitarianism and democracy? How can totalitarianism be defined in opposition to what it originates from? In the second instance, if such an antinomic connection with democracy holds true for communism, certainly that cannot be the case for Nazism, which is ousted from the analytical framework of all these authors. But then the category of totalitarianism loses some of its logical consistency. Already on shaky historical ground, totalitarianism stumbles on philosophical terrain, which had seemed to provide it with its last foothold.

Quite different from the paradigm of totalitarianism, biopolitics is not part of a philosophical presupposition, which is to say, it is derived not from any philosophical form of history but rather from concrete events—not only from facts but also from languages that make it comprehensible. Even before Foucault's analysis of biopolitics there was Nietzsche's genealogy and more precisely his deconstruction of the concept of origin—the origin that the theoreticians of totalitarianism searched for—which must be addressed to uncover the perspective of this new gaze.[6] If a full and absolute origin of the historical process cannot be said to exist—if the origin is never a unity, if it always splits and multiplies into many origins—then not only are such origins not definable—as Nietzsche himself explains, thus radically opposing all forms of philosophical historicism—but the entire historical event *[vicenda]* of the West is destined to assume features that are irreducible to the linearity of a single perspective. The global interpretation of modernity emerges here as profoundly altered. Every possibility of a unified reading of modernity comes to an end in favor of a frame traversed by horizontal and vertical breaks that make it difficult to presuppose continuity. Furthermore, note that what in the preceding paradigm is configured as a completely closed event in the specialist language of politics is now enlarged to include a more complex relation derived from the meeting, conflict, and layering with other disciplinary lexicons that interact and contaminate each other to create new and different effects. The appearance onstage of biological life—which anything but predisposes modern philosophy to a depoliticizing movement as in the Arendtian model—has a disruptive effect that then positions modern philosophy along different vectors of sense, which overlap without coming together in a single line. The force of the biopolitical perspective lies precisely in its capacity to read this interweaving and this conflict, this gap in meaning and what is implied, which is to say the powerful antinomy between intersecting languages that are originally heterogeneous, such as those of politics and biology. What happens when an outside—life—bursts into politics, thereby breaking apart its presumed autonomy, shifting discourse onto a terrain that is irreducible to traditional terms like "democracy," "power," and "ideology"?

This is the frame in which the phenomenon of Nazism is situated and where its radical heterogeneity ought to be interrogated. Without having recourse to more recent interpretations, Ernst Nolte, a witness who cannot be suspected of Gauchist sympathies, recognized the theoretical fallacy of situating an ideology such as communism (however catastrophic in its political consequences) on the same lexical level with something like Nazism; under no circumstance can they be thought as belonging to the same category.[7] This is different from what Arendt thought. Nazism is not an ideology because it belongs to a dimension that is different from and subordinate to that of ideas, from which Marxist communism was born. Nazism is not a markedly different species within the same genus, namely that of totalitarianism, because it is situated outside Western tradition (a tradition that also includes the philosophy of communism among its offspring). It is counter to such a tradition, notwithstanding the differences between communism and liberalism (which fundamentally share a common reference to a universal idea of transcendence), that Nazism elaborates a radically different conception, one that no longer needs to be legitimated in any idea because it finds its essential foundation in its simple material force. It is not the contingent and necessary product of a history that defines the relation between men on the basis of their free decisions or, as communism holds, on the basis of their social conditions; rather it is based on an absolutely natural fact *[dato]* that concerns basic *[nuda]* aspects of biological life. To recognize in Nazism the singular attempt to free the natural features of existence from their historical peculiarity upends the Arendtian thesis of the totalitarian overlap between the philosophy of nature and the philosophy of history. It means instead seeing Arendt's blind spot—that these philosophies cannot be assimilated—and therefore recognizing the philosophical impracticability of the notion of totalitarianism.

Seen from the biopolitical perspective, the twentieth century and indeed the entire course of modernity is not determined or decided by the superficial and contradictory antithesis between totalitarianism and democracy but by that which is much more profound—because it pertains to the preservation of life—between history and nature and the historicization of nature and the naturalization of history. This antithesis is much more profound, I would say, because it is not ascribable to a symmetrical bipolarity from the moment that nature, understood in a biological sense, as Nazism understood it, becomes not an antihistory, or a philosophy, or an ideology that is opposed to history but more precisely a nonphilosophy and a nonideology. Nazism is not a political philosophy but a political biology, a politics of *[di]* life and a politics over *[su]* life transformed into its opposite and for that very reason productive of death. As Emmanuel Levinas wrote in the 1930s, in Nazism the biological, with all the fatality that inheres therein, becomes much more than an object of spiritual life; it becomes its core.[8] It is this immediately biological element of Nazism, the thanatopolitical, regardless of the number of its victims, that makes the category of totalitarianism historically and theoretically unusable.

The question of biopolitics now moves increasingly front and center, in spite of those who mistakenly thought that the related disasters—one by explosion and the other by implosion—of what they referred to as totalitarianisms made possible a return to the old political lexicon that preceded it. Seen from this perspective, the end of the Second World War does not mark the victory of the alliance between democracy and communism on the level of language and material practice but that of a liberalism situated in the same biopolitical regime that, certainly inclined in an opposite direction, resulted in Nazism. In ways more unfamiliar than communism, Nazism emerges decisively defeated from the war both militarily and politically but less so either culturally or linguistically in the sense that the centrality of *bios* as object and subject of politics is reaffirmed, even if it has changed to reflect a liberal influence, namely, in that the appropriation and the possible modification of the body is not on the part of the state but on the part of the individual owner of himself.[9] If man for Nazism is his body and only his body, for liberalism, beginning with Locke, man is the possessor of his own body and therefore can use it, transform it, and sell it, as if the body were a slave. In this sense, liberalism—naturally I am speaking about the underlying category of liberalism—turns the Nazi perspective inside out, transferring the property of the body of the state to the individual, but within the same biopolitical lexicon. Yet it is precisely the biopolitical characterization of liberalism that separates it from democracy.

One could say, with an exaggeration that is not completely unjustified, that we cannot return to liberal democracy after the advent of so-called totalitarianisms because liberal democracy never really existed as such. Just as the assimilation between Nazism and communism in the category of totalitarianism is to be deconstructed, so too is the notion of liberal democracy to be problematized. The ideology of liberalism, in its logic, presuppositions, and conceptual language (antiegalitarian, particularist, and at times also naturalistic) if not opposed to the ideology of democracy is quite different; the latter tends to be universalist and egalitarian, as Carl Schmitt noted in his essay on parlamentarism and democracy.[10] If we stop ourselves from representing modernity as a historicist might, if we reject the idea of a chronological succession between liberal-democratic and totalitarian regimes in favor of a different genealogical or topological representation, we see that the correct and conceptually important distinction is not the vertical one between totalitarianism and liberal democracy but the horizontal and transversal one between democracy and communism on one side—communism as the paroxysmal fulfillment of egalitarian democracy—and biopolitics on the other. Biopolitics in turn breaks off into two antithetical but not unrelated forms: Nazism, the biopolitics of the state, and liberalism, the biopolitics of the individual.

Foucault recognized the biopolitical character of liberalism, locating it on the level of the government of life, in opposition to or at least apart from the universalist procedures of democracy. Democracy—or anyway that which defines itself as such, which is to say that is founded on the primacy of abstract law and the equal rights of individuals who are endowed with reason and will—had already come to an end by the 1920s and 1930s. It is no longer capable of being reconstituted, let alone exported elsewhere. Naturally, if the democratic regime is reduced only to the presence of formal political parties in competition with one another and to the electoral method by which governing majorities are formed, one can always argue, as some do, that the number of democracies in the world continues to grow. But in doing so we lose sight of the radical transformation that swept across them, dragging them down into a semantic domain that cannot be reduced to what the concept of democracy requires as its precondition.

However, we should be careful. In arguing for this thesis, I am not referring to the dysfunctions, defects, limits, and contradictions that are implicit in all political forms, which are necessarily imperfect and incomplete. Rather I am speaking of a profound laceration of the same democratic horizon. This laceration becomes visible the moment we move from the level of form to the level of content, where content is understood as the material of the current biopolitical regime. It is true that democracy as such does not have contents; it is a technology *[tecnica]*, an ensemble of rules directed to distributing power proportionally according to the will of the electors. But it is precisely for this reason that it explodes or implodes; it takes in an element that it can no longer hold without changing into something radically different.

What we are dealing with here is the establishment of the biological life of individuals *[singoli]* and populations as fundamental to all of the most important political decisions of today. Of course, this does not mean that in the engagement and conflict between political forces other possibilities are not also at stake, concerning international relations and internal order, the model of economic development and the definition of civil rights. But the disruptive element with respect to the traditional democratic framework lies in the fact that each of these choices refers to the body of its citizens without mediation. If we consider only the recent example of Italy, where the laws that have chiefly involved the opinions of the public are those that concern highway security, immigration, artificial insemination, and bans on smoking and drugs, we can begin to measure the direction of this paradigm shift; the model of medical care has become not only the privileged object but also the form itself of political life, which is to say a politics that finds its only possible basis of legitimacy in life. This is principally what happens when citizens are continually interpellated or objectively involved with regard to questions that pertain to the preservation of, the limits of, or what is to be excluded from their own bodies. But—and here is the decisive point—when the living or dying body becomes the symbolic and material epicenter of the dynamics of politics as well as its conflicts, we move into a dimension that lies not simply, as we sometimes hear, after or beyond democracy but resolutely outside it—not only removed from its procedures but from its language and conceptual apparatus. Democracy is always directed to a totality of equal subjects, given the fact that they are separated from their own bodies and therefore understood as pure logical atoms endowed with rational will.

This element of abstraction or disembodiment in democracy also echoes in the proposition that places the person at the center of democratic praxis, where *person* is understood, according to the originary meaning of the term, precisely as a disembodied subjectivity—as distinct from that totality of impulses, needs, and desires that are aggregated in the corporeal dimension.[11] When the biopolitical shift *[svolta]* that we are reconstructing takes place, this bodily dimension becomes the real interlocutor of government, at once subject and object. What is emphasized less is the principle of equality, which is inapplicable to something like the body, constituted as it is differently from every other body according to criteria that can be defined and changed over time. What is at stake is not only the principle of equality but also an entire series of distinctions or oppositions on which the conception of modern politics is based and from which democracy is generated: the distinctions between public and private, artificial and natural, and law *[diritto]* and theology. This is because at the moment when the body substitutes or "restores" *[riempie]* the abstract subjectivity of the juridical person it becomes difficult if not impossible to distinguish the concerns of the public sphere from the private as well as what belongs to the natural order and what can be subjected to the intervention of technology *[tecnica]*, with all of the ethical as well as religious questions that this kind of choice raises.

The reason for the indistinction and unresolvable contrasts that inevitably result is that human life is precisely the space in which public and private, natural and artificial, and the political and theological are entwined to such a degree that no decision of the majority can undo it. This is why biopolitics is incompatible with the conceptual lexicon of democracy. Contrary to what we might think, the onset of life into *dispositifs* of power marks the eclipse of democracy, at least democracy as we have imagined it up until now. Of course, this does not mean that another kind of democracy is not imaginable, one that is compatible with the biopolitical shift underway, a shift, we should add, that cannot be reversed. How do we look for and how do we think about a biopolitical democracy or a democratic biopolitics, one that is capable of being used not on bodies but in favor of bodies? These questions are difficult to answer. At present we can glimpse only the outlines of possible responses. What is certain, however, is that to begin thinking in this direction all of the old philosophies of history and all the conceptual paradigms that refer to them must be dismantled *[disfarsi]*.

Notes

This essay was published originally in the journal Critical Inquiry *34 (2008): 633–44.*

1. See Augusto Del Noce, *L'Interpretazione transpolitica della storia contemporanea* (Naples: Guida, 1982).
2. See Hannah Arendt, *The Origins of Totalitarianism* (1957; New York: Harcourt, Brace, Jovanovich, 1973); and Jacob Leib Talmon, *The Origins of Totalitarian Democracy* (London: Secker & Warburg, 1952).
3. See François Furet, *Revolutionary France, 1770–1880*, trans. Antonia Nevill (Oxford: Blackwell, 1992), and *The Passing of an Illusion: The Idea of Communism in the Twentieth Century*, trans. Deborah Furet (Chicago: University of Chicago Press, 1999).
4. See Raymond Aron, *L'Opium des intellectuels* (1955; Paris: Editions Pluriel, 2002); *Democracy and Totalitarianism: A Theory of Political Systems*, ed. Roy Pierce (1968; Ann Arbor: University of Michigan Press, 1990); and *Power, Modernity, and Sociology: Selected Sociological Writings*, trans. Peter Morris, ed. Dominique Schnapper (Brookfield, VT: Gower, 1988).
5. See Marcel Gauchet and Gladys Swain, *Madness and Democracy: The Modern Psychiatric Universe*, trans. Catherine Porter (Princeton, NJ: Princeton University Press, 1999); *La Révolution des pouvoirs: La Souveraineté, le peuple, et la représentation, 1789–1799* (Paris: Gallimard, 1995); and *La Révolution des droits de l'homme* (Paris: Gallimard, 1989); and Claude Lefort, *Writing: The Political Test*, trans. David Ames Curtis (Durham, NC: Duke University Press, 2000); and *La Complication: Retour sur le communisme* (Paris: Fayard, 1999).
6. Esposito here is referring to Foucault's discussion of biopolitics as it appears in the seminar from 1975–76 entitled *"Society Must Be Defended."* Recall that for Foucault, biopolitics names a technology of power that is to be distinguished from the mechanisms of discipline that emerge at the end of the eighteenth century. This new configuration of power aims to take "control of life and the biological processes of man-as-species and of ensuring that they are not disciplined, but regularized." The biopolitical apparatus includes "forecasts, statistical estimates, and overall mea-

sures. . . . In a word, security mechanisms have to be installed around the random element inherent in a population of living beings so as to optimize a state of life." As such, biopolitics is juxtaposed in Foucault's analysis to sovereignty, leading to the important distinction between them: "It is the power to make live. Sovereignty took life and let live. And now we have the emergence of a power that I would call the power of regularization, and it, in contrast, consists in making live and letting die." Biopolitics thus is that which guarantees the continuous living of the human species. Michel Foucault, *"Society Must Be Defended": Lectures at the Collège de France, 1975–76*, trans. David Macey, ed. Mauro Bertani and Alessandro Fontana (New York: Picador, 2003), 246–47). [Translator's note.]

7. See Ernst Nolte, *Der europäische Bürgerkrieg, 1917–1945: Nationalsozialismus und Bolschewismus* (Berlin: Propyläen Verlag, 1987).
8. See Emmanuel Levinas, "Reflections on the Philosophy of Hitlerism," trans. Seán Hand, *Critical Inquiry* 17 (Autumn 1990): 62–71.
9. Esposito is drawing on the distinction in Greek between *zoē* and *bios* as employed most famously by Giorgio Agamben in *Homo Sacer*. Agamben writes: "The Greeks had no single term to express what we mean by the word 'life.' They used two terms that, although traceable to a common etymological root, are semantically and morphologically distinct: *zoē*, which expressed the simple fact of living common to all living beings (animals, men, or gods), and *bios*, which indicated the form or way of life proper to an individual or group." Giorgio Agamben, *Homo Sacer: Sovereign Power and Bare Life*, trans. Daniel Heller-Roazen (Stanford, CA: Stanford University Press, 1998), 1. [Translator's note.]
10. See Carl Schmitt, *Die geistesgeschictliche Lage des heutigen Parlamentarismus* (Munich: Duncker & Humblot, 1923).
11. See Roberto Esposito, *Terza persona: Politica della vita e filosofia dell'impersonale* (Turin: Einaudi, 2007).

CONTRIBUTORS

SAMANTHA BANKSTON is Assistant Professor in the Department of Humanities and Social Sciences at Sierra Nevada College, where she is also the Honors Program Director. Her research is focused in the areas of continental philosophy, feminist philosophy, literature, and modern art. She has published numerous articles on contemporary continental philosophy, and her first authored book, *Deleuze and Becoming(s)*, is forthcoming with Bloomsbury/Continuum; her second book, *Deleuze and Žižek*, will be published in 2016 by Palgrave Macmillan. She is also the translator of Anne Sauvagnargues' *Deleuze and Art* (Bloomsbury/Continuum, 2014).

TIMOTHY CAMPBELL is Professor of Italian in the Department of Romance Studies at Cornell University. In addition to authoring *Wireless Writing in the Age of Marconi* (University of Minnesota Press, 2006) and *Improper Life: Biopolitics and Technology from Heidegger to Agamben* (University of Minnesota Press, 2011), he has just completed a study of cinematic forms of life titled *Holding Patterns: Cinema and the Generous Form of Life*. He has also translated Roberto Esposito's *Bios: Biopolitics and Philosophy* (University of Minnesota Press, 2008), *Communitas: The Origin and Destiny of Community* (Stanford University Press, 2009), and, most recently, Carlo Diano's *Form and Event* (Fordham University Press, forthcoming, 2016). At Cornell he teaches courses on contemporary Italian philosophy, cinema, and literature.

VERNON W. CISNEY is a Visiting Assistant Professor of Philosophy at Gettysburg College. His work is broadly focused on questions of ethics and politics, subjectivity, and thought in the areas of nineteenth through twenty-first century continental philosophy. He is currently working on questions of the relation between politics, aesthetics, and cinema. He is the author of *Derrida's Voice and Phenomenon: An Edinburgh Philosophical Guide* (Edinburgh University Press, 2014), as well as *Derrida and Deleuze: Difference, Immanence, and the Future* (Edinburgh University Press, forthcoming, 2016). He is also the coeditor of *The Way of Nature and the Way of Grace: Philosophical Footholds on Terrence Malick's Tree of Life* (Northwestern University Press, forthcoming,

2016), and *Between Foucault and Derrida* (Edinburgh University Press, forthcoming, 2016). In addition, he is cotranslating *La Monnaie Vivante (Living Currency)* by Pierre Klossowski for Bloomsbury Academic Press.

ROBERTO ESPOSITO is the Vice President of the Italian Institute of Human Sciences and Professor of Philosophy in charge of the PhD Program "Modern and Contemporary Philosophy." He was a member of the scientific board of the International College of Philosophy (Paris), as well as the founder of a Research Center on European Politics. Coeditor of the *Political Philosophy* journal, he is also member of the editorial boards of *Il Mulino*, Liguori Press, and Einaudi Press. He has published several books, including *Bios: Biopolitics and Philosophy* (University of Minnesota Press, 2008) and *Communitas: The Origin and Destiny of Community* (Stanford University Press, 2009).

DIANA GARVIN is currently a doctoral candidate in Italian Studies at Cornell University. She received her AB in Romance Studies (Italian, French, and Spanish) from Harvard University in 2006. *Critical Inquiry* will publish her article "Taylorist Breastfeeding in Rationalist Clinics: Constructing Industrial Motherhood in Fascist Italy" in its Spring 2015 volume.

FRÉDÉRIC GROS is a Professor of Political Philosophy at the Institut d'Etudes Politiques in Paris. He has published extensively on Michel Foucault's writings. His books include *Michel Foucault* (Presses Universitaires de France, 1996) and *Foucault et la folie* (Presses Universitaires de France, 1997).

IAN HACKING is a University Professor Emeritus at the University of Toronto and a Professor Emeritus of the chair of Philosophy and History of Scientific Concepts at the Collège de France. His many books include *The Social Construction of What?* (Harvard University Press, 2000), *Mad Travelers: Reflections on the Reality of Transient Mental Illnesses* (Harvard University Press, 2002), and *Historical Ontology* (Harvard University Press, 2004).

DAVID M. HALPERIN is W. H. Auden Distinguished University Professor of the History and Theory of Sexuality at the University of Michigan, where he is also Professor of English Language and Literature, Women's Studies, Comparative Literature, and Classical Studies. He coedited *Before Sexuality: The Construction of Erotic Experience in the Ancient Greek World* (Princeton University Press, 1990), *The Lesbian and Gay Studies Reader* (Routledge, 1993), and *Gay Shame* (University of Chicago Press, 2009); he also cofounded *GLQ: A Journal of Lesbian and Gay Studies*, which he coedited from 1991 to 2005. He is the author of *Before Pastoral: Theocritus and the Ancient Tradition of Bucolic Poetry* (Yale University Press, 1983), *One Hundred Years of Homosexuality and Other Essays on Greek Love* (Routledge, 1990), *Saint Foucault: Towards a Gay Hagiography* (Oxford University Press, 1995), *How to Do the History of Homosexuality* (University of Chicago Press, 2002), *What Do Gay Men Want? An Essay on Sex, Risk, and Subjectivity* (University of Michigan Press, 2007; rev. ed. 2009), and *How To Be Gay* (Belknap Press of Harvard University Press, 2012).

MARY BETH MADER is Professor of Philosophy at the University of Memphis. Her research focuses on recent and contemporary continental philosophy, feminist philosophy, and the intersections of continental thought and the life sciences. She is the author of *Sleights of Reason: Norm, Bisexuality, Development* (SUNY Press, 2011). Her essays on Foucault include "Foucault and Social Measure," "Foucault's 'Metabody,'" and "Modern Living and Vital Race: Foucault and the Science of Life." She is translating Deleuze's lecture courses on Foucault.

TODD MAY is Class of 1941 Memorial Professor of the Humanities at Clemson University. He has published dozens of articles in various areas of recent French thought, particularly that of Michel Foucault, Gilles Deleuze, and Jacques Rancière. He is the author of many books, including *The Political Philosophy of Poststructuralist Anarchism* (Pennsylvania State University Press, 1994), *Reconsidering Difference: Nancy, Derrida, Levinas, and Deleuze* (Pennsylvania State University Press, 1997), *Gilles Deleuze: An Introduction* (Cambridge University Press, 2005), and *The Philosophy of Foucault* (Acumen Press, 2006). His most recent book is *Contemporary Movements and the Thought of Jacques Rancière: Equality in Action* (Edinburgh University Press, 2010). He is currently writing a book on friendship in the contemporary period.

LADELLE McWHORTER is the author of *Bodies and Pleasures: Foucault and the Politics of Sexual Normalization* (Indiana University Press, 1999), *Racism and Sexual Oppression in Anglo-America: A Genealogy* (Indiana University Press, 2009), and more than two dozen articles on Foucault, Bataille, Irigaray, and race theory. With Gail Stenstad, she edited a revised and greatly expanded second edition of her 1992 anthology *Heidegger and the Earth: Essays in Environmental Philosophy*, which was issued by Toronto University Press in 2009. She holds the James Thomas Chair in Philosophy and is also a Professor of Environmental Studies and Women, Gender, and Sexuality Studies at the University of Richmond in Virginia.

EDUARDO MENDIETA is a Professor in the Philosophy Department at Pennsylvania State University. His areas of specializations are ethics (broadly construed), contemporary European philosophy, Latin American philosophy, critical theory, and race theory. He has published many articles on these topics, along with three monographs: *Global Fragments: Critical Theory, Latin America and Globalizations* (SUNY Press, 2007), *Adventures of Transcendental Philosophy: Karl-Otto Apel's Semiotics and Discourse Ethics* (Rowman and Littlefield, 2002), and *The Philosophical Animal: On Zoopoetics and Interspecies Cosmopolitanism* (SUNY Press, forthcoming).

CATHERINE MILLS is Associate Professor and Australian Research Council Future Fellow in the Centre for Human Bioethics at Monash University. Her areas of research include bioethics, feminist philosophy, and continental philosophy. In addition to publishing numerous articles, she is the author of *Futures of Reproduction: Bioethics and Biopolitics* (Springer, 2011), and *The Philosophy of Agamben* (McGill/Queens University Press, 2008).

NICOLAE MORAR is an Assistant Professor of Philosophy and Environmental Studies and an Associate Member of the Institute of Ecology and Evolution at University of Oregon. He is a member of the Consortium for Socially Relevant Philosophy of/in Science and Engineering (SRPoiSE). Dr. Morar specializes in bioethics (especially biomedical, genethics, environmental, and research ethics), philosophy of biology and ecology, and recent continental philosophy. Morar is the coeditor of *Perspectives in Bioethics, Science, and Public Policy* (Purdue University Press, 2013). Forthcoming publications include the coedited volume *Between Foucault and Derrida* (Edinburgh University Press, 2016). He is also the coeditor of *Between Foucault and Deleuze* and the cotranslator of Pierre Klossowski's *La Monnaie Vivante (Living Currency)* for Bloomsbury Academic Press. Morar is currently completing a monograph titled *Biology, BioEthics, and BioPolitics: How to Think Differently About Human Nature.*

JEFFREY T. NEALON is Edwin Erle Sparks Professor of English and Philosophy at Penn State University. He is author of *Foucault Beyond Foucault: Power and Its Intensifications Since 1984* (Stanford University Press, 2007), and several other monographs on cultural theory, among which are *Alterity Politics: Ethics and Performative Subjectivity* (Duke University Press, 1998) and *Post-Postmodernism; Or, The Cultural Logic of Just-in-Time Capitalism* (Stanford University Press, 2012), which looks at the interconnections of American cultural and economic production since the 1980s. His newest project is *Plant Theory: Biopower and Vegetable Life* (Stanford, 2015).

ANTONIO NEGRI is an ex-Professor in Padua and an independent researcher. His areas of research include globalization, Marxism, and anticapitalism. He has published many books, including *Marx Beyond Marx: Lessons on the Grundrisse* (Autonomedia, 1991), *The Savage Anomaly: The Power of Spinoza's Metaphysics and Politics* (University of Minnesota Press, 1991), *Time for Revolution* (Continuum, 2003), *Subversive Spinoza: (Un)Contemporary Variations* (Manchester University Press, 2004), and *Political Descartes: Reason, Ideology, and the Bourgeois Project* (Verso, 2007). He has coauthored four books with Michael Hardt: *Labor of Dionysus: A Critique of the State-Form* (University of Minnesota Press, 1994), *Empire* (Harvard University Press, 2000), *Multitude: War and Democracy in the Age of Empire* (Penguin, 2004), and *Commonwealth* (Belknap Press of Harvard University Press, 2009).

CARLOS NOVAS is an Associate Professor in the Department of Sociology and Anthropology at Carleton University where he teaches science and technology studies. Carlos obtained his PhD in sociology from Goldsmiths College, University of London. His doctoral research concentrated on the governance of genetic risk. This research focused on the history of the development of a genetic test for Huntington's disease, the forms of advice offered to persons at risk of developing this illness, and how persons at risk interact with one another through the medium of the Internet. After completing his PhD, Carlos was awarded a one-year ESRC Postdoctoral Fellowship where he concentrated on disseminating the results of his PhD research. Carlos was also awarded a Wellcome Trust Postdoctoral Fellowship undertaken at the BIOS Centre for the Study of Bioscience, Biomedicine, Biotechnology and Society, London School of Economics. His current research interests relate to biopolitics, the life sciences, biotechnology firms, patients' organizations, and health-care rationing. A recent publication consists of a coedited volume with Sahra Gibbon, *Biosocialities, Genetics and the Social Sciences: Making Biologies and Identities* (Routledge, 2008).

PAUL PATTON is Scientia Professor of Philosophy at the University of New South Wales in Sydney, Australia. His publications deal with aspects of French poststructuralist philosophy, Nietzsche, and a variety of topics in contemporary political philosophy. He is the author of *Deleuze and the Political* (Routledge, 2000) and *Deleuzian Concepts: Philosophy, Colonization, Politics* (Stanford, 2010). He is editor of *Deleuze: A Critical Reader* (Blackwell 1996), (with Duncan Ivison and Will Sanders) *Political Theory and the Rights of Indigenous Peoples* (Cambridge University Press, 2000), (with John Protevi) *Between Deleuze and Derrida* (Continuum, 2003), (with Simone Bignall) *Deleuze and the Postcolonial* (Edinburgh University Press, 2010), and (with Sean Bowden and Simone Bignall) *Deleuze and Pragmatism* (Routledge, 2015).

CHRISTOPHER PENFIELD is a PhD candidate in the Department of Philosophy at Purdue University, where he recently defended his dissertation, *Foucault, Kant, Deleuze, and the Problem of Political Agency*. His present work is focused on developing a political philosophy of agency, understood as an account of the necessary conditions for collective action to transform the social order. He is the author of "Toward a Theory of Transversal Politics: Deleuze and Foucault's Block of Becoming" in *Foucault Studies* (April 2014); articles on Foucault in *The Cambridge Foucault Lexicon* (Cambridge University Press, 2014) and the *Encyclopedia of Global Justice* (Springer, 2011); a book review on Deleuze and theology in *Notre Dame Philosophical Reviews* (2014); and several forthcoming translations of texts by and about Foucault.

PAUL RABINOW is Professor of Anthropology at the University of California at Berkeley. His work is centered primarily on modernity itself as a problem: a problem for those seeking to live with its diverse forms, and a problem for those seeking to advance or resist modern projects of power and knowledge. He has covered topics as diverse as the problems facing descendants of a Moroccan saint coping with the changes wrought by colonial and postcolonial regimes, to the wide array of knowledge and power relations entailed in the great assemblage of social planning in France, to work in the last decade on molecular biology and genomics. He has written dozens of articles and authored many books, among which are *Essays on the Anthropology of Reason* (Princeton University Press, 1996), *Marking Time: On the Anthropology of the Contemporary* (Princeton University Press, 2007), *Designs for an Anthropology of the Contemporary* (Duke University Press, 2008), (with Hubert Dreyfus) *Foucault: Beyond Structuralism and Hermeneutics* (Chicago University Press, 1983), (with Talia Dan-Cohen) *A Machine to Make a Future: Biotech Chronicles* (Princeton University Press, 2006) and

Anthropos Today: Reflections on Modern Equipment (Princeton University Press, 2003). In addition, he has edited many books, such as *The Essential Works of Michel Foucault, 1954–1984*, vol. 1, *Ethics, Subjectivity and Truth* (New Press, 1997), (with Nikolas Rose) *The Foucault Reader* (Vintage, 1984) and *The Essential Foucault* (New Press, 2003), and (with William M. Sullivan) *Interpretive Social Science: A Second Look* (University of California Press, 1988).

JUDITH REVEL is an Associate Professor in the Foreign Languages and Literature Department at Université Paris 1 Panthéon-Sorbonne. She is a member of the Bureau scientifique du Centre Michel Foucault (The Scientific Committee of the Michel Foucault Center), and she codirects a research seminar at l'Ecole des hautes etudes en sciences sociales on "The History of Intellectual Engagement in the Twentieth Century." She is interested in contemporary Italian and French philosophy, particularly in political theory, biopolitics, and processes of subjectivation. She has published widely on these topics, including in *Foucault. Expériences de la pensée* (Bordas, 2005), *Dictionnaire Foucault* (éditions Ellipses, 2007), and *Qui a peur de la banlieue ?* (Bayard, 2008).

NIKOLAS ROSE is a Professor of Sociology and head of the Department of Social Science, Health and Medicine at King's College London. He was originally trained as a biologist before switching to psychology and then to sociology. His books include *The Psychological Complex: Psychology, Politics and Society in England, 1869–1939* (Routledge, 1984), *Governing the Soul: The Shaping of the Private Self* (Routledge, 1989; 2nd ed., Free Association Press, 1999), *Inventing Our Selves: Psychology, Power and Personhood* (Cambridge University Press, 1996), *Powers of Freedom: Reframing Political Thought* (Cambridge University Press, 1999), *The Politics of Life Itself : Biomedicine, Power, and Subjectivity in the Twenty-first Century* (Princeton University Press, 2006), and (with Peter Miller) *Governing the Present* (Polity Press, 2008). His current work is on the social life of neuroscience, which he writes about (with Joelle Abi-Rached) in *Neuro: The New Brain Sciences and the Remaking of the Human* (Princeton University Press, 2013). He is one of the directors of the joint LSE-Imperial College Centre for Synthetic Biology and Innovation, and a member of numerous advisory committees including the Nuffield Council on Bioethics.

JANA SAWICKI is the Carl W. Vogt '58 Professor of Philosophy at Williams College. Her research interests include recent continental philosophy, feminist theory, critical theory, queer theory, and social and political philosophy. She has published many articles in these areas and is the coeditor of *Foucault Companion* (Wiley-Blackwell, 2013), along with *Foucault Studies: Special Issue on Queer Theory* (September 14, 2012). In addition, she is the author of *Disciplining Foucault: Feminism, Power and the Body* (Routledge, 1991).

ANN LAURA STOLER is Willy Brandt Distinguished University Professor of Anthropology and Historical Studies at the New School for Social Research. She is the founding director of the summer Institute for Critical Social Inquiry at The New School and a founding coeditor of the online journal *Political Concepts: A Critical Lexicon*. Her work is focused in the areas of postcolonial theory, critical race theory, gender studies, and political economy. She is the author of numerous monographs: *Capitalism and Confrontation in Sumatra's Plantation Belt, 1870–1979* (University of Michigan Press, 1995), *Race and the Education of Desire: Foucault's History of Sexuality and the Colonial Order of Things* (Duke University Press, 1995), *Along the Archival Grain: Epistemic Anxieties and Colonial Common Sense* (Princeton University Press, 2009), and *Carnal Knowledge and Imperial Power: Race and the Intimate in Colonial Rule* (University of California Press, 2010). She has also edited or coedited many collections: for example, *Haunted by Empire: Geographies of Intimacy in North American History* (Duke University Press, 2006), *Imperial Formations* (School for Advanced Research Press, 2007), and *Imperial Debris: On Ruins and Ruinations* (Duke University Press, 2013).

MARTINA TAZZIOLI is a PhD candidate in the Philosophy Department at Goldsmiths College. She received her BA and MA degrees in philosophy from University of Pisa (Italy), where she wrote a thesis on Michel Foucault's critical analysis of neoliberal governmentality. She was a fellow at University of Chicago and at Université Paris XII. Her research project concerns a Foucaultian lecture of the European migratory policies, by focusing, on one hand, on the so-called neoliberal advanced techniques of government and, on the other hand, on the forms and practices of resistances of the very migrants. She is a member of the editorial board of Materiali Foucaultiani (www.materialifoucaultiani.org).

INDEX

abnormal/abnormality, 6, 151, 199, 209; biological, 93; exception, 151; versus normal, 96, 249, 253; sexual, 200
abstraction, 121–22, 131–36, 357
activism, 188, 196, 239; patient (group), 183, 184, 192, 323; political, 19
Africa, 255, 332; Jane Alexander, 163; colonialism, 327, 332, 338–39; colored bodies, 327; genomics, 312; HIV (transmission and deaths), 303; migration, 279; North Africa emergency, 276–81, 290; (post)apartheid, 171, 176, 335; technology and social manipulation, 176
African American, 209, 312
Agamben, Giorgio, 148, 150–53, 155–57, 276, 293, 299; bare life, 24, 82–99, 110, 116–17, 293, 302–4, 359; and biopower/biopolitics, 16–18, 82–83; *bios*, 43, 55, 58, 83, 96, 100, 110, 151, 154, 302, 304, 307, 355; *Homo Sacer*, 83–88, 302–4; state of exception, 304; *zoē*, 83–84, 87, 151, 154, 302, 304
animal, 131, 155, 161–65, 168, 170–71, 175–78, 359; animality, 19, 140, 145; animal life, 85, 138, 140–52; animal studies, 138, 140, 144–49, 152–55; becoming animal, 19; biology, 122–31, 143–53; capital, 140–41; factory, 138, 140; forms, 129; human-animal divide, 141, 158–59; living animal, 2, 131; political, 4, 13
Arab uprising, 21, 274, 281–82
Archaeology of Knowledge, The (Foucault), 31, 45, 150, 156, 329, 344, 347
Arendt, Hannah, 87, 289, 293, 350–54, 358; *The Origins of Totalitarianism*, 293, 350–51, 358

barebacking, 209, 221–25; practice, 202–5
Bataille, Georges, 33, 35
beast, 141, 161–62, 168, 170–71, 179; and power, 152; and *Übermensch*, 161–62
bestiary, 141, 143, 159, 161–64, 177, 178
bioethics, 306
biohistory, 23, 176, 179
biophilic, 176
biopolitics/biopolitical, 138–41, 162; affirmative, 92; and animality, 147–56, 176–79; biopolitics of the population, 9, 12, 14, 19, 42, 65, 96, 298, 313; versus biopower, 43, 51, 59; and Judith Butler, 228–37; and capitalism, 265–73; concept, 7, 9, 13, 18, 29–30, 42–45, 48, 82–92, 298–324; critique of, 88; and Holocaust, 91; life and politics, 82, 83, 88, 95–98, 102–8, 112–17, 240; and medicine, 183–92, 196; and neoliberalism, 20; origins of, 17–18, 30, 42–44, 51–57, 67, 259–61, 265–73; overt versus subversive, 67, 75; population management, 248; race, 21, 113, 236, 333–36, 340, 345; resistance, 19, 234; scholarship, 82; sexual practices, 213, 219, 220; and social medicine, 43; as

biopolitics/biopolitical (*continued*)
sovereignty, 51, 84; statistics, 73–78; study of, 17–22; and subjectivity, 45; *thanatopolitics*, 89; "to make live and to let die," 110–11; versus totalitarianism, 21–22, 348–49, 353–59
biopower: and animal studies, 138–41, 144–54; versus biopolitical, 61, 63; and capitalism, 189; concept, 1, 83, 88–90, 102–9, 121, 139, 199–200, 229–30, 269–71, 297–309; and discipline, 105, 138–39; gender, 230–34; genomic medicine, 318–22; and life, 95–97, 307; modernity, 106–7, 112–14, 121; and normalization, 95–96, 98; objections to, 14–17, 299–308; origins, 3–7, 144–45, 302; race/racism, 112–13, 303–4, 309–12, 332–36, 343; reproduction, 313–18; revolutionary, 77–80; and sexuality, 5–6; as sovereignty, 14, 51, 151; versus sovereignty, 3–4, 14–15, 95, 151, 248; subjection to, 86; subjectivity, 176–77, 219; transition to governmentality, 7–14; two poles, anatomo-politics, 4, 65, 95, 260, 298; two poles, biopolitics of the population, 5, 40, 42–43, 95, 248, 260, 264–65, 298
biotechnology, 176; engineering, 164; industry, 183, 185, 190–91
Blanchot, Maurice, 33–36
body/bodies, 40, 92–95, 191–92, 238, 297, 336, 341, 355–58; advisory, 315; anatomo-politics of the, 12–13, 19, 106, 298; in art, 168; and biology, 162; biopolitical, 84, 88; and biopower, 17, 102–10; counting of, 67, 78, 80; and discipline, 9, 43, 103, 139, 189, 247–51, 256, 260–64, 268, 300, 340; and gender, 231; human, 193, 313; immanence, 58, 62–63; living, 87, 122–23; as machine, 4, 65; physiological processes, 132; political, 18, 336; versus population, 102–10, 199; and power, 30, 231; and rights, 230; and sexuality, 229, 338; of the sick, 318; social, 41, 55, 304, 335, 336–37; state/nonstate, 306; of work, 138, 219
bourgeois, 213, 328–30, 336–43, 347; bourgeoisie, 263; subject, 326, 332
Butler, Judith, 20, 228–42, 275, 293, 342
capital/capitalism, 6, 41, 54–56, 63, 77, 247, 270–72, 321, 332–33; animal, 140–41; biocapital, 301, 317; and biopolitics, 265–68; *Capital* (Marx), 71, 264; and discipline, 249–55, 261; human, 271; industrial, 6; neoliberal, 152; production, 49–52, 189
capture, 83, 84, 91, 213, 272, 277, 291–93
census, 66, 74–79
citizen, 67, 192, 269, 270, 272, 303, 356, 357; citizenship, 220, 235, 239, 284, 288–89, 290–92, 299, 306; and colonialism, 330, 332, 335, 336, 345; global, 238; US citizens, 17
class, 50, 69, 262, 323, 333, 337–38; classification, 121–35, 148, 218; lower, 342; middle, 339; power, 41, 268; social, 78, 340; struggle, 49, 55–56, 66; working, 48–50, 263
colonial, 171–72, 176, 334–43, 351; power, 305, 309, 327, 331; studies, 326–27; violence, 171
cosmopolitan/cosmopolitanism, 158, 162–63, 168, 175–78
courses at Collège de France: *The Birth of Biopolitics*, 20, 103, 107, 138, 251, 259–60, 265; *Security, Territory, Population*, 7–9, 102–6, 248; *Society Must Be Defended*, 138–39, 260, 301, 309, 335
Cuvier, Georges, 19, 121–34, 136–37

Deleuze, Gilles, 24, 89, 95, 246, 300, 307, 323; and control, 246, 300; and desire, 152; and the fold, 36–37, 59–61; and life, 85–86; and structuralism, 53–54
democracy, 63, 274–75, 288, 349, 353–58; conquest of, 283; and totalitarianism, 350, 352
Derrida, Jacques, 52, 61–64, 85, 87, 138, 156; animals, 149–50; deconstruction, 59; specter of communism, 63
disciplinary power, 106–9, 113, 139, 264, 332, 336; and criminality, 247; *Discipline and Punish*, 4–5, 261; discipline versus security, 105
discipline, 40, 199, 264, 268–70, 299, 328, 342; as anatomo-politics of the body, 43, 65; and biopolitics, 96, 138–40,

260; birth of, 40; function of, 8–10, 104; as a model of power, 12–13; normalizing, 245–49, 253–55; self-, 339–40
Discipline and Punish (Foucault), 109, 245–51, 260–62, 329
discourse, 13, 30–32, 37–40, 150, 333–36, 339–45; academic, 147; biological, 140, 142, 231, 299; biopolitical, 229, 237; counter-, 218, 230, 234, 239; governmental, 291; nationalistic, 15; order of, 34–35; political, 51; public, 6, 188, 210, 215; and race, 112, 309–10, 327–31; of reason, 145; truth, 21–22, 89, 299, 307
disease, 70, 71, 73, 212, 249, 309, 311, 319; control of, 103, 201, 212; and health, 93; identity as, 216–17; list of, 79, 317, 320; rare, 19, 183–91
dispositif, 52, 54, 57, 60–61, 159, 358; immunitary, 92; technological, 174
domination, 40, 52, 63, 113, 272, 300; biopolitical, 83; colonial, 309; mechanisms of, 15; power as, 2–3; practices of, 42

economic, 8, 291, 305, 319, 337–38, 349; action, 12; analysis, 41, 189–95, 315–17; bioeconomics, 306, 317; and biopower, 138; and capital, 254; classical, 52; crisis, 279–80; development, 48, 263, 320, 356; and political, 39, 54–55, 271–73; power, 114, 301; processes, 6, 150–52; reasons, 6, 183–84, 313; subjects, 114, 211, 269, 282; system, 6, 262, 264
emergent, 275–76, 306, 308, 323; political subject, 10, 299; spatialities, 292; technologies, 190
empire, 62, 67, 326–28, 330–31, 339, 342–43, 346; Ottoman, 332; Soviet, 310
Empire (Hardt and Negri), 300–302, 304
epistemology, 45, 53, 99, 280, 288, 322
errancy, 83, 95, 98
Esposito, Roberto, 16, 18, 21, 83, 90–95; *Bios*, 91, 92, 304
ethics, 107, 167–68, 237–38, 240; and aesthetics of the self, 43, 302; dimensions of, 43, 148–49, 161, 308; and politics, 228; and sacrifice, 140
exploitation, 134, 238, 251, 300, 332, 338; biopolitical, 273; conditions of, 232, 287; processes of, 50–52, 264, 290, 301

feral, 162, 176–77
Fordism, 48, 50, 55
function, 123–28, 130–35, 144, 319–20, 336; biopolitical, 20; economic, 261–64; functionalist, 19, 121, 122; function of, 16, 71, 94, 253, 286; linguistic, 34, 36; of power, 40, 50, 53, 153; regulative, 31, 201, 261; reproductive, 336
Furet, François, 351, 352, 358

gay, 199–212, 214–21; communities, 206; and HIV risks, 201, 205, 207, 209; identity, 215; rights, 199; sexual practices, 207, 216, 217; subjectivity, 199, 200, 201, 210, 214
gender, 228, 230, 299, 336, 338; identification, 231–32; normative, 199, 248; politics of, 234–35; and sex, 231
genocide, 113, 256, 316, 333, 334
genome, 308, 310–11, 320, 325
genomic medicine, 21, 308, 318–19
govern/government, 6, 75, 95, 107, 212, 236, 319, 321–22, 350; art and general problems of, 13; authorities/groups, 184, 186, 188; British, 76–77; colonial, 305; (community of) the governed, 239–40, 271–72; and discipline, 40, 55; and drugs, 186; European, 283; governing class, 68–69; and health, 196; and individuals/individualization, 40, 42–43, 111, 114, 228, 231, 275; Italian, 276, 283–84, 290; and liberalism/neoliberalism, 250–56, 267; and life, 103, 111–12, 265, 303, 316, 356–57; and migrants, 279–84, 286–92; offices of, 71; and power, 102; representational, 2, 112, 356; and scarcity, 10–14, 315; seat of, 8; self-, 216; strategies and techniques of, 97, 103, 329; technology, 190, 221; of words, 37
governmentality, 40; colonial, 329; concept, 7–10, 20, 22, 302; forms of, 103, 113–14, 176; and migrants, 21, 274–76, 279–81, 292; neoliberal, 228, 252; and population, 13–14; and sovereignty, 236; techniques of, 42, 176–77, 275

Haraway, Donna, 148, 154, 163, 176, 178; species chauvinism, 19, 140
Hardt, Michael, 16, 18, 24, 300–301, 304, 324
health technology, 192–93
History of Madness (Foucault), 19, 146
History of Sexuality, The, vol. 1 (Foucault), 138, 140, 184, 245, 298, 328–29, 340, 342; Anglo-American reception, 82; and biopower/biopolitics, 7, 65, 95, 97, 102, 105, 297; and normalization, 247; and racism, 334–35; and the "right of death," 108
History of Sexuality, The, vol. 2 (Foucault), 115, 117, 240, 298
History of Sexuality, The, vol. 3 (Foucault), 240, 298
Hitler, Adolf, 89, 305, 359
HIV, 201–17, 219–21; disclosure, 219; normative taxonomy, 20; risk of transmission, 201, 220; and unsafe sex, 202
Holocaust, 89, 91, 151, 230, 297, 302–4; regime, 230; trauma, 37
homosexual, 206, 235; "invention" of, 142; pathology and, 199–201, 205, 215–18; "scientific" explanations, 20; and subjectivity, 215
human: activity/behavior, 13, 56, 210–11, 250–51, 266; being, 18, 54, 125, 135, 153, 158–59, 253, 298–301; biology, 19, 103–4; body, 4, 65, 106, 193, 298; capacities, 6; capital, 270–71; health, 195; humanism/humanist, 144, 147, 229, 234–35, 240, 311, 321; humanity/humanitarian, 16, 50, 158, 272; life, 3–4, 86, 102–3, 108, 112, 138, 199, 298, 300, 305, 357; nature, 331, 340; and nonhuman animals, 161–77; and norms, 94; populations, 184; and power, 97–98, 103–4; race, 265, 271; rights, 272, 278, 284; sciences, 4, 31, 41, 61; sexual practices, 6; and statistics, 67–68; subjectivity/subjectivation, 41–44, 200–201, 299

immanence, 54–56, 58–63, 85–86, 92–95, 99
immune/immunity, 90, 91, 92, 100, 222
imperial politics, 326, 328, 337
indigenous, 220
interspecies, 162–64, 168, 176, 178
Italy, 53, 280–82, 287, 288, 356; and migration, 276–78; and Tunisian Revolution, 282; Western Marxism, 18, 48–51

language, 171–72, 278, 350, 353, 355; academic departments, 254; of biopower, 7, 13; and bodies, 58; of democracy, 357; "The Discourse on Language," 32, 38; Dutch, 336; European, 339; of governmentality, 13–14; Martin Heidegger, 172; of liberalism, 355; literary, 17, 29–38, 43; object of knowledge, 65, 142; and the outside, 32–38; plant, 148; of race, 343
lesbian, 199, 215, 218
liberalism, 91, 259–60, 306, 332, 339–40, 350, 354–56; and biopolitics, 103, 272–73; and capitalism, 265, 317; classical, 250–51, 268; and democracy, 350; as economic doctrine, 267
life, 68–70, 138–54, 185, 201, 228–40, 248, 251, 297–308; administration of, 184, 187–89, 191, 199–200, 213, 221, 234, 297–98, 314, 319; animal life, 138, 140–41; bare life, 82–88, 110, 302–4; biological, 83–84, 90, 110, 275, 302, 353–54, 356; and biopolitics, 43, 50–51, 54–55, 82, 103, 229–30; and biopower, 14–17, 102–6, 140; concept of life, 82–98, 121–26; conditions of life, 106, 110, 114, 233; Georges Cuvier, 122–25, 130; and discipline, 254; efficiency, 4; expectancy, 105; forms of, 165; Foucault's, 14, 29; functions, 130–31, 133–35; human, 65, 103, 108, 112, 138, 140, 184, 228, 300, 357; immanent, 83; of individuals, 40, 111, 213–15, 283–84, 336; life and death, 2, 69, 86, 98, 109–12, 213, 236, 297; life in general, 6, 102, 105, 236, 303; neoliberal, 270–71; and norms, 213; political life, 357; of population, 299; and power, 1, 4, 6–7, 30, 45, 57–63, 105–6, 139–40, 142, 272, 274, 298, 301–2, 304, 338; quality of life, 139, 317–18, 322; and racism, 230; right to life, 111, 230, 239, 298; science of life,

121; and sexuality, 218–19, 234; social, 300; and sovereignty, 2–3, 106, 109; subjectivity, 261; time of life, 50, 264, 269; urban, 104
life sciences, 89, 177, 235
literary, 29, 33, 35, 39, 42, 44; birth of biopolitics, 30; linguistic decade, 38; speech, 34

Malthus, Thomas, 75
manage/management, 255; child, 341; crisis, 213; of error, 98; forces, 8; of inclusion and exclusion, 300; of individuals, 12, 42, 277–79, 291; of life, 40, 88, 90, 95–96, 297, 307, 309; managerial capitalism, 266–68, 271; of migrants, 21; of populations, 236–37, 248, 317–21; and power, 3; of production, 40, 43; of reproduction, 314; self-, 270; of sexuality, 328, 331, 336; time, 253
Marx, Karl, 55–57, 66, 71, 266, 300, 354; and emancipation, 50; and labor, 49; and Marxism, 18, 48, 52, 53, 56
medicine, 93, 184–85, 301, 302; bioscience and, 83, 88–91, 196, 337; and discipline, 73; funding, 188–90, 192–94; genomic, 21, 308, 318–20; and human capacities, 313; and normalization, 19; and population, 104; and race, 309; social, 43; and split personalities, 80
Mediterranean, 274, 276, 279, 281–83
migrant, 21, 63, 274–92, 293, 294
migration, 250, 277–83, 285–86; African internal, 279; categories, 280; migration governmentality, 274–75, 277–83; practices of, 275, 283, 290–93; Tunisian, 283
militant, 49–50
minority, 202, 215, 216, 225, 284
money, 278; circulation, 248–49, 252–53, 257; and health, 195, 319; research and development, 76
multitude, 18, 56, 61–62, 301, 304

Nazi, 315; anti-Semitism, 351; biology, 352–56; concentration camps, 151, 303; eugenics, 91; Nazism, 89, 91–93, 106, 113, 250, 350–56; regime, 91–92, 236, 303–5, 335
Negri, Antonio, 16–18, 24, 47, 63, 299–301, 304, 323–24; and bios, 43, 55, 58
neoliberalism, 216, 249–52, 254–55, 257, 268, 270
Nietzsche, Friedrich, 327; animal, 158–63, 176–77; and exteriority, 35; genealogy, 234, 353; immanence, 85; materialism, 54; power and knowledge, 40; self-overcoming, 145; subjectivity, 232; vitalism, 57
norm, 216, 246–49, 310; behavioral, 220, 228, 240; and biopolitics, 96, 228–35; developmental, 247; and discipline, 261; Foucault, 151; gender, 230–34; hygienic, 264; as immanent impulse of life, 83, 92–96; national, 282, 290; and practices, 257; the sciences, 4; sexual, 247; social, 94–95, 290; statistics, 18, 107–8, 199, 247
normal/normalization: and biopower, 5–6, 96, 108, 230, 232; capital punishment, 16; concept, 20; discipline, 4, 245, 252–53, 256–57; law, 5; and life, 93–96; ontology, 57; versus pathological, 91–96, 218, 318; and racism, 333, 343; and sexuality, 201, 210, 212, 218–19, 228, 230–37, 247–50, 328; and subjection, 176; technologies of, 19

ontology: biopolitical, 51, 54–55, 57–58; of function, 125; and gender, 231; of nature, 159; temporal, 44
operaismo, 48–49, 51, 54–55, 63, 64
Order of Things, The (Foucault), 35–36, 38–39, 140, 145–47, 152; and Cuvier, 121–26; and *episteme*, 31; and life, 142–43
orphan drug, 19, 183–95, 196, 197, 198
Orphan Drug Act, 183, 189, 196, 198
outside: of biopower, 83; the body, 122–23; of capitalism, 55; the contract, 111; democracy, 357; detainee camps, 276, 278; Europe, 317; exteriority of meaning, 348; the family, 341; the law, 86; life, 353; neoliberal apparatus, 255; of production, 52–53; self-forming outside of norms, 240; of the subject, 233–34; "The Thought of the Outside," 32–37, 40–42, 44; of the United States, 190–91; of the world, 58

pharmaceutical, 183–92, 312; companies, 19, 183, 185, 188–89, 311, 319, 321; groups, 184; industry, 186–88, 190–91; markets, 192, 194, 317; Pharmaceutical Manufacturers Association (PMA), 188; regulation, 185, 196

politics/political, 14, 29, 54, 56, 328; absolutist, 89; activism, 19–20, 82, 183, 239, 270, 274–77; affirmative, 92; agency, 234; animals, 4, 13, 141, 146; art, 168; "autonomy of the political," 53; biology and, 19; and biopolitical, 56, 82, 297–305, 307–10, 312–13, 316–19, 321–23; body, 18; body politic, 65, 336; challenges, 43; change, 239, 262, 275; chronopolitics, 264; concept, 58–62, 121, 162–63; control, 51; debate, 51; and deconstruction, 58–59; discourse, 39–45, 51; economy, 7, 11, 12, 40–41, 236, 251, 271; expediency, 194, 210; events, 20–21, 259; Foucault's political identity, 260; global, 228, 238, 264; hierarchies, 199; imperial, 326, 337, 339; knowledge, 327; life and, 1–2, 7, 21, 82–98, 103, 213, 228–30, 239, 270, 298–300, 338, 353–57; Marxism, 53; migration, 279–92; modern, 82, 92, 102, 138, 140, 261, 356; order, 110; organization, 6, 48; paradigms, 22; parties, 356; versus pastoral power, 107; philosophy/theory, 16, 51, 56, 108, 111–13, 235, 248, 261, 297, 299, 333, 341, 350–51, 354; political imaginary, 163, 176; politics of truth, 8, 240; and population, 10, 12, 105; power, 40, 92, 107–9, 151, 337; and prisons, 262–63; problems, 18; and race, 333–34, 338, 340, 344–45; regulation, 95; representation, 109, 112; resistance/struggle, 8, 21, 41, 43, 56, 188, 192–93, 200, 298; revolts, 263; right, 110–12; rulers, 2; science, 13, 349; and sex/sexuality, 5, 214–18, 234; social reality, 54; society, 111; statistics, 96; strategy, 68; subjects, 1–2, 10, 12, 50, 95, 248, 269–73; *thanatopolitics*, 89–92, 303–4, 355; and technology/technologies, 43, 163, 275–76; will, 267; working-class, 48–51

population, 105, 236, 307; animals, 153; biopolitics of the, 5, 9, 18, 19, 20, 95–96, 102–3, 271, 298, 300, 313; and biopower, 40, 230; concept, 7; discipline, 107; and disease, 103–4; economics, 6, 12, 315; and eugenics, 316–18; exploitation of, 238; gay men, 209; and genocide, 16, 230, 236, 297; global, 313–14; and governmentality, 7, 12–14, 20; and health, 183–96, 299, 301, 318, 320–22; and HIV, 205; and individuals, 9, 12, 96, 104; and life, 65, 105, 199, 221, 299, 356; migrant, 21; neoliberalism, 20; as political subject, 95, 248; and power, 10, 106; and race/racism, 21, 230, 303, 309–12, 330, 334–37; regulatory mechanisms of, 5, 110, 231, 247–51, 260, 264–65, 268; relation to multiplicity, 10, 12; and security, 9–10, 12, 103, 106; set of living beings, 42–43, 298; societal attitudes toward, 6; and statistics, 18, 65–67, 71, 75, 77, 79, 105, 248; United States population, 17; and war, 16, 230

power, 30, 228–29; analytics of, 1–3, 29, 39, 237, 308–9; as assemblage, 40, 358, biopower, 1, 14–17, 43, 61, 63, 77, 102–14, 300–307; capitalist, 50; colonial, 331; counter-, 44; deductive versus productive, 2, 93, 95; disciplinary, 105, 247, 257, 261, 264, 332, 336; economic, 114, 273; emergence of bio-, 3–7; as force, 3, 106; genealogy of, 40; juridical, 230; labor, 49, 51–52, 55–56, 63, 264, 266; of language, 37, 144, 288; over life, 2, 85, 96, 105, 135, 230, 275, 276, 283, 297, 298, 301–2; mechanisms of, 1, 32, 235, 303; neoliberal, 246, 257; and networks, 3, 257; ontology of, 54, 59, 344; pastoral, 13, 177; political, 40, 92, 95, 109, 337; power-knowledge, 42, 65, 66, 141, 177, 230, 234, 237, 245, 257, 322, 326; *puissance* and *pouvoir*, 45, 57, 59, 60; relations of, 38, 40, 41, 59, 184, 231, 233, 247, 271, 292, 326; as security, 12, 249; and sexuality, 328; sovereign, 3, 88, 89, 106, 110–12, 152, 153, 236, 247, 260–61, 268, 297,

304–8; state, 107, 236; strategies of, 10; technologies of, 88, 104, 139, 189
power/knowledge, 230–37, 245, 257
precarious, 50, 242, 287, 289, 290; creatures, 168, 337; *Precarious Life* (Butler), 235, 293; and precariousness, 235–38, 240
psychoanalysis, 200–201, 217, 227, 240
punish/punishment: biopower, 107; capital punishment, 16; and discipline, 247, 261–63; sexual, 216; and sovereignty, 247

queer: culture, 20, 200; people, 199–200, 215, 217–21; project, 19, 200, 235; studies/theory, 199, 229, 231; subjectivity, 20, 199–200

race, 271, 299, 301, 316, 325–30; and (bio) politics, 89, 104, 112–13, 308–10, 340; conceptions of, 312, 332–34, 337–38, 342–45; human, 265, 271; and population, 303; racialized, 171–72, 326, 329, 335, 339; racism, 112–13, 230, 235–36, 265, 303, 309–12, 328–30, 332–40, 343–45, 352; and sexuality, 209–10; superior, 177; war, 15
rare disease, 19, 183–91, 193, 195–97
reproduction, 53, 61, 133, 135, 313–17; and biology, 89, 306, 308–9; definition, 52; and labor, 49–51; of value, 40
resistance, 60, 69, 151, 175, 220, 252, 260, 300; (bio)political, 8, 19–20, 50–51, 272–73; and Butler, 234; modes of, 21, 54–55, 230; movements, 185; practice of, 32, 39–43, 279; strategies of, 275–76, 284–86, 289–90, 292
right, 106, 239, 263, 269, 279, 287–89, 308; biopower as, 108–13; civil, 356; to decide life and death, 297–98, 304–6, 335; discourse, 4; gay, 199; governed's, 271–72; human, 284–85; to life, 230, 236, 239, 298; sovereign, 106, 248, 261; unconditional, 282, 285; violations, 239
risk, 189, 190, 264, 278, 283–84, 320, 341; of circulation, 248–49; health, 185, 317; HIV/AIDS, 20, 220; and life, 86; and sex, 201–20; of social death, 233–34
Roussel, Raymond, 32–34, 45
Saussure, Ferdinand de, 32–37
scarcity, 10–12
science, 245; academic, 188; applied sciences, 254; biomedical sciences, 92, 192; bioscience, 83, 90, 306; European, 348; history of science, 129; human genome, 311; human sciences, 31, 41, 61, 142–45, 149–51, 199; of life/life sciences, 19, 89, 121–22, 126, 142, 177, 235; "Life: Experience and Science" (Foucault), 85, 95; linguistics, 32; medical science, 219; and norms, 4; policy, 319; and political philosophy, 85; political science, 13, 62, 349; and power, 40; psychological sciences, 232; and race, 337; statistics as "moral science," 67–68, 73, 213, 338; of wealth, 265
security, 50; apparatuses of, 236, 248–50, 252; and biopolitics, 96; concept, 8–10; disease, 68; financial, 254; and governmentality, 7; and health, 68, 307; highway, 357; and multiplicity, 10–13; national, 238, 302, 306; and population, 10–13, 236; security mechanisms, 103, 105–8, 113–14; of society, 16, 103; and sovereign, 2
sex/sexuality, 297; and biopower/biopolitics, 5–8, 67, 105, 151, 153, 229–30, 298, 313; and census, 75, 78; and children, 5; and colonialism, 327–43; and education of desire, 21; and gender, 228–40; individual practices, 6, 152; and life, 40, 65, 138–39; and medicine, 19; and normalization, 247–48; perversion, 6, 19, 199–201; and politics, 43; and power, 326; and race, 21, 326, 327–32, 334–35; regime of sexuality, 231–34; relations, 8; and reproduction, 313; safe sex, 202–9, 215, 219–21, 224, 225, 226; sex organs, 164; sex selection, 317; sexual minorities, 235; and women, 5–6
Sloterdijk, Peter, 172, 176, 178
sovereign/sovereignty, 65, 256, 304–8; Giorgio Agamben, 84–85, 88–89, 151–53, 302–5; versus biopower, 1–4, 14–16, 95, 106, 108–14, 152, 229–30, 248, 260–61, 265, 268, 297; Judith

sovereign/sovereignty (*continued*)
Butler, 236–39; and census, 75; versus discipline, 247; versus Empire, 304–5; and freedom, 14–15, 213, 272; and governmentality, 12–14; as juridical, 40–41; modernity, 84–85, 88–89, 151–53; national, 62, 286; as power, 49, 51, 257; and rights, 272; techniques of, 103; and territory, 8–10
Stalin, Joseph, 89, 106, 305, 335, 351
state, 13–16, 192, 302–9, 355–56; apparatus, 41, 185, 264–65, 305, 306; biopolitical, 83, 98, 102–14, 333, 340; colonial, 328–29, 339; constitution of, 83–84; of exception, 302, 303; and life, 90–91; modern, 74–75, 102, 328, 334; nation-state, 62, 304, 306, 316, 351; power, 236, 239, 250; racism, 265, 333, 335–36, 343–45; sovereign, 286, 297, 303; welfare, 256, 332
statist, 185, 332, 337, 343
statistics, 8, 72–75, 79, 96, 254, 338; of AIDS, 206; bureaucratic, 68, 70; of deviance, 73; as moral science, 67–68; and poverty, 5; of sexual practices, 212, 220; Victorian, 65
subject/subjectivity, 37–45, 91; "American," 176; appropriation, 39; biopower/biopolitical, 139, 230, 322, 340; bourgeois, 328, 332; and capitalism, 6; Cartesian, 41; and class, 56; colonial, 331, 336; constitution/production of, 30, 40–43, 229, 231; creative/productive, 18, 41, 43, 57–59, 61–63, 322; death/disappearance of, 34, 38, 42, 200; and democracy, 357; and desire, 56, 332, 336; and discipline, 103, 249, 261; and discourse, 37, 41; economic, 269; ethical, 228, 302; gay, 19–20, 199–201, 210–19; history as subject of philosophy, 349; history of subjectivity, 39; human, 299; and life, 17, 95, 111–12, 300, 302–3, 322; and migration, 276, 279, 281, 286–89; and the multitude, 61–63; neoliberal, 213, 216, 249, 253; and object, 7, 12, 37–38; objectivation of, 41–42; political, 1–3, 12, 84–85, 95, 248, 264, 355; and population, 10, 95, 248; and power, 18, 30, 40–41, 95, 307, 318–19; of practice, 39; "process without a subject" (Louis Althusser), 53; and resistance, 42; revolutionary, 49; and rights, 272, 288; self-relation of, 42; and sexuality, 231–40, 248; and sovereign, 1–2, 9–10, 14, 75, 109–14, 297, 304; and structuralism, 54; "The Subject and Power" (Foucault), 107; subjected/subjecting to, 16, 29, 42, 65, 86, 103, 188, 246–47, 256, 268; subjectification, 21, 89, 107, 299, 302, 307, 313–14, 322–23; subjection, 20, 41, 176–77, 231–35; subjectivation, 3, 18, 39, 41–44, 230, 234, 240; subjective, 32, 34–35, 86, 92, 139, 292; subject to, 56, 69, 71, 112, 153, 256, 315, 323, 330; thinking, 38; Western, 260

Talmon, Jacob, 350–52, 358
taxation, 66–67, 74, 77
taxonomy, 20, 128, 134; Borgesian, 38; classical, 126, 131; functional, 124; racial, 338
technology, 163, 168, 171–72, 298, 306, 349, 356–57; bio-, 176, 183, 185, 190–91; of (bio)power, 5–6, 95, 104, 106, 108–9, 112–13, 139, 230, 343; disciplinary, 104; of normalization, 236; Office for Health Technology Assessment, 192–93; of sex, 336–37; of state, 77, 190; political, 275–76
totalitarianism, 21–22, 100, 349–55; *The Origins of Totalitarianism* (Arendt), 293, 350–51, 358
transcendental, 60, 142–43, 144, 152; approach, 56–57; arguments, 234; concepts, 53; historico-, 150; normativism, 92–93

Übermensch, 158, 176
unsafe sex, 20, 201–2, 215–16, 218

visibility, 135–36, 143, 249; and Cuvier, 121–22; versus invisibility, 133, 269; and natural history, 126–27; strategy of, 285
vitalism, 57
vulnerability, 168, 170, 235, 237–38

wealth, 13, 56, 103, 261, 266–68, 315; analysis of, 41; and capital, 49, 265; creation of, 109, 247